Aspergillus and *Penicillium* in the Post-genomic Era

Edited by

Ronald P. de Vries,[1] Isabelle Benoit Gelber[1] and
Mikael Rørdam Andersen[2]

[1]Fungal Physiology
CBS-KNAW Fungal Biodiversity Centre; and
Fungal Molecular Physiology
Utrecht University
Utrecht
The Netherlands

[2]Department of Systems Biology
Technical University of Denmark
Kongens Lyngby
Denmark

Caister Academic Press

Caister Academic Press
Norfolk, UK

www.caister.com

British Library Cataloguing-in-Publication Data
A catalogue record for this book is available from the British Library

ISBN: 978-1-910190-39-5 (paperback)
ISBN: 978-1-910190-40-1 (ebook)

Cover design adapted from image courtesy of Jan Dijksterhuis, Ad Wiebenga and Ronald P. de Vries (*Aspergillus niger* grown on wheat bran).

Contents

Contributors v

Preface xi

1 Taxonomy of *Aspergillus*, *Penicillium* and *Talaromyces* and its Significance for Biotechnology 1
Jos Houbraken, Robert A. Samson and Neriman Yilmaz

2 Comparative Genomics, Resequencing and Fast Forward Genetics in *Aspergillus* and *Penicillium* 17
Scott E. Baker and Erin L. Bredeweg

3 Diversity and Mechanisms of Genomic Adaptation in *Penicillium* 27
Jeanne Ropars, Ricardo C. Rodríguez de la Vega, Manuela López-Villavicencio, Joëlle Dupont, Dominique Swennen, Emilie Dumas, Tatiana Giraud and Antoine Branca

4 Approaches for Comparative Genomics in *Aspergillus* and *Penicillium* 43
Jane L. Nybo, Sebastian Theobald, Julian Brandl, Tammi C. Vesth and Mikael R. Andersen

5 Blue Mould to Genomics and Beyond: Insights into the Biology and Virulence of Phytopathogenic *Penicillium* Species 75
Wayne M. Jurick II, Jiujiang Yu and Joan W. Bennett

6 Post-genomic Approaches to Dissect Carbon Starvation Responses in Aspergilli 89
Jolanda M. van Munster, Anne-Marie Burggraaf, István Pócsi, Melinda Szilágyi, Tamás Emri and Arthur F.J. Ram

7 Genetics and Physiology of Sulfur Metabolism in *Aspergillus* 113
Andrzej Paszewski, Jerzy Brzywczy, Marzena Sieńko and Sebastian Piłsyk

8 Production of Feruloyl Esterases by *Aspergillus* Species 129

Miia R. Mäkelä, Luis Alexis Jiménez Barboza, Ronald P. de Vries and Kristiina S. Hildén

9 Secondary Metabolite Formation by the Filamentous Fungus *Penicillium chrysogenum* in the Post-genomic Era 145

Marta M. Samol, Oleksandr Salo, Peter Lankhorst, Roel A.L. Bovenberg and Arnold J.M. Driessen

10 pH Modulation by Fungal Secreted Effecting Molecules: A Mechanism Affecting Pathogenicity and Mycotoxin Accumulation During Colonization by *Penicillium expansum* 173

Dov Prusky, Shiri Barad, Nofar Glam, Nancy Keller and Amir Sherman

11 Evolutionary Adaptation as a Tool to Generate Targeted Mutant Strains as Evidenced by Increased Inulinase Production in *Aspergillus oryzae* 189

Helena Culleton, Eline Majoor, Vincent A. McKie and Ronald P. de Vries

Index 197

Contributors

Mikael Rørdam Andersen
Department of Systems Biology
Technical University of Denmark
Kongens Lyngby
Denmark

mr@bio.dtu.dk

Scott E. Baker
US Department of Energy Environmental Molecular Sciences Laboratory
Pacific Northwest National Laboratory; and
US Department of Energy Joint BioEnergy Institute
Richland, WA
USA

scott.baker@pnnl.gov

Shiri Barad
Department of Postharvest Science of Fresh Produce
Agricultural Research Organization (ARO)
Volcani Center
Bet Dagan
Israel

barad.shiri@gmail.com

Luis Alexis Jiménez Barboza
Department of Food and Environmental Sciences
Division of Microbiology and Biotechnology
Viikki Biocenter 1
University of Helsinki
Helsinki
Finland

luajimbar@gmail.com

Joan W. Bennett
Department of Plant Biology and Pathology
Rutgers University
New Brunswick, NJ
USA

profmycogirl@yahoo.com

Isabelle Benoit Gelber
Fungal Physiology
CBS-KNAW Fungal Biodiversity Centre; and
Fungal Molecular Physiology
Utrecht University
Utrecht
The Netherlands

i.benoit@cbs.knaw.nl

Roel A.L. Bovenberg
Synthetic Biology and Cell Engineering
Groningen Biomolecular Sciences and Biotechnology Institute
University of Groningen
Groningen; and
DSM Biotechnology Centre
Delft
The Netherlands

roel.bovenberg@dsm.com

Antoine Branca
Laboratory of Ecology, Systematics and Evolution (ESE)
CNRS-UMR8079
Paris-Sud University
Orsay
France

antoine.branca@u-psud.fr

Julian Brandl
Department of Systems Biology
Technical University of Denmark
Kongens Lyngby
Denmark

jubra@bio.dtu.dk

Erin L. Bredeweg
US Department of Energy Environmental Molecular Sciences Laboratory
Pacific Northwest National Laboratory
Richland, WA
USA

erin.bredeweg@pnnl.gov

Jerzy Brzywczy
Institute of Biochemistry and Biophysics
Polish Academy of Sciences
Warsaw
Poland

jurek.ibb.waw.pl

Anne-Marie Burggraaf
Molecular Microbiology and Biotechnology
Institute of Biology Leiden
Leiden University
Leiden
The Netherlands

m.a.burggraaf@outlook.com

Helena Culleton
Fungal Molecular Physiology
Utrecht University; and
Fungal Physiology
CBS-KNAW Fungal Biodiversity Centre
Utrecht
The Netherlands; and
Megazyme International Ireland
Bray
Ireland

helena@megazyme.com

Arnold J.M. Driessen
Molecular Microbiology
Groningen Biomolecular Sciences and Biotechnology Institute
University of Groningen
Groningen; and
Kluyver Centre for Genomics of Industrial Fermentations
Delft
The Netherlands

a.j.m.driessen@rug.nl

Emilie Dumas
Laboratory of Ecology, Systematics and Evolution (ESE)
CNRS-UMR8079
Paris-Sud University
Orsay
France

emilie.dumas@u-psud.fr

Joëlle Dupont
Institute of Systematics, Evolution, Biodiversity
UMR7205 CNRS-MNHN-UPMC-EPH
Sorbonne University
Paris
France

jdupont@mnhn.fr

Tamás Emri
Department of Biotechnology and Microbiology
Faculty of Science and Technology
University of Debrecen
Debrecen
Hungary

emri.tamas@science.unideb.hu

Tatiana Giraud
Laboratory of Ecology, Systematics and Evolution (ESE)
CNRS-UMR8079
Paris-Sud University
Orsay
France

tatiana.giraud@u-psud.fr

Nofar Glam
Department of Postharvest Science of Fresh Produce
Agricultural Research Organization (ARO)
Volcani Center
Bet Dagan
Israel

nofar.glam@mail.huji.ac.il

Kristiina S. Hildén
Department of Food and Environmental Sciences
Division of Microbiology and Biotechnology
Viikki Biocenter 1
University of Helsinki
Helsinki
Finland

kristiina.s.hilden@helsinki.fi

Jos Houbraken
CBS-KNAW Fungal Biodiversity Centre
Utrecht
The Netherlands

j.houbraken@cbs.knaw.nl

Wayne M. Jurick II
USDA-ARS
Food Quality Laboratory
Beltsville, MD
USA

wayne.jurick@ars.usda.gov

Nancy Keller
Department of Medical Microbiology and Immunology and Bacteriology
University of Wisconsin
Madison, WI
USA

npkeller@wisc.edu

Peter Lankhorst
DSM Biotechnology Centre
Delft
The Netherlands

peter.lankhorst@dsm.com

Manuela López-Villavicencio
Institute of Systematics, Evolution, Biodiversity
UMR7205 CNRS-MNHN-UPMC-EPH
Sorbonne University
Paris
France

mlopez@mnhn.fr

Vincent A. McKie
Megazyme International Ireland
Bray
Ireland

vincent@megazyme.com

Eline Majoor
Fungal Molecular Physiology
Utrecht University; and
Fungal Physiology
CBS-KNAW Fungal Biodiversity Centre
Utrecht
The Netherlands

eline_majoor@hotmail.com

Miia R. Mäkelä
Department of Food and Environmental Sciences
Division of Microbiology and Biotechnology
Viikki Biocenter 1
University of Helsinki
Helsinki
Finland

miia.r.makela@helsinki.fi

Jolanda M. van Munster
School of Life Sciences
University of Nottingham
Nottingham
UK

jolanda.van_munster@nottingham.ac.uk

Jane L. Nybo
Department of Systems Biology
Technical University of Denmark
Kongens Lyngby
Denmark

jlnr@bio.dtu.dk

Andrzej Paszewski
Institute of Biochemistry and Biophysics
Polish Academy of Sciences
Warsaw
Poland

apasz@ibb.waw.pl

Sebastian Piłsyk
Institute of Biochemistry and Biophysics
Polish Academy of Sciences
Warsaw
Poland

seba@ibb.waw.pl

István Pócsi
Department of Biotechnology and Microbiology
Faculty of Science and Technology
University of Debrecen
Debrecen
Hungary

pocsi.istvan@science.unideb.hu

Dov Prusky
Department of Postharvest Science of Fresh Produce
Agricultural Research Organization (ARO)
Volcani Center
Bet Dagan
Israel

dovprusk@volcani.agri.gov.il

Arthur F.J. Ram
Molecular Microbiology and Biotechnology
Institute of Biology Leiden
Leiden University
Leiden
The Netherlands

a.f.j.ram@biology.leidenuniv.nl

Ricardo C. Rodríguez de la Vega
Laboratory of Ecology, Systematics and Evolution (ESE)
CNRS-UMR8079
Paris-Sud University
Orsay
France

ricardo.rodriguezdelavega@gmail.com

Jeanne Ropars
Laboratory of Ecology, Systematics and Evolution (ESE)
CNRS-UMR8079
Paris-Sud University
Orsay
France

jeanne.ropars@gmail.com

Oleksandr Salo
Molecular Microbiology
Groningen Biomolecular Sciences and Biotechnology Institute
University of Groningen
Groningen
The Netherlands

o.v.salo@rug.nl

Marta M. Samol
Molecular Microbiology
Groningen Biomolecular Sciences and Biotechnology Institute
University of Groningen
Groningen
The Netherlands

marta.m.samol@gmail.com

Robert A. Samson
CBS-KNAW Fungal Biodiversity Centre
Utrecht
The Netherlands

r.samson@cbs.knaw.nl

Amir Sherman
Department of Postharvest Science of Fresh Produce
Agricultural Research Organization (ARO)
Volcani Center
Bet Dagan
Israel

asherman@volcani.agri.gov.il

Marzena Sieńko
Institute of Biochemistry and Biophysics
Polish Academy of Sciences
Warsaw
Poland

marsi@ibb.waw.pl

Dominique Swennen
INRA; and
AgroParisTech
UMR1319 Micalis
Thiverval-Grignon
France

dominique.swennen@grignon.inra.fr

Melinda Szilágyi
Department of Human Genetics
Faculty of Medicine
University of Debrecen
Debrecen
Hungary

szilagyi.melinda@med.unideb.hu

Sebastian Theobald
Department of Systems Biology
Technical University of Denmark
Kongens Lyngby
Denmark

setd@bio.dtu.dk

Tammi C. Vesth
Department of Systems Biology
Technical University of Denmark
Kongens Lyngby
Denmark

tcve@bio.dtu.dk

Ronald P. de Vries
Fungal Physiology
CBS-KNAW Fungal Biodiversity Centre; and
Fungal Molecular Physiology
Utrecht University
Utrecht
The Netherlands

r.devries@cbs.knaw.nl

Neriman Yilmaz
CBS-KNAW Fungal Biodiversity Centre
Utrecht
The Netherlands; and
Biodiversity (Mycology and Microbiology)
Agriculture and Agri-Food Canada
Ottawa, ON
Canada

nerimanyilmaz82@gmail.com

Jiujiang Yu
USDA-ARS
Food Quality Laboratory
Beltsville, MD
USA

jiujiang.yu@ars.usda.gov

Current Books of Interest

Chloroplasts: Current Research and Future Trends	2016
Microbial Biodegradation: From Omics to Function and Application	2016
Influenza: Current Research	2016
MALDI-TOF Mass Spectrometry in Microbiology	2016
The Bacteriocins: Current Knowledge and Future Prospects	2016
Omics in Plant Disease Resistance	2016
Acidophiles: Life in Extremely Acidic Environments	2016
Climate Change and Microbial Ecology: Current Research and Future Trends	2016
Biofilms in Bioremediation: Current Research and Emerging Technologies	2016
Microalgae: Current Research and Applications	2016
Gas Plasma Sterilization in Microbiology: Theory, Applications, Pitfalls and New Perspectives	2016
Virus Evolution: Current Research and Future Directions	2016
Arboviruses: Molecular Biology, Evolution and Control	2016
Shigella: Molecular and Cellular Biology	2016
Aquatic Biofilms: Ecology, Water Quality and Wastewater Treatment	2016
Alphaviruses: Current Biology	2016
Thermophilic Microorganisms	2015
Flow Cytometry in Microbiology: Technology and Applications	2015
Probiotics and Prebiotics: Current Research and Future Trends	2015
Epigenetics: Current Research and Emerging Trends	2015
Corynebacterium glutamicum: From Systems Biology to Biotechnological Applications	2015
Advanced Vaccine Research Methods for the Decade of Vaccines	2015
Antifungals: From Genomics to Resistance and the Development of Novel Agents	2015
Bacteria-Plant Interactions: Advanced Research and Future Trends	2015
Aeromonas	2015
Antibiotics: Current Innovations and Future Trends	2015
Leishmania: Current Biology and Control	2015
Acanthamoeba: Biology and Pathogenesis (2nd edition)	2015
Microarrays: Current Technology, Innovations and Applications	2014

Full details at www.caister.com

Preface

The ascomycete genera *Aspergillus* and *Penicillium* are among the most widely studied filamentous fungi. They have had a profound impact on human society and continue to do so. The antibiotic penicillin, produced by *Penicillium rubens*, has been instrumental in treating patients with infections during and after World War II and *Penicillium* species are still actively mined for novel antibiotics. *Aspergillus* species are better known for the industrial production of enzymes and metabolites (e.g. organic acids) that have found applications in various industrial sectors, such as food & feed, paper & pulp, textiles, beverages, wine, beer, pharmaceuticals, and biofuels & biochemicals. Several *Aspergillus* species also have a darker side, being able to cause invasive infections in immune-compromised patients, leading still to a high mortality rate.

Genome sequencing has affected studies into the biology of all classes of organisms and this is certainly true for filamentous fungi. The level with which biological systems can be studied since the availability of genomes and postgenomic technologies is beyond what most people could have imagined previously. *Aspergillus* and *Penicillium* are at the forefront of fungal genomics with many genome sequences available and a whole genus genome sequencing project in progress for *Aspergillus*. Genomic and post-genomic analysis has both broadened and changed our understanding of fungal biology, in particular with respect to the complexity of fungal biology and the high diversity amongst fungal species.

In this book we showcase the impact of genomics on studies in *Aspergillus* and *Penicillium*. The book starts with an overview of the taxonomy these two genera and its sister genus *Talaromyces*, which has recently been re-visited and updated. It then continues with three chapters on techniques currently used in relation to genomics.

Next several chapters present a genomic look on a variety of biological processes in these fungi: pathogenicity, carbon starvation, sulfur metabolism, feruloyl esterases, secondary metabolism and pH modulation. The final chapter presents a novel approach to generating targeted mutants that in combination with genomics can help to gain more insights into the mechanism underlying enzyme production.

While this book only provides a small selection of the broad range of topics that are actively studied in these fungi by many scientists around the world, its contents provide a clear picture of the influence of genomics on fungal biology. It also demonstrates the areas that require further development and as such can be a reference not only for scientists working with these genera but for fungal biology as a whole.

We are very grateful for the many authors who contributed to the book and provided us with high quality chapters sharing their considerable expertise. It is those contributions that provide the quality to this book and we enjoyed our interaction with them very much.

Finally, we would also like to thank Annette Griffin and the other staff at Caister Academic Press, who have made this a smooth and efficient process and were also available for any queries we have had.

Ronald P. de Vries
Isabelle Benoit Gelber
Mikael Rørdam Andersen

Taxonomy of *Aspergillus*, *Penicillium* and *Talaromyces* and its Significance for Biotechnology

1

Jos Houbraken, Robert A. Samson and Neriman Yilmaz

Abstract

Various fungi are used in biotechnology for their ability to produce a variety of small molecules and enzymes. The order *Eurotiales* contains the species-rich genera *Aspergillus, Penicillium* and *Talaromyces* and some species belonging to those genera are utilized for biotechnology. Application of the single name nomenclature has led to numerous name changes for many fungi. This chapter will provide an overview of important name changes for the genera *Aspergillus, Penicillium, Talaromyces* and other related genera. The number of newly described species has also increased significantly in the last decade. A sequence-based approach is currently recommended to correctly identify species of these genera. This chapter will also provide an overview of molecular identification techniques for isolates belonging to these genera.

Introduction

The genera *Aspergillus, Penicillium* and *Talaromyces* are phylogenetically related, belonging to the order *Eurotiales*. The species belonging to these genera are commonly encountered microfungi with diverse physiological properties. Some species grow at low (psychrotolerant) and others at high (thermophilic) temperatures, low pH and/or low oxygen levels, and others can grow at low water activities. Many species are well known and can have a positive and/or negative impact (Houbraken and Samson, 2011). Negative impacts include the spoilage of foods and feed, and the production of mycotoxins (e.g. patulin, aflatoxins) in those products. Additionally, some species are opportunistic human pathogens (e.g. *Aspergillus fumigatus, Aspergillus flavus, Aspergillus terreus,* and *Talaromyces* (= *Penicillium*) *marneffei*) (De Hoog *et al.*, 2000) and some species can cause allergies in individuals following exposure to fragments of those micro-organisms. On a positive side, some species are used in food fermentation processes, in biotechnology for the production of enzymes, organic acids, and for the manufacturing of drugs. Examples include *Penicillium roqueforti* and *Penicillium camemberti,* which are used in the production of Roquefort and Camembert cheese respectively, *Aspergillus oryzae,* which is used to produce soy sauce, and *Penicillium rubens,* which is used to produce penicillin (Houbraken *et al.*, 2011b; Samson *et al.*, 2010).

As these species can have both positive and negative impacts, their correct identification to species level is an important process in a number of fields, including food and medical mycology and biotechnology. Ideally, identification should be simple, accurate, unequivocal and immutable. Until recently, fungal identification was primarily based on phenotypic characteristics however this method is difficult for most researchers. Sequencing of specific genes for species identification is now commonly applied. In the last decade, these new techniques have helped re-classify well-known species into species complexes. Correct identification is functionally important as many species belonging to these complexes have different and unique properties. Examples of such properties include the ability to produce certain enzymes and/or enzyme profiles, mycotoxins and higher resistance to certain anti-fungal pesticides and therapeutics. An example of this are species belonging to *Aspergillus* sect. *Nigri*. Species in this section have been the subject of various taxonomic studies (Hong *et al.*, 2013; Jurjević *et al.*, 2012; Varga *et al.*, 2011) and 26 species are currently accepted (Houbraken *et al.*, 2014). These species are difficult to separate based on phenotypic characteristics; however, their correct species identification is required as each species in this section can have different properties. An example of this is their ability to form different mycotoxins. *Aspergillus niger*, *Aspergillus welwitschiae* and *Aspergillus carbonarius* can produce ochratoxin A, while *A. niger* and *A. welwitschiae* produce fumonisins. Furthermore, Meijer *et al.* (2011) showed that most species of the section *Nigri* produce specific enzyme profiles. Therefore a sequence based identification approach is recommended to identify species. The choice of the target gene and use of reference databases for identification are also crucial for correct species identification.

In this chapter an overview of the current phylogenetic relationships between *Aspergillus*, *Penicillium*, *Talaromyces* and related genera is presented. The changes over time to the classification schemes of *Aspergillus*, *Penicillium* and *Talaromyces* species are also given. The recent changes in nomenclature (single versus dual nomenclature), the discovery of new species and the elucidation of species complexes led to an enormous increase of species (names). The impact of these changes is also discussed.

Phylogeny of *Eurotiales*

The order *Eurotiales* is divided into the families *Aspergillaceae*, *Thermoascaceae* and *Trichocomaceae*. The *Elaphomycetaceae* probably represents a fourth family in this order and this family is phylogenetically most closely related to the *Trichocomaceae* (Quandt *et al.*, 2015) (Fig. 1.1). Most members of these families produce anamorphs with complex, branched conidiophores with monophialides and conidia borne in chains.

The *Aspergillaceae* are characterized by the production of asci in cleistothecia or surrounded by Hülle cells, and mainly have ascospores with a furrow or slit. This family contains two species-rich genera, namely *Aspergillus* and *Penicillium*. These genera are, together with *Monascus*, well known for their application in biotechnology and/or food fermentations. The position of *Monascus* (and the *Monascaceae*) has been the subject of discussion in various papers and it was often placed outside the order *Eurotiales*. The anamorph of *Monascus* differs from most other members of the order *Eurotiales*. Conidiogenous structures are Basipetospora-like (producing aleurioconidia from undifferentiated conidiogenous cells) and the conidia resemble chlamydospores. Using a four-gene phylogeny, Houbraken and Samson (2011) placed this genus in the *Aspergillaeceae*. Other genera in this family are *Hamigera*, *Leiothecium*, *Penicilliopsis*, *Phialomyces*, *Sclerocleista*, *Warcupiella*, *Xerochrysium* and *Xeromyces* (Fig. 1.1).

The *Trichocomaceae* as defined by Malloch (1985a,b) largely corresponds with Houbraken and Samson's (2011) circumscription of the *Aspergillaceae* and *Trichocomaceae*.

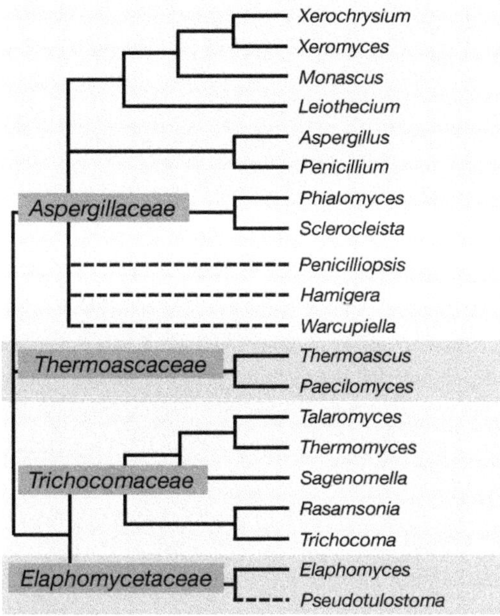

Figure 1.1 Phylogram based on the current state of knowledge regarding phylogenetic relationships within *Eurotiales*. Genera of uncertain or poorly supported phylogenetic positions are represented as dashed lines. The phylogram is based on the studies of Geiser *et al*. (2015), Houbraken and Samson (2011), Pitt *et al*. (2013) and Quandt *et al*. (2015).

Houbraken and Samson (2011) characterized the *Trichocomaceae* as having asci borne in a tuft or layer of loose hyphae, and ascospores lacking slits or furrows. This family includes the genera *Rasamsonia, Sagenomella, Talaromyces, Thermomyces* and *Trichocoma*. Several *Talaromyces* species (e.g. *T. funiculosus, T. pinophilus*) are known as cellulose producers and others are able to produce azaphilone polyketide pigments. The genera *Rasamsonia* and *Thermomyces* contain thermophilic species (e.g. *Rasamsonia emersonii, Thermomyces dupontii, Thermomyces lanuginosus*) and are also used in biotechnology.

The genera *Byssochlamys* and *Thermoascus* belong in the family *Thermoascaceae*. The relationship between these two genera is moderately supported in the study of Houbraken and Samson (2011) (0.99 posterior probability (pp); 67% bootstrap support (bs)). These genera are also phenotypically distinct. *Byssochlamys* produces naked ascomata and *Thermoascus* forms well-defined firm ascomata. However, they share similar characteristics such as the production of asci in croziers and the formation of smooth or finely roughened ascospores lacking a furrow or slit (Houbraken and Samson, 2011).

The genera *Elaphomyces* and *Pseudotulostoma* are positioned in the *Elaphomycetaceae*. These are the only genera in the *Eurotiales* that include ectomycorrhizal fungi. Interestingly, *Elaphomycetaceae* represents one of the few independent origins of the mycorrhizal symbiosis in *Ascomycota* (Tedersoo *et al.*, 2010). *Elaphomyces* species produce subglobose, hypogeous 'truffle' fruiting bodies, which have an organized outer layer of tissue (peridium) that encloses the gleba or spore-bearing tissue (Trappe, 1979). The evolutionary relationship of *Elaphomyces* and *Pseudotulostoma* within the *Ascomycota* was unclear for a long time, but genome analysis reliably placed these genera in *Eurotiales* (Quandt *et al.*, 2015). An anamorph for *Elaphomyces* or *Pseudotulostoma* has not been observed, and *Elaphomyces* grows only very slowly in culture (Miller and Miller 1984). One uniting character of the *Elaphomycetaceae* with other families of the *Eurotiales* is the production of cleistothecia (completely enclosed ascomata), although there are a few exceptions (e.g. *Trichocoma* spp.) (Quandt *et al.*, 2015).

Species and species concepts

Many *Aspergillus* and *Penicillium* species were described in the nineteenth century, and most of them remain unidentifiable today. *Aspergillus candidus, Aspergillus glaucus, Aspergillus flavus* and *Penicillium expansum* are the oldest currently accepted species in *Aspergillus* and *Penicillium* and these species were introduced by Link in 1809. *Talaromyces* is younger and the history of this genus starts with Benjamin (1955), who introduced it as a teleomorphic genus for *Penicillium* species.

Traditional taxonomic studies primarily relied on phenotypic characteristics (Pitt 1980; Raper and Fennell, 1965; Raper and Thom, 1949). Thom (1910), Biourge (1923) and Thom and Church (1926) made great contributions to *Aspergillus* and *Penicillium* taxonomy. In these classical monographs overviews of the accepted species of that time were given and identification keys were provided. These monographs were also of importance as they were using standardized media for species comparisons. Although the types of media used have since changed, comparison of growth patterns on standard media is still used for present day mycology. Phenotype based classifications were the standard in *Aspergillus, Penicillium* and *Talaromyces* taxonomy until the early 1990s.

During the 1990s new classification schemes evolved from the traditional morphology based schemes to include data from extrolite profiles, ubiquinones and DNA sequence data (Frisvad *et al.*, 1990; Kuraishi *et al.*, 1991; Yaguchi *et al.*, 1996). Colwell (1970) introduced the term 'polyphasic taxonomy' to reflect these additional classification characteristics for her taxonomic description of the prokaryote *Vibrio*. The aim of polyphasic taxonomy is to integrate different kinds of data and information on microorganisms, such as phenotypic, genotypic and phylogenetic data, to form a consensus type of taxonomy. The polyphasic approach has become standard in *Aspergillus, Penicillium* and *Talaromyces* taxonomy (Boysen *et al.*, 1996; Christensen *et al.*, 2000; Frisvad and Samson, 2004; Houbraken *et al.*, 2011a, 2014; Varga *et al.*, 2011). Characteristics commonly incorporated in a polyphasic approach are phenotypic (micro- and macromorphology) and physiological (e.g. growth on different media at different temperatures and water activities), extrolite profiles and multigene phylogenies.

The extrolite profiles are a key component in the polyphasic species concept. In general, members of the *Eurotiales* produce a large number of (bioactive) extrolites (secondary metabolites), including several mycotoxins. Many studies have shown that extrolite profiles are species-specific and usually consistently expressed in different isolates of the same species, regardless of geographic origin or habitat (Frisvad and Filtenborg, 1989). Samson *et al.* (2011) compared extrolites of *Aspergillus, Penicillium* and *Talaromyces* and found that *Aspergillus* and *Penicillium* share 91 biosynthetic families, *Penicillium* and *Talaromyces* share 9, and *Aspergillus* and *Talaromyces* share 11, demonstrating that *Talaromyces* is more distant from these two genera.

The first phylogenetic studies with sequence data date back from the late 1980s, using rRNA sequences (e.g. Dupont *et al.*, 1990; Edman *et al.*, 1988; Logrieco *et al.*, 1990). Taylor *et al.* (2000) endorsed the genealogical concordance phylogenetic species recognition (GCPSR) as the method to identify the species, which incorporates the analyses of multiple gene sequences for species delineation. Currently, GCPSR analysis is commonly used to determine species limited in *Aspergillus, Penicillium* and *Talaromyces.* Usually two- to four loci are analysed and the most commonly used genes are ITS, beta tubulin (*BenA*), calmodulin (*CaM*), and RNA polymerase II second largest subunit (*RPB2*). Other genes such as RNA polymerase II largest subunit (*RPB1*), actin (*Act*) or translation elongation factor 1 alpha (*TEF1a*) are also used, but their application in GCPSR analysis is hampered by a low number of reference sequences in public databases.

Single name nomenclature and its impact on applied mycology

The International Code of Nomenclature for Algae, Fungi and Plants (ICN) governs the naming of fungi (McNeill *et al.*, 2012). For a long time, dual nomenclature (Art. 59) was used for pleomorphic fungi, meaning that one species can have a name for the sexual morph and another for the asexual morph, with priority given to the sexual morph name. After its introduction, dual nomenclature was not immediately applied to *Aspergillus* and *Penicillium* taxonomy. Major taxonomic treatments of these genera (e.g. Raper and Fennell, 1965; Raper and Thom, 1949; Thom and Raper, 1945) ignored this rule and instead used a single name nomenclature. In those monographs the generic names *Aspergillus* and *Penicillium* were applied irrespective of their reproductive structures. In the early 1970s, the single name nomenclature was less commonly used and dual nomenclature was also accepted for *Aspergillus* and *Penicillium* taxonomy. As a consequence, teleomorph genera were revised or established and a new combination was proposed (e.g. Malloch and Cain, 1972; Pitt, 1980; Stolk and Samson, 1972; Subramanian, 1972). Dual nomenclature was commonly applied for approximately four decades until the recent changes in the ICN. One of the most significant changes to the ICN was the abandonment of Art. 59 leading the way for a move to single name nomenclature (McNeil *et al.*, 2012; Norvell 2011), allowing one name for one species even when they display different morphs. Application of the single name nomenclature and recent changes in *Aspergillus* and *Penicillium* taxonomy led to numerous name changes. An overview of important changes is given for the genera *Aspergillus, Penicillium, Talaromyces* and related genera is given in this chapter.

Identification

DNA sequence data quickly changed the way fungal taxonomy was approached. Today it surpasses morphology as the main criterion for characterizing species. DNA sequencing became one of the most important tools for species identification and the DNA barcoding initiative is based on this principle (Blaxter 2003; Hebert *et al.*, 2003). The ITS rDNA region was accepted as the official DNA barcode for fungi (Schoch *et al.*, 2012). It was chosen because of the universal primer sets available for PCR amplification and it is the most widely sequenced gene for fungi. However this barcode has its limitations and is unable to distinguish all species, especially in genera such as *Aspergillus, Penicillium* and *Talaromyces.* As a result, a secondary marker is needed for species identification. The identification of a universal secondary marker is under investigation (Stielow *et al.*, 2015). However, this is a difficult task as different genes have variable performance between genera. Currently, each genus has a specific secondary marker. For isolate identification purposes, this means that ITS is able to place a strain into a species complex, sometimes giving a correct identification, and that a secondary marker will then be used for a final identification. On the other hand, if the genus is already known (e.g. by a phenotype based identification), then the secondary marker can be sequenced directly. For reliable species identification, sequencing of the calmodulin gene is recommended for *Aspergillus*, and partial beta-tubulin gene sequencing is preferred in *Penicillium* and *Talaromyces* (Samson *et al.*, 2010, 2014; Visagie *et al.*, 2014a; Yilmaz *et al.*, 2014).

Even though many sequence projects have been undertaken to barcode fungi, very few comprehensive data sets are available due the large volume of taxonomic work necessary to label barcodes with the correct name. In addition, even though barcoding is not the same as taxonomy, its success is reliant on well-defined taxonomies and thus reliable reference sequences often originate from taxonomic studies. A high-quality reference sequence database is important for correct species identification. Initiatives such as RefSeq have been launched (Schoch *et al.*, 2014) to address these issues. Specific lists with accepted *Aspergillus*, *Penicillium* and *Talaromyces* species linked to GenBank accession numbers of ITS, *BenA*, *CaM* and *RPB2* sequences are also published, and can serve as a starting point to develop curated identification databases. These lists might contribute to solving problems concerning identification of biotechnologically important strains. Species names that are no longer taxonomically correct will hopefully not be used in the future. Examples of incorrect species names and their current, correct names are listed in Table 1.1.

New species, new names

On 25 August 2015, 344 *Aspergillus*, 363 *Penicillium* and 98 *Talaromyces* species were accepted (J. Houbraken, CBS database), resulting in a total number of 805 species. The number of accepted species in these genera has dramatically increased in the last decade. Fig. 1.2 shows the cumulative number of accepted species plotted against the year of publication. This line can be divided into three parts, reflecting three eras: 1809–1900, 1900–2004 and 2004–present. The increase of the cumulative number of current accepted species from the year 1809 (four species) until 1900 (18 species) is low. Many new *Aspergillus* and *Penicillium* species (> 175) were described in that time period, although only 18 of those are currently in use. Material of some species that were described in that period is present in (European) herbaria. Old herbarium specimens that are non-viable and unrecognizable using traditional techniques might still contain DNA of analysable quality. Analysis of this material could show that these species also represent currently accepted species. It is undesirable to revive these old names in case they predate current (well-known) species. This will lead to confusion and may necessitate giving up well-characterized and well-known names (Pitt and Samson, 2007).

From 1900 onwards, the number of currently accepted *Aspergillus* and *Penicillium* species steadily increased. This can be attributed to two main factors. Firstly, living cultures were distributed as type material within the mycological community, making it possible for taxonomists during this period (and still today) to study these species. Secondly, *Penicillium* (and to a lesser extent *Aspergillus*) attracted much attention due to its economic significance.

Table 1.1 Overview of invalid names used in biotechnological literature and their current correct name

Incorrect name	Correct name	Reference
Acremonium cellulolyticus	*Talaromyces pinophilus*	Yilmaz *et al*., 2014
Aspergillus acidus	*Aspergillus luchuensis*	Hong *et al*., 2013
Aspergillus awamori sensu Nakazawa	Unknown identity; either *A. niger* or *A. luchuensis*	Yamada *et al*., 2011; Hong *et al*., 2013; Houbraken *et al*., 2014
Aspergillus awamori sensu Perrone *et al*. (2011)	*Aspergillus welwitschiae*	Hong *et al*., 2013
Aspergillus citricus	*Aspergillus niger*	Frisvad *et al*., 2011
Aspergillus ficuum	*Aspergillus welwitschiae*	Houbraken *et al*., 2014
Aspergillus foetidus	*Aspergillus niger*	Varga *et al*., 2011a
Aspergillus foetidus var. *acidus*	*Aspergillus luchuensis*	Hong *et al*., 2013
Aspergillus kawachii nom. inval.	*Aspergillus luchuensis*	Hong *et al*., 2013
Aspergillus phoenicis	Representative strains of *A. phoenicis* belong to *A. niger* and *A. tubingensis*	Frisvad *et al*., 1990; Kozakiewicz *et al*., 1992; Houbraken *et al*., 2014
Aspergillus saitoi nom. inval.	*Aspergillus tubingensis*	Houbraken *et al*., 2014
Aspergillus usamii	*Aspergillus niger*	Yamada *et al*., 2011
Aspergillus violaceofuscus	Exact identity of this species remains unsolved	Houbraken *et al*., 2014
Emericella nidulans	*Aspergillus nidulans*	Samson *et al*., 2014
Geosmithia argillacea	*Rasamsonia argillacea*	Houbraken *et al*., 2012
Penicillium funiculosum	*Talaromyces funiculosus*	Samson *et al*., 2011
Penicillium griseoroseum	*Penicillium chrysogenum* sensu lato	Houbraken *et al*., 2012
Penicillium purpurogenum	*Talaromyces purpurogenus*	Samson *et al*., 2011
Talaromyces emersonii	*Rasamsonia emersonii*	Houbraken *et al*., 2012
Talaromyces thermophilus	*Thermomyces dupontii*	Houbraken *et al*., 2014

This is marked by the pioneering studies of Dierckx (1901), Thom (1910), Westling (1911), Biourge (1923), Zaleski (1927) and Thom (1930). Many of the species described in these studies are still accepted today (Fig. 1.2).

In 2003, the total number of currently accepted species was 459, which is an average increase of 4.2 species per year from 1900 to 2004. This average increased to 28.8 species per year in the time period 2004 to 2015. This can be attributed to several reasons. Firstly, the use of Sanger sequencing was introduced in many laboratories world-wide. Phenotype based classifications require a high level of expertise but sequence based approaches enabled a broader scientific community to study relationships and describe new species. Sequencing clarified relationships among *Aspergillus, Penicillium, Talaromyces* and related genera. This resulted in modified generic descriptions that also included deviating structures (Houbraken and Samson, 2011; Samson *et al.*, 2014; Yilmaz *et al.*, 2014). Genera such as *Thysanophora, Torulomyces* and *Eladia* were synonymized with *Penicillium,* and *Phialosimplex*

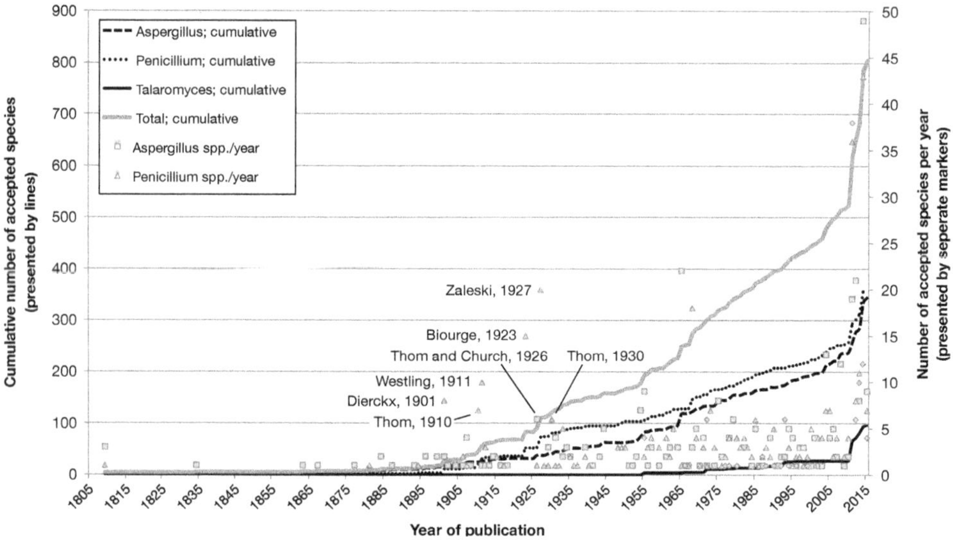

Figure 1.2 Overview of the number of currently accepted *Aspergillus*, *Penicillium* and *Talaromyces* species plotted against the year of publication. This graph is based on the studies of Samson *et al.* (2014), Yilmaz *et al.* (2014), and Visagie *et al.* (2014), and is supplemented with the most recent insights on the taxonomy of those genera.

and *Polypaecilum* with *Aspergillus* (Houbraken and Samson, 2011; Samson *et al.*, 2014). Furthermore, species complexes were unravelled. Species that are phenotypically well-defined proved to represent a number of species that are genetically and evolutionary distinct. Another main factor was the discontinuation of dual nomenclature and the introduction of the single name nomenclature system. These changes had the largest impact in *Talaromyces*, followed by *Aspergillus* and *Penicillium*. In 2011, Samson *et al.* redefined the genus *Talaromyces* and included anamorph and teleomorph characters. In total, 38 species previously belonging to *Penicillium* were combined in *Talaromyces*. Another five anamorphic species belonging to various other genera (*Geosmithia, Paecilomyces, Sagenomella*) were transferred to the redefined genus *Talaromyces* (Samson *et al.*, 2011; Yilmaz *et al.*, 2014). In *Aspergillus*, 18 species known under their teleomorph name and belonging to the genera *Cristaspora, Neosartorya* and *Emericella* were combined in *Aspergillus* (Samson *et al.*, 2014), and five teleomorph species belonging to *Eupenicillium, Chromocleista, Hemicarpenteles,* were transferred to *Penicillium* (Houbraken *et al.*, 2010; Houbraken and Samson, 2011; Visagie *et al.*, 2014a)

The steep increase in the number of accepted *Aspergillus, Penicillium* and *Talaromyces* species in the last decade (Fig. 1.2) will probably continue into the next. Many new species are still being discovered and are waiting to be described. Biodiversity studies on specific substrates or habitats also revealed the presence of many new species. For example, 18 new *Aspergillus, Penicillium* and *Talaromyces* species were described from house dust samples (Visagie *et al.*, 2014b) and many new species were and are yet to be described from soils from South Africa and marine environments in Korea (Park *et al.*, 2014; Visagie 2012). It is likely that when more biodiversity studies are performed, many more new *Aspergillus, Penicillium* and *Talaromyces* species will be discovered.

Aspergillus, *Penicillium* and *Talaromyces* and related genera

Several genera related to *Aspergillus, Penicillium* and *Talaromyces* are of importance in biotechnology, food mycology and/or medical mycology. The most well-known are *Hamigera, Monascus, Rasamsonia* and *Thermomyces* (Fig. 1.1). Data on the taxonomy and occurrence of these genera are given below.

Aspergillus

The genus *Aspergillus* represents a highly diverse group of fungi, however, genetic studies tend to concentrate on four species: *A. nidulans, A. niger, A. oryzae* and *A. fumigatus*. The success of aspergilli is partly due to the fact that they can grow on a wide variety of substrates and under a wide variety of conditions. *Aspergillus* species have also been used in biotechnology for a long time. For example, the use of one of the most commonly utilized food additives, citric acid (E330) (produced by *A. niger*), dates back to 1917. Currently, four subgenera and 19 sections are accepted in *Aspergillus* and this classification largely corresponds with that of Raper and Fennell (1965), who divided the genus into 18 groups.

The conidiophore structure is an important microscopic character used in *Aspergillus* taxonomy. Most aspergilli produce conidiophores that consist of a foot cell at the base, long stipes (usually non-septate) that end in a swollen apex, called a vesicle. Vesicles are decorated with phialides (uniseriate) or with metulae and phialides (biseriate), which are borne simultaneously. Other characters used in phenotype based identification are the colour of the conidia, growth rate on agar media, and physiological characteristics (growth temperature, water activity).

The introduction of the single name nomenclature has had a great impact on *Aspergillus*. Two papers on the application of the single nomenclature in *Aspergillus* were published (Pitt and Taylor, 2014; Samson *et al.*, 2014). Samson *et al.* (2014) used a broad concept of *Aspergillus*, and the majority of the *Aspergillus* names, including their teleomorphs, were maintained under their *Aspergillus* name. This option had the highest support in the International Commission of *Penicillium* and *Aspergillus* (ICPA) taxonomy (Samson *et al.*, 2014). On the other hand, Pitt and Taylor (2014) advocated applying the sexual names to clades within *Aspergillus*. It was proposed to split *Aspergillus* in *Eurotium, Neosartorya, Emericella* and *Chaetosartorya* and keep the *Aspergillus* names for species belonging to section *Circumdati*. In this system the genus name would be more informative because physiological characteristics such as xerophily and thermophily would be linked to a genus name. However, Pitt and Taylor's (2014) proposal would imply that *Aspergillus* should have neotypification of *A. niger* in place of *A. glaucus*. A review of the most recent publications dealing with *Aspergillus* taxonomy show that the nomenclature as proposed by Samson *et al.* (2014) has received more acceptance than that of Pitt and Taylor (2014).

Hamigera

Stolk and Samson (1971) introduced *Hamigera* for *Talaromyces* species that produce single asci, limiting *Talaromyces* to species producing asci in chains (Stolk and Samson, 1972). Although Pitt (1980) considered *Hamigera* synonymous with *Talaromyces*, Houbraken and Samson (2011) showed that it is a distinct genus closely related to *Warcupiella*. In addition, Houbraken *et al.* (2012) showed that *T. leycettanus* is unrelated to *Talaromyces* and belongs in the *Hamigera/Warcupiella* clade, noting that the exact position of this species within the clade should be further investigated.

Monascus

The genus *Monascus* belongs to the family *Aspergillaceae* and is phylogenetically related to *Aspergillus* and *Penicillium*. Species belonging to this genus produce aleurioconidia from undifferentiated conidiogenous cells and can form stalked ascomata. These features set this genus apart from *Aspergillus* and *Penicillium*. *Monascus* species are used in fungal fermented foods and are of special interest because of their production of secondary metabolites. Red rice (ang-kak, angka, 'red kojic rice') is produced by *Monascus* species and is of interest for its health promoting effects. However, *Monascus* species can also produce the mycotoxin citrinin, and the presence of this mycotoxin in food should be avoided.

Penicillium

Penicillium species are commonly occurring and, similar to *Aspergillus*, have diverse properties. Several species are known as food spoilage agents; however, the majority of species are soil-borne. Some species are also applied in biotechnology, e.g. *Penicillium rubens* is used in the production of the antibiotic penicillin, *P. roqueforti* and *P. camemberti* in cheese fermentations and *P. brevicompactum* in the production of the immunosuppressant drug mycophenolic acid. The genus name *Penicillium* was introduced by Link (1809) and is derived from '*penicillus*', which is Latin for little brush. A '*penicillus*' consists of a well-defined cluster of phialides that are either attached to a stipe directly, or through one or more stages of branching (Pitt and Hocking, 2009). *Penicillium* is phylogenetically closely related to *Aspergillus*, but can easily be phenotypically differentiated. For example, the phialides are born successively and not simultaneously as in *Aspergillus*. Furthermore, *Aspergillus* conidiophores have a foot cell and often non-septate stipes, while penicillia lack a foot cell and have septated stipes.

In anticipation of the new International Code of Nomenclature of algae, fungi and plants (ICN), Houbraken and Samson (2011) redefined the genus *Penicillium*. Teleomorph genera, such as *Eupenicillium*, *Chromocleista* and *Hemicarpenteles*, are now included in the redefined genus. Additionally, the generic description of *Penicillium* was changed. Sexual states were included, but species with acerose phialides, usually with symmetrical branched conidiophores, were excluded. Species with the latter features are now included in *Talaromyces*.

Rasamsonia

The genus *Rasamsonia* was introduced for thermotolerant and thermophilic *Penicillium*-like species, which form a monophyletic clade distinct from *Penicillium* and *Talaromyces*. The biotechnologically important species formerly known as *Talaromyces emersonii* was renamed as *Rasamsonia emersonii* (Houbraken *et al.*, 2012). This species produces thermostable enzymes and is genome sequenced. Currently, this genus encompasses ten species: *R. aegroticola*, *R. argillacea* (syn. *Geosmithia argillacea*), *R. brevistipitata*, *R. byssochlamydoides* (syn. *Talaromyces byssochlamydoides*), *R. composticola*, *R. cylindrospora*, *R. eburnea* (syn. *Talaromyces eburneus*, *Geosmithia eburnea*), *R. emersonii* (syn. *Talaromyces emersonii*), *R. piperina* and *R. pulvericola*. Several *Rasamsonia* species are of medical importance and have been isolated from animals and humans, including *R. argillacea* (Doyon *et al.*, 2013; Giraud *et al.*, 2013; Salgüero *et al.*, 2013), *R. cylindrospora*, *R. aegroticola*, *R. eburnea* and *R. piperina* (Houbraken *et al.*, 2013).

Talaromyces

Talaromyces was initially described as a teleomorphic genus by Benjamin in 1955. It was introduced as a sexual morph for *Penicillium* species producing generally white or yellow, soft-walled ascomata. Up until the late 1980s the classification of *Talaromyces* was based on morphology, and at that time several anamorph genera such as *Geosmithia, Merimbla, Paecilomyces* and *Penicillium* (Pitt *et al.*, 2000) were linked to *Talaromyces*. However, sequence data showed that only a few species from these genera belong in the modern concept of *Talaromyces,* with the remaining taxa not related to *Talaromyces*.

The ability to produce enzymes and soluble pigments make *Talaromyces* an important genus in biotechnology. Both *T. funiculosus* and *T. pinophilus* are good cellulase producers and *T. rugulosus* is reported to produce inulinase (Brown *et al.*, 1987; Maeda *et al.*, 2011, 2013; Mansouri *et al.*, 2013; Mishra *et al.*, 1984; Visser *et al.*, 2013; Wood and McCrae, 1986). Furthermore, *Talaromyces flavus* is used as a bio-control agent of soil-borne pathogens such as *Verticillium dahliae, V. albo-atrum, Rhizoctonia solani* and *Sclerotinia sclerotiorum* (Brunner *et al.*, 2005; Gohel *et al.*, 2006; Marois *et al.*, 1984; Punja 2001). *Talaromyces* species generally produce yellow, orange or red pigments in their mycelia or diffusing pigments. Of the pigment-producing *Talaromyces* species, *T. atroroseus* is of interest as this species consistently produces the azaphilone biosynthetic families mitorubrins and *Monascus* red pigments without any mycotoxins (Frisvad *et al.*, 2013).

Thermomyces

Currently, two species are accepted in the genus *Thermomyces*: *Th. lanuginosus* and *Th. dupontii*. In literature, various synonyms are used. For example, *Humicola lanuginosa* is used for *Th. lanuginosus* (e.g. Bokhari *et al.*, 2008) and synonyms of *Th. dupontii* include *Talaromyces thermophilus, Talaromyces dupontii* and *Penicillium dupontii* (e.g. Niu *et al.*, 2014; Zhang *et al.*, 2015). *Thermomyces lanuginosus* and *Th. dupontii* are able to grow at high temperatures and are used for the production of enzymes (Houbraken *et al.*, 2014).

Conclusion

Owing to the application of single name nomenclature, *Aspergillus, Penicillium* and *Talaromyces* now contain sexual and asexual species. The generic descriptions of these genera were adjusted to incorporate these states and as a consequence new combinations were made. The number of new combinations or name changes is rather low because *Aspergillus, Penicillium* or *Talaromyces* names were for many species already available. Nevertheless, in some cases names of well-known species have changed e.g. *Talaromyces* (*Penicillium*) *marneffei, Talaromyces* (*Penicillium*) *funiculosus, Aspergillus* (*Neosartorya*) *fischeri* and *Aspergillus glaucus* (*Eurotium herbariorum*).

Previous classifications based on morphology alone resulted in numerous misidentifications. In the modern era, a polyphasic approach was adopted in taxonomic studies and resulted in more stable species definitions. With the steep increase in the number of newly described species and the changes in nomenclature, correct species identification might appear more difficult. However, accurate species identification is possible with the use of sequence data. However sequence based identification is still problematic, mainly due to the lack of curated databases for the many sequences published on *Aspergillus, Penicillium* and *Talaromyces*. Lists with accepted species that are linked to GenBank accession numbers for ITS, *BenA*, *CaM* and *RPB2* sequences have been compiled and published (Samson *et al.*, 2014; Yilmaz *et al.*, 2014; Visagie *et al.*, 2014a). These data sets well help the mycology community to correctly identify species in the future.

Acknowledgements

The authors thank Jessica Talbot and Ronald de Vries for their useful suggestions and proof-reading the manuscript.

References

Benjamin, C.R. (1955). Ascocarps of *Aspergillus* and *Penicillium*. Mycologia *47*, 669–687.

Biourge, P. (1923). Les moisissures de groupe *Penicillium* Link. Cellule *33*, 7–331.

Blaxter, M. (2003). Counting angels with DNA. Nature *421*, 1–3.

Bokhari, S.A., Latif, F., and Rajoka, M.I. (2008). Kinetics of high-level of β-glucosidase production by a 2-deoxyglucose-resistant mutant of *Humicola lanuginosa* in submerged fermentation. Braz. J. Microbiol. *39*, 724–733.

Boysen, M., Skouboe, P., Frisvad, J.C., and Rossen, L. (1996). Reclassification of the *Penicillium roqueforti* group into three species on the basis of molecular genetic and biochemical profiles. Microbiology *142*, 542–549.

Brown, J.A., Collin, S.A., and Wood, T.M. (1987). Enhanced enzyme production by the cellulolytic fungus *Penicillium pinophilum*, mutant strain NTG III/6. Enzyme Microb. Techn. *9*, 176–180.

Brunner, K., Zeilinger, S., Ciliento, R., Woo, S.L., Lorito, M., Kubicek, C.P., and Mach, R.L. (2005). Improvement of the fungal biocontrol agent *Trichoderma atroviride* to enhance both antagonism and induction of plant systemic disease resistance. Appl. Environm. Microb. *71*, 3959–3965.

Christensen, M., Frisvad, J.C., and Tuthill, D.E. (2000). *Penicillium* species diversity in soil and some taxonomic and ecological notes. In Integration of Modern Taxonomic Methods for *Penicillium* and *Aspergillus* Classification, R.A. Samson, and J.I. Pitt, eds. (Harwood Academic Publishers, Amsterdam, the Netherlands), pp. 309–320.

Colwell, R.R. (1970). Polyphasic taxonomy of the genus *Vibrio*: numerical taxonomy of *Vibrio cholerae*, *Vibrio parahaemolyticus*, and related *Vibrio* species. J. Bacteriol. *104*, 410–433.

De Hoog, G.S., Guarro, J., Gené, J., and Figueras, M.J. (2000). Atlas of clinical fungi, 2nd edn. (Centraalbureau voor Schimmelcultures, Utrecht, The Netherlands).

Dierckx, R.P. (1901). Un essai de revision du genre *Penicillium* Link. Note preliminaire. Ann. Soc. Sci. Brux. *25*, 83–89.

Doyon, J.B., Sutton, D.A., Theodore, P., Dhillon, G., Jones, K.D., Thompson, E.H., Fu, J., Wickes, B.L., Koehler, J.E., and Schwartz, B.S. (2013). *Rasamsonia argillacea* pulmonary and aortic graft infection in an immune-competent patient. J. Clin. Microb. *51*, 719–722.

Dupont, J.A. (1990). Molecular assessment of the position of *Stilbothamnium* in the genus *Aspergillus*. In Modern Concepts in *Penicillium* and *Aspergillus* Systematics, R.A. Samson, and J.I. Pitt, eds. (Plenum Press, New York, NY, USA), pp. 335–342.

Edman, J.C., Kovacs, J.A., Masur, H., Santi, D.V., Elwood, H.J., and Sogin, M.L. (1988). Ribosomal RNA sequence shows *Pneumocystis carinii* to be a member of the fungi. Nature *334*, 519–522.

Frisvad, J.C., and Filtenborg, O. (1989). Terverticillate Penicillia: chemotaxonomy and mycotoxin production. Mycologia *81*, 837–861.

Frisvad, J.C., Filtenborg, O., Samson, R.A., and Stolk, A.C. (1990). Chemotaxonomy of the genus *Talaromyces*. Antonie van Leeuwenhoek *57*, 179–189.

Frisvad, J.C., Larsen, T.O., Thrane, U., Meijer, M., Varga, J., Samson, R.A., and Nielsen, K.F. (2011). Fumonisin and ochratoxin production in industrial *Aspergillus niger* strains. PLoS One *6*, e23496.

Frisvad, J.C., and Samson, R.A. (2004). Polyphasic taxonomy of *Penicillium* subgenus *Penicillium*. A guide to identification of food and air-borne terverticillate Penicillia and their mycotoxins. Stud. Mycol. *49*, 1–174.

Frisvad, J.C., Yilmaz, N., Thrane, U., Rasmussen, K.B., Houbraken, J., and Samson, R.A. (2013). *Talaromyces atroroseus*, a new species efficiently producing industrially relevant red pigments. PLoS One *8*, e84102.

Geiser, G.M., Lobuglio, K.F., and Gueidan, C. (2015). *Pezizomycotina*: *Eurotiomycetes*. In The Mycota, Systematics and Evolution, part B, Esser, K., Spatafora, J.W., and McLaughlin, D.J., eds. (Springer-Verlag Berlin, Heidelberg, Germany), pp. 121–142.

Giraud, S., Favennec, L., Bougnoux, M.E., and Bouchara, J.P. (2013). *Rasamsonia argillacea* species complex: taxonomy, pathogenesis and clinical relevance. Future Microbiol. *8*, 967–978.

Hebert, P.D.N., Ratnasingham, S., and deWaard, J.R. (2003). Barcoding animal life: cytochrome c oxidase subunit 1 divergences among closely related species. P. Roy. Soc. Lond. B *270*(Suppl. 1), S96–S99.

Hong, S.-B., Lee, M., Kim, D.-H., Varga, J., Frisvad, J.C., Perrone, G., Gomi, K., Yamada, O., Machida, M., Houbraken, J., *et al.* (2013). *Aspergillus luchuensis*, an industrially important black *Aspergillus* in East Asia. PLoS One *8*, e63769.

Houbraken, J., Frisvad, J.C., and Samson, R.A. (2011a). Taxonomy of *Penicillium* section *Citrina*. Stud. Mycol. *70*, 53–138.

Houbraken, J., Frisvad, J.C., and Samson, R.A. (2011b). Fleming's penicillin producing strain is not *Penicillium chrysogenum* but *P. rubens*. IMA Fungus *2*, 87–92.

Houbraken, J., Giraud, S., Meijer, M., Bertout, S., Frisvad, J.C., Meis, J.F., Bouchara, J.P., and Samson, R.A. (2013). Taxonomy and antifungal susceptibility of clinically important *Rasamsonia* species. J. Clin. Microbiol. *31*, 22–30.

Houbraken, J., and Samson, R.A. (2011). Phylogeny of *Penicillium* and the segregation of *Trichocomaceae* into three families. Stud. Mycol. *70*, 1–51.

Houbraken, J., Spierenburg, H., and Frisvad, J.C. (2012). *Rasamsonia*, a new genus comprising thermotolerant and thermophillic *Talaromyces* and *Geosmithia* species. Antonie van Leeuwenhoek *101*, 403–421.

Houbraken, J., Visagie, C.M., Meijer, M., Frisvad, J.C., Busby, P.E., Pitt, J.I., Seifert, K.A., Louis-Seize, G., Demirel, R., Yilmaz, N., *et al.* (2014). A taxonomic and phylogenetic revision of *Penicillium* section *Aspergilloides*. Stud. Mycol. *78*, 373–451.

Houbraken, J., Vries, R.P. de, and Samson, R.A. (2014). Modern taxonomy of biotechnologically important *Aspergillus* and *Penicillium* species. Adv. Appl. Microbiol. *86*, 199–249.

Jurjević, Ž., Peterson, S.W., Stea, G., Solfrizzo, M., Varga, J., Hubka, V., and Perrone, G. (2012). Two novel species of *Aspergillus* section *Nigri* from indoor air. IMA Fungus *3*, 159–173.

Kozakiewicz, Z., Frisvad, J.C., Hawksworth, D.L., Pitt, J.I., Samson, R.A., and Stolk, A.C. (1992). Proposals for nomina specifica conservanda and rejicienda in *Aspergillus* and *Penicillium* (Fungi). Taxon *41*, 109–113.

Kuraishi, H., Aoki, M., and Itoh, M. (1991). Distribution of ubiquinones in *Penicillium* and related genera. Mycol. Res. *95*, 705–711.

Link, H.F. (1809). Observationes in Ordines plantarum naturales, Dissertatio 1ma. Magazin der Gesellschaft Naturforschenden Freunde Berlin *3*, 3–42.

Logrieco, A., Peterson, S.W., and Wicklow, D.T. (1990). Ribosomal RNA comparisons among taxa of the terverticilliate penicillia. In Modern Concepts in *Penicillium* and *Aspergillus* Systematics, R.A. Samson, and J.I. Pitt, eds. (Plenum Press, New York, NY, USA), pp. 343–355.

McNeill, J., Barrie, F.R., Buck, W.R., Demoulin, V., Greuter, W., Hawksworth, D.L., Herendeen, P.S., Knapp, S., Marhold, K., Prado, J., *et al.* (2012). International Code of Nomenclature for algae, fungi, and plants (Melbourne Code) adopted by the Eighteenth International Botanical Congress Melbourne, Australia, July 2011. Regnum Vegetabile 154. (A.R.G. Gantner Verlag, Ruggell, Liechtenstein).

Maeda, R.N., Barcelos, C.A., Anna, L.M.M.S., and Pereira, N. (2013). Cellulase production by *Penicillium funiculosum* and its application in the hydrolysis of sugar cane bagasse for second generation ethanol production by fed batch operation. J. Biotechnol. *163*, 38–44.

Maeda, R.N., Serpa, V.I., Rocha, V.A.L., Mesquita, R.A.A., Anna, L.M.M.S., Castro, A.M., Driemeier, C.E., Pereira, N., and Polikarpov, I. (2011). Enzymatic hydrolysis of pretreated sugar cane bagasse using *Penicillium funiculosum* and *Trichoderma harzianum* cellulases. Process Biochem. *46*, 1196–1201.

Malloch, D. (1985a). Taxonomy of *Trichocomaceae*. In Filamentous Microorganisms, T. Arai, ed. (Japan Scientific Society Press, Tokyo, Japan), pp. 37–45.

Malloch, D. (1985b). The *Trichocomaceae*: relationships with other *Ascomycetes*. In Advances in *Penicillium* and *Aspergillus* Systematics, R.A. Samson, and J.I. Pitt, eds (Plenum Press, New York, NY, USA), pp. 365–382.

Malloch, D., and Cain, R.F. (1972). New species and combinations in cleistothecial Ascomycetes. Can. J. Bot. *50*, 61–72.

Mansouri, S., Houbraken, J., Samson, R.A., Frisvad, J.C., Christensen, M., Tuthill, D.E., Koutaniemi, S., Hatakka, A., and Lankinen, P. (2013). *Penicillium subrubescens*, a new species efficiently producing inulinase. Antonie van Leeuwenhoek *103*, 1343–1357.
Marois, J.J., Fravel, D.R., and Papavizas, G.C. (1984). Ability of *Talaromyces flavus* to occupy the rhizosphere. Soil Biol. Biochem. *16*, 387–390.
Meijer, M., Houbraken, J., Dalhuijsen, S., Samson, R.A., and de Vries, R.P. (2011). Growth and hydrolase profiles can be used as characteristics to distinguish *Aspergillus niger* and other black aspergilli. Stud. Mycol. *69*, 19–30.
Mishra, C., Rao, M., Seeta, R., Srinivasan, M.C., and Deshpande, V. (1984). Hydrolysis of lignocelluloses by *Penicillium funiculosum* cellulase. Biotechnol. Bioeng. *26*, 370–373.
Niu, X., Chen, L., Yue, Q., Wang, B., Zhang, J., Zhu, C., Zhang, K., Bills, G.F., and An, Z. (2014). Characterization of thermolide biosynthetic genes and a new thermolide from sister thermophilic fungi. Org Lett. *16*, 3744–3747.
Norvell, L.L. (2011). Fungal nomenclature. 1. Melbourne approves a new code. Mycotaxon *116*, 481–490.
Pitt, J.I. (1980). The genus *Penicillium* and its teleomorphic states *Eupenicillium* and *Talaromyces*. (Academic Press, London, UK).
Pitt, J.I., and Hocking, A.D. (2009). Fungi and food spoilage, 3rd edition (Springer, New York, NY, USA).
Pitt, J.I., Lantz, H., Pettersson, O.V., and Leong, S.L. (2013). *Xerochrysium* gen. nov. and *Bettsia*, genera encompassing xerophilic species of *Chrysosporium*. IMA Fungus *4*, 229–241.
Pitt, J.I., and Samson, R.A. (2007). Nomenclatural considerations in naming species of *Aspergillus* and its teleomorphs. Stud. Mycol. *59*, 67–70.
Pitt, J.I., Samson, R.A., and Frisvad, J.C. (2000). List of accepted species and their synonyms in the family *Trichocomaceae*. In Integration of Modern Taxonomic Methods for *Penicillium* and *Aspergillus* Classification, R.A. Samson, and J.I. Pitt, eds. (Harwood Academic Publishers, Amsterdam, the Netherlands), pp. 9–79.
Pitt, J.I., and Taylor, J.W. (2014). *Aspergillus*, its sexual states and the new International Code of Nomenclature. Mycologia *106*, 1051–1062.
Punja, Z.K. (2001). Genetic engineering of plants to enhance resistance to fungal pathogens – a review of progress and future prospects. Can. J. Plant Pathol. *23*, 216–235.
Quandt, C.A., Kohler, A., Hesse, C.N., Sharpton, T.J., Martin, F., and Spatafora, J.W. (2015). Metagenome sequence of *Elaphomyces granulatus* from sporocarp tissue reveals *Ascomycota* ectomycorrhizal fingerprints of genome expansion and a *Proteobacteria*-rich microbiome. Environ. Microbiol. *17*, 2952–2968.
Raper, K.B., and Fennell, D.I. (1965). The genus *Aspergillus*. (Williams & Wilkins Co., Baltimore, MD, USA).
Raper, K.B., and Thom, C. (1949). A manual of the Penicillia. (The Williams & Wilkins Company, Baltimore, MD, USA).
Salgüero, R., Borman, A.M., Herrtage, M., Benchekroun, G., Abbondati, E., Piola, V., and Vanhaesebrouck, A. (2013). *Rasamsonia argillacea* mycosis in a dog: first case in Europe. Vet. Rec. Case Rep. *1*, e101373.
Samson, R.A., Houbraken, J., Thrane, U., Frisvad, J.C., and Andersen, B. (2010). Food and Indoor Fungi, CBS laboratory manual series 2. (CBS-Fungal Biodiversity Centre, Utrecht, The Netherlands).
Samson, R.A., Visagie, C.M., Houbraken, J., Hong, S.B., Hubka, V., Klaassen, C.H., Perrone, G., Seifert, K.A., Susca, A., Tanney, J.B., *et al.* (2014). Phylogeny, identification and nomenclature of the genus *Aspergillus*. Stud. Mycol. *78*, 141–173.
Samson, R.A., Yilmaz, N., Houbraken, J., Spierenburg, H., Seifert, K.A., Peterson, S.W., Varga, J., and Frisvad, J.C. (2011). Phylogeny and nomenclature of the genus *Talaromyces* and taxa accommodated in *Penicillium* subgenus *Biverticillium*. Stud. Mycol. *70*, 159–183.
Santos, P.E., Piontelli, E., Shea, Y.R., Galluzzo, M.L., Holland, S.M., Zelazko, M.E., and Rosenzweig, S.D. (2006). *Penicillium piceum* infection: diagnosis and successful treatment in chronic granulomatous disease. Med. Mycol. *44*, 749–753.
Schoch, C.L., Robbertse, B., Robert, V., Vu, D., Cardinali, G., Irinyi, L., Meyer, W., Nilsson, R.H., Hughes, K., Miller, A.N., *et al.* (2014). Finding needles in haystacks: linking scientific names, reference specimens and molecular data for Fungi. Database *2014*, 1–21.
Schoch, C.L., Seifert, K.A., Huhndorf, S., Robert, V., Spouge, J.L., Levesque, C.A., Chen, W., and Fungal Barcoding Consortium (2012). Nuclear ribosomal internal transcribed spacer (ITS) region as a universal DNA barcode marker for Fungi. Proc. Natl. Acad. Sci. U.S.A. *109*, 6241–6246.

Stielow, J.B., Lévesque, C.A., Seifert, K.A., Meyer, W., Irinyi, L., Smits, D., Renfurm, R., Verkley, G.J.M., Groenewald, M., Chaduli, D., *et al.* (2015). One fungus, which genes? Development and assessment of universal primers for potential secondary fungal DNA barcodes. Persoonia 35, 242–263.

Stolk, A.C., and Samson, R.A. (1971). Studies on *Talaromyces* and related genera I. *Hamigera* gen. nov. and *Byssochlamys*. Persoonia 6, 341–357.

Stolk, A.C., and Samson, R.A. (1972). The genus *Talaromyces* – studies on *Talaromyces* and related genera II. Stud. Mycol. 2, 1–65.

Subramanian, C.V. (1972). The perfect states of *Aspergillus*. Curr. Sci. *41*, 755–761.

Taylor, J.W., Jacobson, D.J., Kroken, S., Kasuga, T., Geiser, D.M., Hibbett, D.S., and Fisher, M.C. (2000). Phylogenetic species recognition and species concepts in fungi. Fungal Genet. Biol. *31*, 21–32.

Tedersoo, L., May, T.W., and Smith, M.E. (2010). Ectomycorrhizal lifestyle in fungi: global diversity, distribution, and evolution of phylogenetic lineages. Myocorrhiza *20*, 217–263.

Thom, C. (1910). Cultural studies of species of *Penicillium*. Bureau of Animal Industry, US Department of Agriculture *118*, 1–109.

Thom, C. (1930). The Penicillia. (The Williams & Wilkins Company, Baltimore, MD, USA).

Thom, C., and Church, M.B. (1926). The Aspergilli. (Williams & Wilkins Company, Baltimore, MD, USA).

Thom, C., and Raper, K.B. (1945). Manual of the Aspergilli. (Williams & Wilkins Company, Baltimore, MD, USA).

Trappe, J.M. (1979). The orders, families, and genera of hypogeous *Ascomycotina* (truffles and their relatives). Mycotaxon 9, 297–340.

Varga, J., Frisvad, J.C., Kocsubé, S., Brankovics, B., Tóth, B., Szigeti, G., and Samson, R.A. (2011). New and revisited species in *Aspergillus* section *Nigri*. Stud. Mycol. *69*, 1–17.

Visagie, C.M. (2012). The polyphasic taxonomy of *Penicillium* and *Talaromyces* spp. isolated from the diverse Fynbos biome. PhD Dissertation, University of Stellenbosch, Stellenbosch, South Africa.

Visagie, C.M., Houbraken, J., Frisvad, J.C., Hong, S.B., Klaassen, C.H., Perrone, G., Seifert, K.A., Varga, J., Yaguchi, T., and Samson, R.A. (2014a). Identification and nomenclature of the genus *Penicillium*. Stud. Mycol. *78*, 343–371.

Visagie, C.M., Hirooka, Y., Tanney, J.B., Whitfield, E., Mwange, K., Meijer, M., Amend, A.S., Seifert, K.A., and Samson, R.A. (2014b). *Aspergillus, Penicillium* and *Talaromyces* isolated from house dust samples collected around the world. Stud. Mycol. *78*, 63–139.

Visser, E.M., Falkoski, D.L., de Almeida, M.N., Maitan-Alfenas, G.P., and Guimarães, V.M. (2013). Production and application of an enzyme blend from *Chyrsoporthe cubensis* and *Penicillium pinophilum* with potential for hydrolysis of sugarcane bagasse. Bioresour. Technol. *144*, 587–594.

Westling, R. (1911). Über die Grünen Spezies der Gattung *Penicillium*. Arkiv før Botanik *11*, 1–156.

Wood, T.M., and McCrae, S.I. (1986). The cellulase *Penicillium pinophilum*. Synergism between enzyme components in solubilizing cellulose with special reference to the involvement of two immunologically distinct cellobiohydrolases. Biochem. J. *234*, 93–99.

Yaguchi, T., Someya, A., and Udagawa, S. (1996). A reappraisal of intrageneric classification of *Talaromyces* based on the ubiquinone systems. Mycoscience 37, 55–60.

Yamada, O., Takara, R., Hamada, R., Hayashi, R., Tsukahara, M., and Mikami, S. (2011). Molecular biological researches of Kuro-Koji molds, their classification and safety. J. Biosci. Bioeng. *112*, 233–237.

Yilmaz, N., Visagie, C.M., Houbraken, J., Frisvad, J.C., and Samson, R.A. (2014). Polyphasic taxonomy of the genus *Talaromyces*. Stud. Mycol. *78*, 175–341.

Yilmaz, N., Visagie, C.M., Frisvad, J.C., Houbraken, J., Jacobs, K., and Samson, R.A. (2016). Taxonomic re-evaluation of species in *Talaromyces* section *Islandici*, using a polyphasic approach. Persoonia *36*, 37–56.

Zaleski, K.M. (1927). Über die in Polen gefundenen Arten der Gruppe *Penicillium* Link. I, II and III Teil. Bulletin de l'Académie Polonaise des Sciences et des Lettres, Classe des Sciences Mathématiques et Naturelles – Série B: Sciences Naturelles, 417–563, pls 36–44 (printed in 1928).

Zhang, X., Li, X., and Xia, L. (2015). Heterologous expression of an alkali and thermotolerant lipase from *Talaromyces thermophilus* in *Trichoderma reesei*. Appl. Biochem. Biotechnol. *176*, 1722–1735.

Comparative Genomics, Resequencing and Fast Forward Genetics in *Aspergillus* and *Penicillium*

2

Scott E. Baker and Erin L. Bredeweg

Abstract

New methods have accelerated the pace of DNA sequencing for comparative genomics and genetics of fungi. High throughput genome sequencing enables comparative analysis of multiple strains of the same species and mutagenized strain lineages with interesting phenotypes. With their compact genomes, species from the genera *Aspergillus* and *Penicillium* are ideal for these multi-strain genome analyses that connect genotypes with phenotypes of interest to fungal biology and biotechnology.

Introduction

Among fungi, *Aspergillus* and *Penicillium* are two important groups of fungi in both basic and applied research. *Aspergillus* has a long history intertwined with humans (for example reviewed in Baker and Bennett, 2008; Bennett, 2009). Aspergilli koji moulds been used in Asian food production for well over 1000 years. Since the turn of the nineteenth century aspergilli have been used in industrial processes such as enzyme and organic acid production. Species of *Aspergillus* are also known for their detrimental effects on human and plant health, as pathogens and mycotoxin producers. Like *Aspergillus, Penicillium* is another genus that has had a profound impact on human health and well-being. This genus is best known for its ability to produce the antibiotic penicillin. Discovered by Alexander Fleming, penicillin proved difficult to produce in large amounts for several years (Fleming, 1929). Efforts to increase production of penicillin during the Second World War led to the identification of a *Penicillium chrysogenum* strain producing higher levels of penicillin, isolated at the USDA laboratory in Peoria by 'Moldy Mary' (Mary Hunt) from a moldy cantaloupe (Bennett and Chung, 2001). Continued strain development using mutagenesis and screening has led to the development of higher and higher productivity. It should be noted that although the Fleming and USDA isolated strains continue to be called *P. chrysogenum,* they have been taxonomically reclassified as *Penicillium rubens* (Houbraken *et al.,* 2011, 2012).

For over a century genetic analysis has provided a wealth of information about biological processes – penicillin production in *P. chrysogenum* is an example – across a diverse range of organisms. Ascomycete fungi are a key class of organisms for genetic analysis in that they produce discrete numbers of offspring in contained structures, and can be screened for altered outward appearances, and tolerance or susceptibility to different stresses or nutrients. Mutant screens in fungi have led to the identification of genes whose protein products play roles in circadian rhythms, central metabolism, morphology, adhesion, secondary metabolism and invasiveness against plant, animal, and even fungal hosts. A good mutant screen has the potential to identify multiple genes involved in the biological process of interest. However, it is quite laborious to connect mutant phenotypes with physical loci. While genetic mapping studies can narrow down the chromosome or area where a mutant gene is located, a variety of molecular biology techniques must be used to definitively identify the gene whose mutation leads to the phenotype in question. A crucial tool set in connecting phenotypes with genotypes are reference genomic sequences for fungi used in genetic studies.

Less than a decade ago the cost of sequencing a fungal genome was high; for this reason a limited number of fungi were sequenced. Now high-throughput DNA sequencing technology makes it possible to think about generating reference genome sequences for hundreds, if not thousands, of fungi (Grigoriev *et al.*, 2011; Martin *et al.*, 2011; Baker *et al.*, 2008). In fact, over the last two decades tools have been developed that have moved DNA sequencing, assembly, gene modelling and analysis from high labour tasks to automated pipelines (Grigoriev *et al.*, 2014). As a consequence, genome sequencing centres such as the US Department of Energy joint genome Institute have embarked on huge sequencing projects such as the 1000 fungal genomes project (Grigoriev *et al.*, 2014). Not only has this made rapid comparison of multiple mutants to an unmutated parent possible, but novel species identification and analysis provides an ideal environment for discovery of new pathways, new genes in closely related species, and a world of evolutionary insight.

As genomes are being produced at such a rapid pace, advances in industry and research in cell biology have created the need to associate biological behaviours or characteristics with genes. The ability to perform a robust genetic analysis including linkage mapping greatly facilitates the association of phenotypes and genotypes. For example, genetics played a key role in the elucidation of the biosynthetic pathway for production of the economically important mycotoxin, aflatoxin, a food and feed contaminant, in *Aspergillus flavus* and *Aspergillus parasiticus* (Bennett and Goldblatt, 1973). In order to identify the genes involved in production of aflatoxin and its precursor sterigmatocystin, it took many years of work beyond the identification of mutants in the related model genetic organism *Aspergillus nidulans* (Brown et al., 1996; Yu et al., 1995). One of the reasons for completing this work in *A. nidulans* is that it has been exploited for both 'classical' and 'forward' genetics research in order to make it into a tractable 'model genetic organism'. But not every organism has a robust forward genetic toolbox, which would allow prediction of a set of genes responsible for a particular characteristic or process. For example, *Penicillium chrysogenum*, the penicillin production organism, has only recent had its sexual cycle characterized and the important enzyme and organic acid producer *Aspergillus niger* has yet to have an identified sexual cycle. For both of these organism, low resolution genetic mapping and analysis was performed using non-meiotic or parasexual genetics (Pontecorvo, 1952; Pontecorvo and Sermonti, 1954).

Genetic analysis including random mutagenesis is clearly a powerful tool for understanding biological processes and *Aspergillus* and *Penicillium*. Over several decades the ability to associate genes and phenotypes improved gradually and with much effort. With the latest advance of genomic sequencing continuing to become cheaper and faster, the possibility of using this tool to identify mutations associated with phenotypes in specific strains is tractable and economical. In this chapter we will discuss examples of how resequencing has been used to associate phenotypes and biological processes with genotypes.

Resequencing and fast forward genetics

The first resequencing study in fungi was performed for *Pichia stipitis* and showed the power of next-generation sequencing methods for identifying sequencing errors in reference genomes and finding mutations (Smith *et al.*, 2008). Another early fungal genome resequencing study focused on the industrial cellulase producer, *Trichoderma reesei*. In this study, two *T. reesei* strains, NG-14 and RUT-C30, of the important lineage derived from the wild type QM6a strain, were resequenced using Illumina technology. The findings indicated that while a significant number of genes were mutated, mutagenesis and screening does not lead to an overabundance of mutation; the classes of proteins encoded by the mutated genes could be reasonably connected to the phenotypes of the mutant strains (Le Crom *et al.*, 2009). A similar study using microarrays on additional *Trichoderma reesei* strains showed similar classes of mutated genes (Vitikainen *et al.*, 2010). Indeed, analyses have identified that a 85 kb deletion in RUT-C30 and other strains containing the same phenotype of increased cellulase production can be tied to the same underpinnings of regulation (Seidl *et al.*, 2008). One issue with the *Trichoderma* studies was the fact that until recently, methods for standard genetic crosses to clean up collateral damage caused by mutagenesis were not available, resulting in an accumulation of mutations. Genetic crossing for *T. reesei* is now available so that future mutagenesis and screening can be followed by genetic mapping and/or chromosomal clean up (Seidl *et al.*, 2009).

As we accumulate new species, and resolve asexual and sexual stages, fungal genomic sequencing, mutation, and resequencing will multiply the genetic potential for discovery. The potential of moving entire antibiotic biosynthetic pathways from a different organism into a tractable *Aspergillus* host for efficient production are growing. For example, the relatively recently discovered genes encoding biosynthetic enzymes for itaconic acid production in *Aspergillus terreus* have been moved into *Aspergillus niger* (Blumhoff *et al.*, 2013; Li *et al.*, 2012, 2013; van der Straat *et al.*, 2013, 2014).

For further insight into industrial strains and their potential, strain depositories are a wonderful resource (McCluskey *et al.*, 2014). A laboratory organism which has moved into the realm of discovery is *Neurospora crassa*, another well studied filamentous fungal model genetic organism. It has been used for decades as a system to better understand a variety of biological processes. Beadle and Tatum won the Nobel Prize using this organism to demonstrate the 'one gene-one enzyme' hypothesis by generating a vitamin B_6 auxotroph mutant (aka 'pyroxidenless') and showing that it segregated as a single gene locus (Beadle and Tatum, 1941). Several hundred forward mutagenesis-derived strains of *N. crassa* have been deposited at the Fungal Genetics Stock Center with many having mutant genes associated with phenotypes that have mapped genetically (McCluskey, 2011). A pilot study in 2011 showed the feasibility of using next-generation DNA sequencing to rapidly identify mutations in multiple strains that were associated with a wide range of mutant phenotypes (McCluskey *et al.*, 2011). Several hundred mutant strains of *N. crassa* are being resequenced by the DOE Joint BioEnergy Institute (JBEI) in collaboration with the DOE JGI. Independent studies of phenotype–genotype utilizing bulk segregant analysis have also come out of the Neurospora community. One example is identification of ndc-1, a temperature-dependent cell cycle arrest mutant, as dependent on spermidine (Pomraning *et al.*, 2011). The interest in a mutation such as this is to understand the unique inner workings between different organisms. A diversity of fungal species is needed to study the breadth for biological processes; for example, *N. crassa* lacks synchronous cell cycles while in *Aspergillus* several species are synchronous.

Fungi are ideal organisms for comparison of different strains of the same organism as well as fast forward genetics on mutagenesis derived strains. Indeed, multiple species of *Aspergillus* and *Penicillium* have been the subject of more recent resequencing efforts, some of which are detailed below.

The aspergilli

Because it has so many important organisms for research, medicine, and industry, *Aspergillus* was one of the earliest fungal genera for which multiple species' genomes were sequenced. References sequences for *Aspergillus fumigatus*, *Aspergillus nidulans* and *Aspergillus oryzae* were published in Nature in 2005 (Galagan *et al.*, 2005; Machida *et al.*, 2005; Nierman *et al.*, 2005). Genome sequences for two strains *of Aspergillus niger* followed (Andersen *et al.*, 2011; Pel *et al.*, 2007).

Aspergillus fumigatus

One of the first *Aspergillus* 'resequencing' studies was published in 2008, when a second strain (Af1163) of *Aspergillus fumigatus* was compared to the reference sequenced strain (Af293) as well as two other related *Aspergillus* strains (Fedorova *et al.*, 2008). This study found that the two strains of *A. fumigatus*, a species found as a pathogen in the lungs of immunocompromised patients, were 99.8% identical at the DNA level with 100% amino acid identity in predicted protein sequences. However, in the genome of each strain there were strain-specific genes clustered in ~10–400 kb genomic 'blocks'. These blocks contained repetitive sequences and pseudogenes as well as one secondary metabolite cluster.

More recently, *A. fumigatus* isolates have been sequenced using next-generation sequencing methods. The selective pressure on these organisms as they grow within human hosts is significant and is known to lead to mutations that lead to azole resistance (Diaz-Guerra *et al.*, 2003; Mann *et al.*, 2003; Nascimento *et al.*, 2003). One study of azole-resistant *A. fumigatus* strains led to the discovery of a gene that was previously not known to play a role in azole resistance, hapE (Camps *et al.*, 2012). Other studies have examined serially isolated *A. fumigatus* strains from single patients in order to look at a progression of aspergillosis (Hagiwara *et al.*, 2014) and at collections of patient isolated strains to in order to examine population genetics of azole resistance (Abdolrasouli *et al.*, 2015).

Aspergillus niger

The *A. niger* citric acid process was initially patented and developed in the early part of the twentieth century ((Currie, 1917). By the late 1920s, it had displaced citrus fruit as the source of citric acid. Glucoamylase is also produced by *A. niger*. The strains that produce citric acid are distinct from those that produce glucoamylase. A reference genome for *A. niger* was initially generated for a mutant strain, CBS 513.88, that produces significant amounts of glucoamylase, an enzyme used in industrial processes for starch saccharification (Pel *et al.*, 2007). A second *A. niger* strain, ATCC 1015, the wild type precursor of a citric acid producing strain (ATCC 11414) was sequenced and compared to CBS 513.88 (Andersen *et al.*, 2011). This comparison showed indicated that the two *A. niger* strains were overall very similar but had differences (clustered strain-specific genes) similar those found between *A. fumigatus* strains Af293 and Af1163. Interestingly, a gene cluster predicted to be involved in ochratoxin A (OTA) production was identified in CBS 513.88 but deleted in ATCC 1015, correlating with their respective OTA production profiles (Andersen *et al.*, 2011; Frisvad *et al.*, 2011). Ochratoxin is carcinogenic and nephrotoxic in animal models. It is produced by fungal secondary metabolite clusters, detailed below, and is typically found in contaminated cereal grains with different species specializing in field versus storage mycotoxin production. It has been found in cereals from Eastern Europe and the Mediterranean, spices from Turkey, and some foodstuffs from Italy (Klaric *et al.*, 2013).

Aspergillus carbonarius

Aspergillus carbonarius is an important producer of the potent kidney-targeting mycotoxin ochratoxin A (OTA) (Battilani *et al.*, 2006). A reference genome sequence was produced by the JGI. Analysis of the genome led to the identification and characterization of genes encoding a polyketide synthase (PKS) and non-ribosomal peptide synthetase (NRPS) involved in OTA production (Gallo *et al.*, 2012, 2014). Using ion torrent technology the genome of a low/non-OTA producing strain of *A. carbonarius* was generated. Their analysis of the resulting sequence indicated a high level of mutations in PKS- and NRPS-encoding genes, including those previously shown to be involved in OTA production and showed that next-generation sequencing methods are an important tool for uncovering the genetics underlying population differences in mycotoxin production (Cabanes *et al.*, 2015).

Aspergillus oryzae

Aspergillus oryzae is an economically important *Aspergillus* species involved in food and beverage production from soy and grain products, such as soy sauce and sake. The reference sequence of *A. oryzae* was generated for strain RIB40 in 2005 (Machida *et al.*, 2005) and since that time, additional strains have been sequenced. One study focused on a comparison to an additional Japanese strain, RIB 326. The major differences that were found were located in non-syntenic blocks (NSBs) that were initially described as species specific genomic regions when the reference sequences of *Aspergillus nidulans*, *A. fumigatus* and *A. oryzae* were compared in 2005. The authors go on to hypothesize that the genes contained in the NSBs probably play a role in the strain-specific flavour characteristics of koji foods (Umemura *et al.*, 2012). Another group looked at the difference between Chinese and Japanese strains and found small numbers of strain-specific genes, speculating that these may play some role in differences in physiology between the strains (Zhao *et al.*, 2013a,b). Mutant 100-8, derived from strain 3.042, shows a difference in protease secretion; strain 100-8 has been sequenced and, while mutations have been identified, they have yet to be definitively linked to the different protease profiles of the two strains (Zhao *et al.*, 2014a,b).

Penicillium

Like *Aspergillus*, *Penicillium* is a significant genus containing species involved in food production, medicine and industrial applications. The history of penicillin, from its discovery to its production during World War II, is fascinating.

Penicillium chrysogenum (*Penicillium rubens*)

Over half a century has elapsed since the Nobel Prize was awarded for penicillin discovery and production. Generation of a reference genome sequence for the organism at the centre of the story, *Penicillium chrysogenum*, was published in 2008 for strain Wisconsin54-1255, a mutant that produced increased levels of penicillin without replication of the biosynthetic cluster (van den Berg *et al.*, 2008). Since then, wild type strains and strains isolated by other means have been sequenced. It should be also be noted that although the Fleming and USDA isolated strains continue to be called *P. chrysogenum*, have been reclassified as *Penicillium rubens* (Houbraken *et al.*, 2011, 2012).

One study compared the genome structure of WI54-1255 with a wild type, UBC (PC0184C) (Wong *et al.*, 2014). One of the major conclusions of this study was that changes in the genome not associated with transposons were due to natural variation in the population of *P. chrysogenum* and that these differences may be associated with altered levels of penicillin production that occur across the natural population. Another comparison of WI54-1255 with strain KF-25 also showed evidence for structural variation and strain-specific genes across the *P. chrysogenum* population (Peng *et al.*, 2014). The strain NCPC10086 is a high producer of penicillin and was recently sequenced and compared with the WI54-1255 reference sequence (Wang *et al.*, 2014). This study found differences in metabolic genes that may play roles in production of penicillin precursors as well as finding a 7-fold amplification of the penicillin biosynthetic cluster. A direct derivative of WI54-1255, P2niaD18, a high penicillin producer and nitrate reductase mutant, was sequenced and compared to the reference strain; the penicillin cluster was duplicated and chromosomal mutations could be associated with the niaD phenotype (Bohm *et al.*, 2015; Specht *et al.*, 2014).

Penicillium decumbens

Penicillium decumbens has been used as a cellulase producer in China. The reference genome sequence for *P. decumbens* strain JU-A10-T was published in 2013 and was shown to encode a diverse set of secreted enzymes predicted to play roles in biomass deconstruction (Liu *et al.*, 2013b). A hyper-producer of secreted biomass deconstruction enzymes, *P. decumbens* strain 114-2 derived from *P. decumbens* strain JU-A10-T was sequenced and found to have a mutation in cre1 (Liu *et al.*, 2013a), a transcription factor involved in carbon catabolite repression also implicated in high levels of cellulase production in mutated strains of *Trichoderma reesei* (Ilmen *et al.*, 1996).

Conclusions

It is clear that the world of genetics has been revolutionized by next-generation sequencing methods. Genetics has been transformed from a discipline that required significant time to identify mutant genes and phenotypes and subsequently associate them with genotypes to one wherein mutants are identified by traditional forward mutagenesis and then only weeks later genomic sequence is generated that allows rapid association of phenotype with the genotype. This transformative technology and method has been applied across a variety of areas from basic to applied industrial and medical fungal biology research challenges, and continues to provide hypotheses for future study.

Future trends

Penicillium and *Aspergillus* genera are populated by many important species involved in basic research, human health, plant health, mycotoxin production and industrial production of organic acids and enzymes. Indeed there are a number of wild type strains with associated mutant lineages that would make attractive targets for resequencing and fast forward genetics. For example, two additional wild type strains of *Aspergillus niger* strains, ATCC 13496 and ATCC 13157, have been sequenced and are available via the JGI Mycocosm platform (Grigoriev *et al.*, 2014). Mutant lineages from these strains with increased secretion of glucoamylase have been sequenced and are currently being analysed (Baker, Magnuson and Gladden, unpublished). Harnessing the power of genetics to make industrial products and perform energetically costly synthetic chemical production is a movement that can be fuelled by further sequencing and gene discovery.

Like *N. crassa*, *A. nidulans* has long been a genetic model organisms. Furthermore, at the Fungal Genetics Stock Center are deposited significant number of classical forward genetic *A. nidulans* mutant strains. In addition, there are classical mutants in the N400 lineage of *A. niger* that could be sequenced. The resequencing of collections of model fungal mutants, as well as industrial and agriculturally relevant strains of fungi has the potential to provide incredible insight into the associated cell biology and characteristics of these organisms.

The future is bright for fast forward genetics.

Acknowledgements

S.E.B. dedicates this chapter to the memory of R.M. Holmboe, who passed away during its writing. The Joint BioEnergy Institute (JBEI) is supported by the Office of Science, Office of Biological and Environmental Research, of the U.S. Department of Energy under Contract No. DE-AC02-05CH11231. Located at the Pacific Northwest National Laboratory, the Environmental Molecular Sciences Laboratory (EMSL) is a DOE Office of Science National User Facility sponsored by the Office of Biological and Environmental Research under Contract DE-AC05-76RL01830.

Web resources

DOE Joint Genome Institute Mycocosm portal to fungal genomes – http://genome.jgi-psf.org/programs/fungi/index.jsf

FungiDB – http://fungidb.org/fungidb/

References

Abdolrasouli, A., Rhodes, J., Beale, M.A., Hagen, F., Rogers, T.R., Chowdhary, A., Meis, J.F., Armstrong-James, D., and Fisher, M.C. (2015). Genomic context of azole resistance mutations in *Aspergillus fumigatus* determined using whole-genome sequencing. MBio *6*.

Andersen, M.R., Salazar, M.P., Schaap, P.J., van de Vondervoort, P.J., Culley, D., Thykaer, J., Frisvad, J.C., Nielsen, K.F., Albang, R., Albermann, K., *et al.* (2011). Comparative genomics of citric-acid-producing *Aspergillus niger* ATCC 1015 versus enzyme-producing CBS 513.88. Genome Res. *21*, 885–897.

Baker, S., and Bennett, J. (2008). An overview of the genus *Aspergillus*. Mycology Series *26*, 3.

Baker, S.E., Thykaer, J., Adney, W.S., Brettin, T.S., Brockman, F.J., D'haeseleer, P., Martinez, A.D., Miller, R.M., Rokhsar, D.S., Schadt, C.W., *et al.* (2008). Fungal genome sequencing and bioenergy. Fungal Biol. Rev. 22, 1–5.

Battilani, P., Magan, N., and Logrieco, A. (2006). European research on ochratoxin A in grapes and wine. Int. J. Food Microbiol. *111*(Suppl. 1), S2–4.

Beadle, G.W., and Tatum, E.L. (1941). Genetic control of biochemical reactions in Neurospora. Proc. Natl. Acad. Sci. U.S.A. *27*, 499–506.

Bennett, J.W. (2009). Aspergillus: a primer for the novice. Med. Mycol. *47*(Suppl. 1), S5–12.
Bennett, J.W., and Chung, K.-T. (2001). Alexander Fleming and the discovery of penicillin. Adv. Appl. Microbiol. *49*, 163–184.
Bennett, J.W., and Goldblatt, L.A. (1973). The isolation of mutants of *Aspergillus flavus* and *A. parasiticus* with altered aflatoxin producing ability. Sabouraudia *11*, 235–241.
Blumhoff, M.L., Steiger, M.G., Mattanovich, D., and Sauer, M. (2013). Targeting enzymes to the right compartment: metabolic engineering for itaconic acid production by *Aspergillus niger*. Metab. Eng. *19*, 26–32.
Bohm, J., Dahlmann, T.A., Gumuser, H., and Kuck, U. (2015). A MAT1-2 wild-type strain from *Penicillium chrysogenum*: functional mating-type locus characterization, genome sequencing and mating with an industrial penicillin-producing strain. Mol. Microbiol. *95*, 859–874.
Brown, D.W., Yu, J.H., Kelkar, H.S., Fernandes, M., Nesbitt, T.C., Keller, N.P., Adams, T.H., and Leonard, T.J. (1996). Twenty-five coregulated transcripts define a sterigmatocystin gene cluster in *Aspergillus nidulans*. Proc. Natl. Acad. Sci. U.S.A. *93*, 1418–1422.
Cabanes, F.J., Sanseverino, W., Castella, G., Bragulat, M.R., Cigliano, R.A., and Sanchez, A. (2015). Rapid genome resequencing of an atoxigenic strain of *Aspergillus carbonarius*. Sci. Rep. *5*, 9086.
Camps, S.M., Dutilh, B.E., Arendrup, M.C., Rijs, A.J., Snelders, E., Huynen, M.A., Verweij, P.E., and Melchers, W.J. (2012). Discovery of a HapE mutation that causes azole resistance in *Aspergillus fumigatus* through whole genome sequencing and sexual crossing. PLoS One *7*, e50034.
Currie, J.N. (1917). The citric acid fermentation of *Aspergillus niger*. J. Biol. Chem. *31*, 15–37.
Diaz-Guerra, T.M., Mellado, E., Cuenca-Estrella, M., and Rodriguez-Tudela, J.L. (2003). A point mutation in the 14alpha-sterol demethylase gene cyp51A contributes to itraconazole resistance in *Aspergillus fumigatus*. Antimicrob. Agents Chemother. *47*, 1120–1124.
Fedorova, N.D., Khaldi, N., Joardar, V.S., Maiti, R., Amedeo, P., Anderson, M.J., Crabtree, J., Silva, J.C., Badger, J.H., Albarraq, A., *et al.* (2008). Genomic islands in the pathogenic filamentous fungus *Aspergillus fumigatus*. PLoS Genet. *4*, e1000046.
Fleming, A. (1929). On the antibacterial action of cultures of a *Penicillium*, with special reference to their use in the isolation of *B. influenzæ*. Br. J. Exp. Pathol. *10*, 226–236.
Frisvad, J.C., Larsen, T.O., Thrane, U., Meijer, M., Varga, J., Samson, R.A., and Nielsen, K.F. (2011). Fumonisin and ochratoxin production in industrial *Aspergillus niger* strains. PLoS One *6*, e23496.
Galagan, J.E., Calvo, S.E., Cuomo, C., Ma, L.J., Wortman, J.R., Batzoglou, S., Lee, S.I., Basturkmen, M., Spevak, C.C., Clutterbuck, J., *et al.* (2005). Sequencing of *Aspergillus nidulans* and comparative analysis with *A. fumigatus* and *A. oryzae*. Nature *438*, 1105–1115.
Gallo, A., Bruno, K.S., Solfrizzo, M., Perrone, G., Mule, G., Visconti, A., and Baker, S.E. (2012). New insight into the ochratoxin A biosynthetic pathway through deletion of a nonribosomal peptide synthetase gene in *Aspergillus carbonarius*. Appl. Environ. Microbiol. *78*, 8208–8218.
Gallo, A., Knox, B.P., Bruno, K.S., Solfrizzo, M., Baker, S.E., and Perrone, G. (2014). Identification and characterization of the polyketide synthase involved in ochratoxin A biosynthesis in *Aspergillus carbonarius*. Int. J. Food Microbiol. *179*, 10–17.
Grigoriev, I.V., Cullen, D., Goodwin, S.B., Hibbett, D., Jeffries, T.W., Kubicek, C.P., Kuske, C., Magnuson, J.K., Martin, F., Spatafora, J.W., *et al.* (2011). Fueling the future with fungal genomics. Mycology *2*, 192–209.
Grigoriev, I.V., Nikitin, R., Haridas, S., Kuo, A., Ohm, R., Otillar, R., Riley, R., Salamov, A., Zhao, X., Korzeniewski, F., *et al.* (2014). MycoCosm portal: gearing up for 1000 fungal genomes. Nucleic Acids Res. *42*, D699–704.
Hagiwara, D., Takahashi, H., Watanabe, A., Takahashi-Nakaguchi, A., Kawamoto, S., Kamei, K., and Gonoi, T. (2014). Whole-genome comparison of *Aspergillus fumigatus* strains serially isolated from patients with aspergillosis. J. Clin. Microbiol. *52*, 4202–4209.
Houbraken, J., Frisvad, J.C., and Samson, R.A. (2011). Fleming's penicillin producing strain is not *Penicillium chrysogenum* but *P. rubens*. IMA Fungus *2*, 87–95.
Houbraken, J., Frisvad, J.C., Seifert, K.A., Overy, D.P., Tuthill, D.M., Valdez, J.G., and Samson, R.A. (2012). New penicillin-producing *Penicillium* species and an overview of section Chrysogena. Persoonia *29*, 78–100.
Ilmen, M., Thrane, C., and Penttila, M. (1996). The glucose repressor gene cre1 of Trichoderma: isolation and expression of a full-length and a truncated mutant form. Mol. Gen. Genet. *251*, 451–460.
Klaric, M.S., Rasic, D., and Peraica, M. (2013). Deleterious effects of mycotoxin combinations involving ochratoxin A. Toxins (Basel) *5*, 1965–1987.

Le Crom, S., Schackwitz, W., Pennacchio, L., Magnuson, J.K., Culley, D.E., Collett, J.R., Martin, J., Druzhinina, I.S., Mathis, H., Monot, F., *et al.* (2009). Tracking the roots of cellulase hyperproduction by the fungus *Trichoderma reesei* using massively parallel DNA sequencing. Proc. Natl. Acad. Sci. U.S.A. *106*, 16151–16156.

Li, A., Pfelzer, N., Zuijderwijk, R., Brickwedde, A., van Zeijl, C., and Punt, P. (2013). Reduced by-product formation and modified oxygen availability improve itaconic acid production in *Aspergillus niger*. Appl. Microbiol. Biotechnol. 97, 3901–3911.

Li, A., Pfelzer, N., Zuijderwijk, R., and Punt, P. (2012). Enhanced itaconic acid production in *Aspergillus niger* using genetic modification and medium optimization. BMC Biotechnol. *12*, 57.

Liu, G., Zhang, L., Qin, Y., Zou, G., Li, Z., Yan, X., Wei, X., Chen, M., Chen, L., Zheng, K., *et al.* (2013a). Long-term strain improvements accumulate mutations in regulatory elements responsible for hyper-production of cellulolytic enzymes. Sci. Rep. 3, 1569.

Liu, G., Zhang, L., Wei, X., Zou, G., Qin, Y., Ma, L., Li, J., Zheng, H., Wang, S., Wang, C., *et al.* (2013b). Genomic and secretomic analyses reveal unique features of the lignocellulolytic enzyme system of *Penicillium decumbens*. PLoS One *8*, e55185.

Machida, M., Asai, K., Sano, M., Tanaka, T., Kumagai, T., Terai, G., Kusumoto, K., Arima, T., Akita, O., Kashiwagi, Y., *et al.* (2005). Genome sequencing and analysis of *Aspergillus oryzae*. Nature *438*, 1157–1161.

Mann, P.A., Parmegiani, R.M., Wei, S.Q., Mendrick, C.A., Li, X., Loebenberg, D., DiDomenico, B., Hare, R.S., Walker, S.S., and McNicholas, P.M. (2003). Mutations in *Aspergillus fumigatus* resulting in reduced susceptibility to posaconazole appear to be restricted to a single amino acid in the cytochrome P450 14alpha-demethylase. Antimicrob. Agents Chemother. *47*, 577–581.

Martin, F., Cullen, D., Hibbett, D., Pisabarro, A., Spatafora, J.W., Baker, S.E., and Grigoriev, I.V. (2011). Sequencing the fungal tree of life. New Phytol. *190*, 818–821.

McCluskey, K. (2011). From genetics to genomics: fungal collections at the Fungal Genetics Stock Center. Mycology 2, 161–168.

McCluskey, K., Wiest, A., and Boundy-Mills, K. (2014). Genome data drives change at culture collections, fungal genomics (Springer Berlin Heidelberg, Germany), pp. 81–96.

McCluskey, K., Wiest, A.E., Grigoriev, I.V., Lipzen, A., Martin, J., Schackwitz, W., and Baker, S.E. (2011). Rediscovery by whole genome sequencing: classical mutations and genome polymorphisms in *Neurospora crassa*. G3 (Bethesda) *1*, 303–316.

Nascimento, A.M., Goldman, G.H., Park, S., Marras, S.A., Delmas, G., Oza, U., Lolans, K., Dudley, M.N., Mann, P.A., and Perlin, D.S. (2003). Multiple resistance mechanisms among *Aspergillus fumigatus* mutants with high-level resistance to itraconazole. Antimicrob. Agents Chemother. *47*, 1719–1726.

Nierman, W.C., Pain, A., Anderson, M.J., Wortman, J.R., Kim, H.S., Arroyo, J., Berriman, M., Abe, K., Archer, D.B., Bermejo, C., *et al.* (2005). Genomic sequence of the pathogenic and allergenic filamentous fungus *Aspergillus fumigatus*. Nature *438*, 1151–1156.

Pel, H.J., de Winde, J.H., Archer, D.B., Dyer, P.S., Hofmann, G., Schaap, P.J., Turner, G., de Vries, R.P., Albang, R., Albermann, K., *et al.* (2007). Genome sequencing and analysis of the versatile cell factory *Aspergillus niger* CBS 513.88. Nat. Biotechnol. *25*, 221–231.

Peng, Q., Yuan, Y., Gao, M., Chen, X., Liu, B., Liu, P., Wu, Y., and Wu, D. (2014). Genomic characteristics and comparative genomics analysis of *Penicillium chrysogenum* KF-25. BMC Genomics *15*, 144.

Pomraning, K.R., Smith, K.M., and Freitag, M. (2011). Bulk segregant analysis followed by high-throughput sequencing reveals the Neurospora cell cycle gene, ndc-1, to be allelic with the gene for ornithine decarboxylase, spe-1. Eukaryot. Cell *10*, 724–733.

Pontecorvo, G. (1952). Non-random distribution of multiple mitotic crossing-over among nuclei of heterozygous diploid *Aspergillus*. Nature *170*, 204–205.

Pontecorvo, G., and Sermonti, G. (1954). Parasexual recombination in *Penicillium chrysogenum*. J. Gen. Microbiol. *11*, 94–104.

Seidl, V., Gamauf, C., Druzhinina, I.S., Seiboth, B., Hartl, L., and Kubicek, C.P. (2008). The Hypocrea jecorina (Trichoderma reesei) hypercellulolytic mutant RUT C30 lacks a 85 kb (29 gene-encoding) region of the wild-type genome. BMC Genomics *9*, 327.

Seidl, V., Seibel, C., Kubicek, C.P., and Schmoll, M. (2009). Sexual development in the industrial workhorse *Trichoderma reesei*. Proc. Natl. Acad. Sci. U.S.A. *106*, 13909–13914.

Smith, D.R., Quinlan, A.R., Peckham, H.E., Makowsky, K., Tao, W., Woolf, B., Shen, L., Donahue, W.F., Tusneem, N., Stromberg, M.P., *et al.* (2008). Rapid whole-genome mutational profiling using next-generation sequencing technologies. Genome Res. *18*, 1638–1642.

Specht, T., Dahlmann, T.A., Zadra, I., Kurnsteiner, H., and Kuck, U. (2014). Complete sequencing and chromosome-scale genome assembly of the industrial progenitor strain P2niaD18 from the penicillin producer *Penicillium chrysogenum*. Genome Announc. *2*.

Umemura, M., Koike, H., Yamane, N., Koyama, Y., Satou, Y., Kikuzato, I., Teruya, M., Tsukahara, M., Imada, Y., Wachi, Y., *et al.* (2012). Comparative genome analysis between *Aspergillus oryzae* strains reveals close relationship between sites of mutation localization and regions of highly divergent genes among *Aspergillus* species. DNA Res. *19*, 375–382.

van den Berg, M.A., Albang, R., Albermann, K., Badger, J.H., Daran, J.M., Driessen, A.J., Garcia-Estrada, C., Fedorova, N.D., Harris, D.M., Heijne, W.H., *et al.* (2008). Genome sequencing and analysis of the filamentous fungus *Penicillium chrysogenum*. Nat. Biotechnol. *26*, 1161–1168.

van der Straat, L., Tamayo-Ramos, J.A., Schonewille, T., and de Graaff, L.H. (2013). Overexpression of a modified 6-phosphofructo-1-kinase results in an increased itaconic acid productivity in *Aspergillus niger*. AMB Express *3*, 57.

van der Straat, L., Vernooij, M., Lammers, M., van den Berg, W., Schonewille, T., Cordewener, J., van der Meer, I., Koops, A., and de Graaff, L.H. (2014). Expression of the *Aspergillus terreus* itaconic acid biosynthesis cluster in *Aspergillus niger*. Microb. Cell Fact. *13*, 11.

Vitikainen, M., Arvas, M., Pakula, T., Oja, M., Penttila, M., and Saloheimo, M. (2010). Array comparative genomic hybridization analysis of *Trichoderma reesei* strains with enhanced cellulase production properties. BMC Genomics *11*, 441.

Wang, F.Q., Zhong, J., Zhao, Y., Xiao, J., Liu, J., Dai, M., Zheng, G., Zhang, L., Yu, J., Wu, J., *et al.* (2014). Genome sequencing of high-penicillin producing industrial strain of *Penicillium chrysogenum*. BMC Genomics *15*(Suppl. 1), S11.

Wong, V.L., Ellison, C.E., Eisen, M.B., Pachter, L., and Brem, R.B. (2014). Structural variation among wild and industrial strains of *Penicillium chrysogenum*. PLoS One *9*, e96784.

Yu, J., Chang, P.K., Cary, J.W., Wright, M., Bhatnagar, D., Cleveland, T.E., Payne, G.A., and Linz, J.E. (1995). Comparative mapping of aflatoxin pathway gene clusters in *Aspergillus parasiticus* and *Aspergillus flavus*. Appl. Environ. Microbiol. *61*, 2365–2371.

Zhao, G., Yao, Y., Chen, W., and Cao, X. (2013a). Comparison and analysis of the genomes of two *Aspergillus oryzae* strains. J. Agric. Food. Chem. *61*, 7805–7809.

Zhao, G., Yao, Y., Hou, L., Wang, C., and Cao, X. (2014a). Comparison of the genomes and transcriptomes associated with the different protease secretions of *Aspergillus oryzae* 100-8 and 3.042. Biotechnol. Lett. *36*, 2053–2058.

Zhao, G., Yao, Y., Hou, L., Wang, C., and Cao, X. (2014b). Draft genome sequence of *Aspergillus oryzae* 100-8, an increased acid protease production strain. Genome Announc. *2*.

Zhao, G., Yao, Y., Wang, C., Hou, L., and Cao, X. (2013b). Comparative genomic analysis of *Aspergillus oryzae* strains 3.042 and RIB40 for soy sauce fermentation. Int. J. Food Microbiol. *164*, 148–154.

3 Diversity and Mechanisms of Genomic Adaptation in *Penicillium*

Jeanne Ropars, Ricardo C. Rodríguez de la Vega, Manuela López-Villavicencio, Joëlle Dupont, Dominique Swennen, Emilie Dumas, Tatiana Giraud and Antoine Branca

Abstract

Penicillium is a diverse fungal genus with hundreds of species occurring worldwide in various substrates, from soil to food, and with various lifestyles, from necrotrophic pathogenicity to endophytic mutualism. Several species are important for human affairs, being widely used in industry, such as the penicillin-producer *P. rubens*, the two cheese starters *P. camemberti* and *P. roqueforti*, and the mould used for fermenting sausages, *P. nalgiovense*. Other species are food spoilers that produce harmful mycotoxins or cause damages in fruit crops. Currently, 30 genomes of *Penicillium* belonging to 18 species are available. In this chapter, we reconstruct a phylogenetic tree based on available *Penicillium* genomes and outline the main features of the genomes, such as gene and transposable element content. We then review the recent advances that the available genomic and transcriptomic resources in the *Penicillium* genus have allowed regarding our understanding of the genomic processes of adaptation, including changes in gene content, expression and strikingly frequent and recent horizontal gene transfers. In addition, we summarize recent studies using genetic markers on the level of genetic diversity, mode of reproduction and population structure within *Penicillium* species. Overall, the *Penicillium* genus appears a highly suitable model for studying the mechanisms of adaptation.

Introduction

Penicillium is a diverse fungal genus with 354 accepted species today (Visagie *et al.*, 2014b), occurring worldwide in various substrates, from soil to food. Their lifestyles also cover a broad range, from necrotrophic pathogenicity to endophytic mutualism, although most are saprotrophs. As a consequence of these ecological niches, many *Penicillium* species have important economic and social relevance for human populations. Several species are widely used in industry, such as the penicillin-producer *P. rubens*, the two cheese starters *P. camemberti* and *P. roqueforti*, and the mould used for fermenting sausages, *P. nalgiovense* (Bernáldez *et al.*, 2013). Other species cause damages and yield loss in fruit crops, e.g. *P. digitatum* and *P. italicum*, while others are a concern for food safety because of their production of mycotoxins, such as patulin (Eckert and Eaks, 1989; McCallum *et al.*, 2002).

In addition to their economic significance, *Penicillium* moulds also provide a tractable model for understanding the genetic and genomic processes underlying adaptation, due to the diversity of their ecological niches, their small genomes, their long haploid phase, their short generation time, and their easy manipulations in laboratory. Therefore, they can help addressing the current key challenges in evolutionary biology, including the identification of the genes involved in ecologically relevant traits as well as the understanding of the nature, time course, and architecture of the genomic changes involved in the origin and processes of population adaptation and divergence (Gladieux *et al.*, 2014).

The *Penicillium* species used in industry (e.g. for cheese maturation, sausage fermentation or for penicillin production) represent particularly well-suited organisms for studying domestication, a selection-based process studied since the dawn of evolutionary thinking as a model of rapid adaptation and diversification (Darwin, 1868). Yet, the process of domestication has been much less studied in eukaryote microorganisms than in plants or animals. Several traits are typically modified in domesticated fungi, such as colour of colonies, growth rate, thallus density, length of conidiophores, rate and rapidity of spore germination (Eichler, 1968; Moreau, 1979). In the baker's yeast *Saccharomyces cerevisiae*, the genomic processes associated with domestication include large-scale duplications leading to genome size expansions (Liti and Louis, 2005; Machida *et al.*, 2005), the acquisition of new traits by horizontal gene transfers (HGT) (Hall *et al.*, 2005; Khaldi and Wolfe, 2008; Khaldi *et al.*, 2008; Novo *et al.*, 2009) and hybridization (Liti *et al.*, 2006). These mechanisms have been suggested as ways for fungi to increase their biochemical repertoire and their ability to adapt to new ecological niches, but its generality remains to be assessed (Friesen *et al.*, 2006; Khaldi and Wolfe, 2008; Khaldi *et al.*, 2008; Rosewich and Kistler, 2000; Wisecaver *et al.*, 2014).

In this chapter, we aimed at summarizing the insights that the available genomic resources in the *Penicillium* genus have contributed to our understanding of the genomic processes of adaptation. In addition, we review recent studies using genetic markers on the level of genetic diversity, mode of reproduction and population structure within *Penicillium* species. We start with a section describing the *Penicillium* species for which genomes are available, with special emphasis on their ecological niches and life history traits. Second, we reconstruct a phylogenetic tree based on available *Penicillium* genomes and outline the main features of the genomes, such as gene and transposable element content. Third, we summarize recent findings on strikingly frequent and recent horizontal gene transfers. Fourth, we review some of the recent transcriptomic studies performed in *Penicillium* fungi. We finally review recent investigations on genetic diversity and population structure within *Penicillium* species.

Ecological niches and life history traits

Currently, 30 genomes of *Penicillium* belonging to 18 species are available in public databases (Table 3.1); these species are necrotrophic plant pathogens (*P. digitatum*, *P. expansum*, *P. italicum*), common food spoilers (*P. biforme*, *P. fuscoglaucum*, *P. carneum*, *P. paneum*), or key industrial species for food production (*P. camemberti*, *P. roqueforti*, *P. nalgiovense*), pharmaceutical industry (*P. rubens*) or biorefinery (*P. decumbens*). The genome described as *P. aurantiogriseum* in the databases (Yang *et al.*, 2014) most likely belongs to *P. expansum* (Ballester *et al.*, 2015). We detail below some of the specific traits and lifestyles of the *Penicillium* species with a sequenced genome.

The two species used for cheese production, *P. camemberti* and *P. roqueforti*, though sharing the same nutrient-rich ecological niche, are not closely related and have different domestication histories, thus providing ideal models to study parallel adaptation (Elmer and Meyer, 2011). The fungus *P. camemberti*, used for the maturation of soft cheeses like Brie and Camembert, is the result of selection programmes aiming at improving the texture of the colony, the colour of the conidia and physiological characteristics of mycelia. This human-created white species is thought to be derived from a single clone of *P. commune* (Pitt *et al.*, 1986), a species complex split into *P. biforme* and *P. fuscoglaucum* (Giraud *et al.*, 2010). *Penicillium camemberti* has never been isolated from other substrates than dairy products, in contrast to *P. fuscoglaucum* and *P. biforme*, which are considered as contaminants by stakeholders because of their blue-grey colour even if used as a starter culture for the production of some cheeses (Frisvad and Samson, 2004). *Penicillium roqueforti*, used as a starter culture in the production of blue veined-cheeses, is in contrast widespread in food environments and has also been isolated from plant environments such as silage or wood (Frisvad and Samson, 2004).

Penicillium carneum and *P. paneum*, two sister species of *P. roqueforti*, are common food spoilers responsible for the production of harmful mycotoxins (O'Brien *et al.*, 2006; Petersson and Schnürer, 1999). *Penicillium nalgiovense* is used in the food industry as an inoculum on fermented dry sausages in Italy, Spain and France (Bernáldez *et al.*, 2013). It contributes to the taste of sausages and it helps preventing desiccation and protecting them from undesirable microorganisms through antibacterial and antifungal activities (Lücke, 1997; Lücke and Hechelmann, 1987).

The discovery by Alexander Fleming and subsequent mass production of β-lactam antibiotics (including penicillin) have revolutionized medicine and greatly contributed to reduce the mortality due to infectious bacterial diseases in the world (Fleming, 1929; Hersbach *et al.*, 1984). The production of antibiotics in large amounts has been made possible through strain improvement in *P. rubens*. Industrial strains are all derived from a single strain, NRRL1951, isolated from a spoiled cantaloupe during the Second World War (Raper *et al.*, 1944). Until recently, *P. rubens* was considered as a synonym of *P. chrysogenum*, but it is now accepted as a closely related but separated species (Houbraken *et al.*, 2011). *Penicillium solitum* is a common food spoiler (in cheeses and dry meats such as sausages and salami), a pomaceous fruit pathogen (Frisvad, 1981; Pitt and Leistner, 1991; Sanderson and Spotts, 1995) and it has also been isolated from apple orchards' soil and house dust (Frisvad and Samson, 2004; Papagianni *et al.*, 2007; Visagie *et al.*, 2014a). It is also used for production of compactin, a cholesterol-lowering agent with also an antifungal effect (Frisvad and Samson, 2004).

Penicillium digitatum is a necrotrophic pathogen responsible for up to 90% post-harvest losses in citrus storage, particularly in arid and subtropical climates (Eckert and Eaks, 1989). In contrast to most other necrotrophic fungi, *P. digitatum* seems highly specialized as it has never been collected in any other substrate than citrus (Barkai-Golan, 2001). It is therefore a model for the study of specialization in the necrotrophic lifestyle. *Penicillium italicum* presents a very similar lifestyle on the same substrate (Palou, 2014).

Penicillium expansum is famous for being the first described *Penicillium* species; it is an important post-harvest spoilage agent that can cause great losses in apple storage facilities (Jurick *et al.*, 2011). In contrast to *P. digitatum*, *P. expansum* can be found in a wide array of other substrates and is a concern for health care as it can produce patulin, a highly toxic

Table 3.1 Summary of available *Penicillium* genomes

Species	Common synonym	Strain	Environment	Sequencing level	Assembly	Contig/scaffold	References	Size (Mb)	GC%	Gene
P. paxilli		ATCC 26601	Insect-damaged pecans	118 scaffolds	GCA_000347475.1	KB644791–KB644908	Berry *et al.*, 2015	34.8	47.9	-
P. decumbens	*P. oxalicum*	114-2	Industrialized strain derived from decayed straw-covered soil	9 scaffolds	GCA_000346795.1	KB644408–KB644416	Liu *et al.* 2013	30.18	50.7	10013
P. decumbens	*P. oxalicum*	JU-A10-T	Industrialized strain derived from decayed straw-covered soil	21 scaffolds	GCA_000383025.1	KB908888–KB908908	Liu *et al.* 2013	30.68	50.5	10473
P. capsulatum		LiaoWQ-2011	*Homo sapiens* lung	62 contigs	GCA_000943765.1	JPLR01000001–JPLR01000062	Li *et al.* (unpublished, 2014)	34.34	49.10	-
P. capsulatum		ATCC 48735	Exposed canvas	65 contigs	GCA_000943775.1	JPLQ01000001–JPLQ01000065	Li *et al.* (unpublished, 2014)	34.37	49.10	-
P. carneum		LCP05634T	Mouldy rye bread	2090 contigs	GCA_000577495.1	HG816029–HG818118	Ropars, J. *et al.* 2015	25.9	48.73	13412
P. roqueforti		FM164	Cheese	48 scaffolds	GCA_000513255.1	HG792015–HG792062	Cheeseman *et al.* 2014	29.01	48.64	12319
P. roqueforti		UASWS P1	Bread	428 contigs	GCA_000737485.1	JNNS01000001–JNNS01000428	Crovadore *et al.* (unpublished, 2014)	27.93	48.2	-
P. paneum		FM227	Cheese	224 scaffolds	GCA_000577715.1	HG813308–HG813531	Ropars *et al.* 2015	26.58	49.01	12201
P. verrucosum		BFE808	Wheat	1632 contigs	GCA_000970515.1	LAKW01000001–LAKW01004207	Stoll *et al.* (unpublished, 2015)	30.16	48.06	-
P. nordicum		UASWS BFE487	Unknown	915 contigs	GCA_000733025.1	JNNR01000001–JNNR01000915	Crovadore *et al.* (unpublished, 2014)	30.42	47.7	-
P. solitum		RS1	Apple	703 contigs	GCA_000952775.1	JYNM01000001–JYNM01000703	Yu *et al.* (unpublished, 2015)	32.34	48.42	-
P. fuscoglaucum	*P. commune*	FM041	Cheese	953 scaffolds	GCA_000576735.1	HG814183–HG815135	Ropars *et al.* 2015	36.09	47.74	15615
P. camemberti		FM013	Cheese	180 scaffolds	GCA_000513335.1	HG793134–HG793313	Cheeseman *et al.* 2014	35.01	48.18	14525
P. biforme	*P. commune*	FM169	Cheese	582 scaffolds	GCA_000577785.1	HG813601–HG814182	Ropars *et al.* 2015	34.87	48.09	15246

P. digitatum		Pd1	Grapefruit	54 scaffolds	GCA_000315645.2	JH993634–JH993686	Marcet-Houben *et al*. 2012	26.08	48.9	-
P. digitatum		PHI26	Orange	101 scaffolds	GCA_000315665.1	JH993687–JH993786	Marcet-Houben *et al*. 2012	26.00	48.9	9153
P. digitatum		Pd01-ZJU	Citrus	1816 contigs	GCA_000485865.1	ANGJ01000001–ANGJ01001816	Sun *et al*. 2013	25.01	48.92	-
P. italicum		PHI-1	Citrus	1632 contigs	GCA_000769765.1	JQGA01000001–JQGA01001632	Stoll *et al*. (unpublished, 2015)	28.85	47.3	9996
P. expansum		R19	Apple	1231 contigs	GCA_000688875.1	JHUC01000001–JHUC01001231	Yu, *et al*. 2014	31.41	48.24	-
P. expansum		CMP-1	Apple	1723 contigs	GCA_000769755.1	JQFX01000001–JQFX01001723	Stoll *et al*. (unpublished, 2015)	31.09	48.2	10683
P. expansum		d1	Apple	270 contigs	GCA_000769735.1	JQFY01000001–JQFY01000270	Stoll *et al*. (unpublished, 2015)	31.43	47.6	11048
P. expansum		MD8	Apple	382 contigs	GCA_000769745.1	JQFZ01000001–JQFZ01000382	Stoll *et al*. (unpublished, 2015)	32.36	47.6	11070
P. expansum		NRRL 62431	Endophyte hazel	4775 contigs	GCA_000584915.1	ALJY01000001–ALJY01004775	Yang *et al*. 2014	31.55	48.5	-
P. rubens	*P. chrysogenum*	Wisconsin 54-1255	Penicillin producer, industrialized strain derived from isolate of mouldy cantaloupe	49 scaffolds	GCA_000226395.1	AM920416–AM920464	Van den Berg *et al*., 2008	32.19	48.9	12943
P. rubens	*P. chrysogenum*	IB 08/921	Unknown, descendant of a wild-type isolate	1151 scaffolds	GCA_000801355.1	JPDR01000001–JPDR01001151	Böhm *et al*. 2015	32.24	49.0	-
P. rubens	*P. chrysogenum*	Wisconsin P2niaD18	Penicillin producer, industrialized strain derived from isolate of mouldy cantaloupe	5 chromosomes	GCA_000710275.1	CM002798–CM002802	Specht *et al*. (unpublished, 2014)	32.53	48.9	-
P. rubens	*P. chrysogenum*	NCPC10086	Penicillin producer, industrialized strain derived from isolate of mouldy cantaloupe	32 scaffolds	GCA_000523475.1	KI963886–KI963917	Wang *et al*. 2014	32.3	48.9	-
P. chrysogenum		KF-25	Soil	87 scaffolds	GCA_000816005.1	KN715891–KN715977	Peng *et al*. (unpublished, 2013)	29.92	49.0	-
P. nalgiovense		FM193	Cheese	869 scaffolds	GCA_000577395.1	HG815136–HG815228, HG815290–HG816004	Ropars *et al*. 2015	31.94	48.48	14503

mycotoxin (McCallum *et al.*, 2002). The role of patulin in the capacity of *P. expansum* to colonize fruits and diverse ecological niches remains debated (Ballester *et al.*, 2015).

Penicillium capsulatum and *P. decumbens* both produce secondary metabolites useful for the industry, in particular highly efficient enzymes for degrading cellulose; *P. decumbens* is also used in biorefinery as a renewable source for oil production (Li *et al.*, 2010). *Penicillium verrucosum* and *P. nordicum* are both known for producing ochratoxin A, one of the most common mycotoxins in spoiled food (Castella *et al.*, 2002). *Penicillium paxilli* is the species which the potent tremor-inducing blocker of calcium-activated potassium channels paxillin was originally isolated from (Berry *et al.*, 2015).

Phylogenetic relationships, genome size and content, and changes in gene content

Despite the economically and ecologically important role of *Penicillium* fungi, their phylogenetic relationships still remain unclear, with only a few phylogenetic trees published, based on single genes (Houbraken and Samson, 2011; Samson *et al.*, 2004; Seifert *et al.*, 2007). We therefore constructed a phylogenetic tree, including all *Penicillium* strains with an available sequenced genome and four *Aspergillus* species as outgroups, based on 3,986 shared single-copy orthologues, corresponding to a concatenated alignment of 1,198,500 bp (Fig. 3.1). In the *Penicillium* clade, some internal nodes remained poorly supported; this may reflect a rapid radiation in this clade, leading to incomplete lineage sorting, and/or hybridizations or recurrent horizontal gene transfers (Cheeseman *et al.*, 2014; Ropars *et al.*, 2015).

Genomes in the *Penicillium* genus appear highly dynamic, with estimated genome sizes ranging from 25 Mb to 36 Mb (Table 3.1). A relationship between host range and genome size has been proposed for fruit pathogens (Ballester *et al.*, 2015; Marcet-Houben *et al.*, 2012). For example, *P. digitatum*, which presents the smallest genome (25.7Mb in average), is only able to infect citrus fruits, whereas the generalist *P. expansum* (pathogen of pome and stone fruits) has the largest genome among fruit pathogens (ca 31 Mb); *P. italicum* presents intermediate host range (mainly citrus fruits) and genome size (ca 29 Mb) (Table 3.1).

We analysed the repeat content of all available genome using RepeatMasker (Smit *et al.*, 2013). Overall, *Penicillium* genomes have a low proportion of interspersed repeats, ranging from 0.32% to 1.71% of total genome assembly lengths, and yet all classes of transposable elements (TEs) are represented (Fig. 3.2). The 10 most abundant types of TEs include four non-LTR *i* retroelements (I-1_AO, I-4_AO, I-5_AO and I-6_AO), three *mariner* DNA transposons (Mariner-6_AN, MarinerL-1_AO and Mariner-1_AF), one *hAT* DNA transposon (hAT-1_AN), one *gypsy* LTR retrotransposon (Gypsy1-I_AO) and one *R1* non-LTR retroelement (RTAg4). *Gypsy* elements are by far the most abundant TE class found in *Penicillium* genomes as they account for 20% of all TEs; *mariner* elements constitute the second most abundant class (13%).

Penicillium biforme, *P. camemberti* and *P. fuscoglaucum*, three closely related species occurring in cheese, show the largest genome sizes among available *Penicillium* genomes (ca. 35–36 Mb, Table 3.1), which is not associated with a particularly high TE content (Fig. 3.2). Interestingly, a large expansion of the proteome seems to have occurred in the ancestor of this clade, indicating that the increase in genome size might be correlated with the

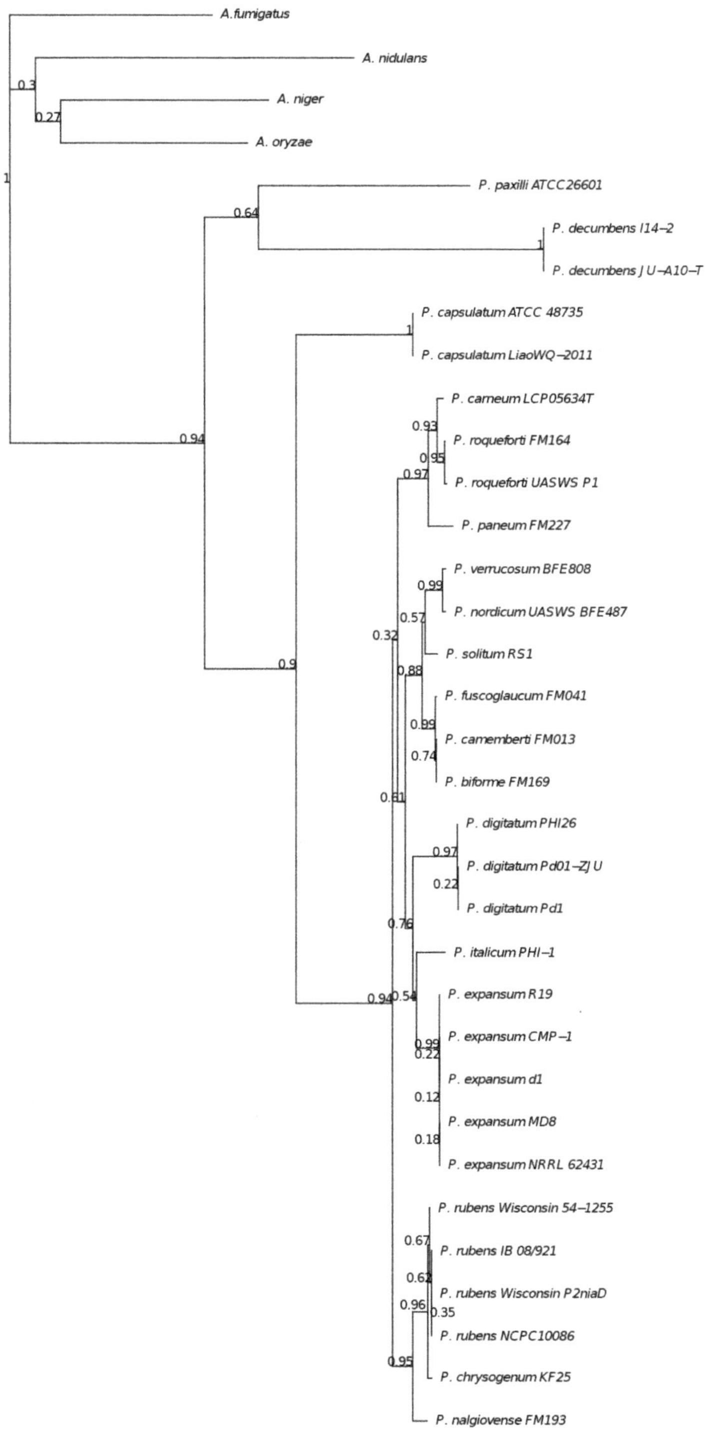

Figure 3.1 Maximum likelihood tree of genome sequenced *Penicillium* based on concatenated alignment of 3986 single copy orthologues using RaxML (Stamatakis, 2014). Node labels correspond to the proportion of gene trees supporting the node.

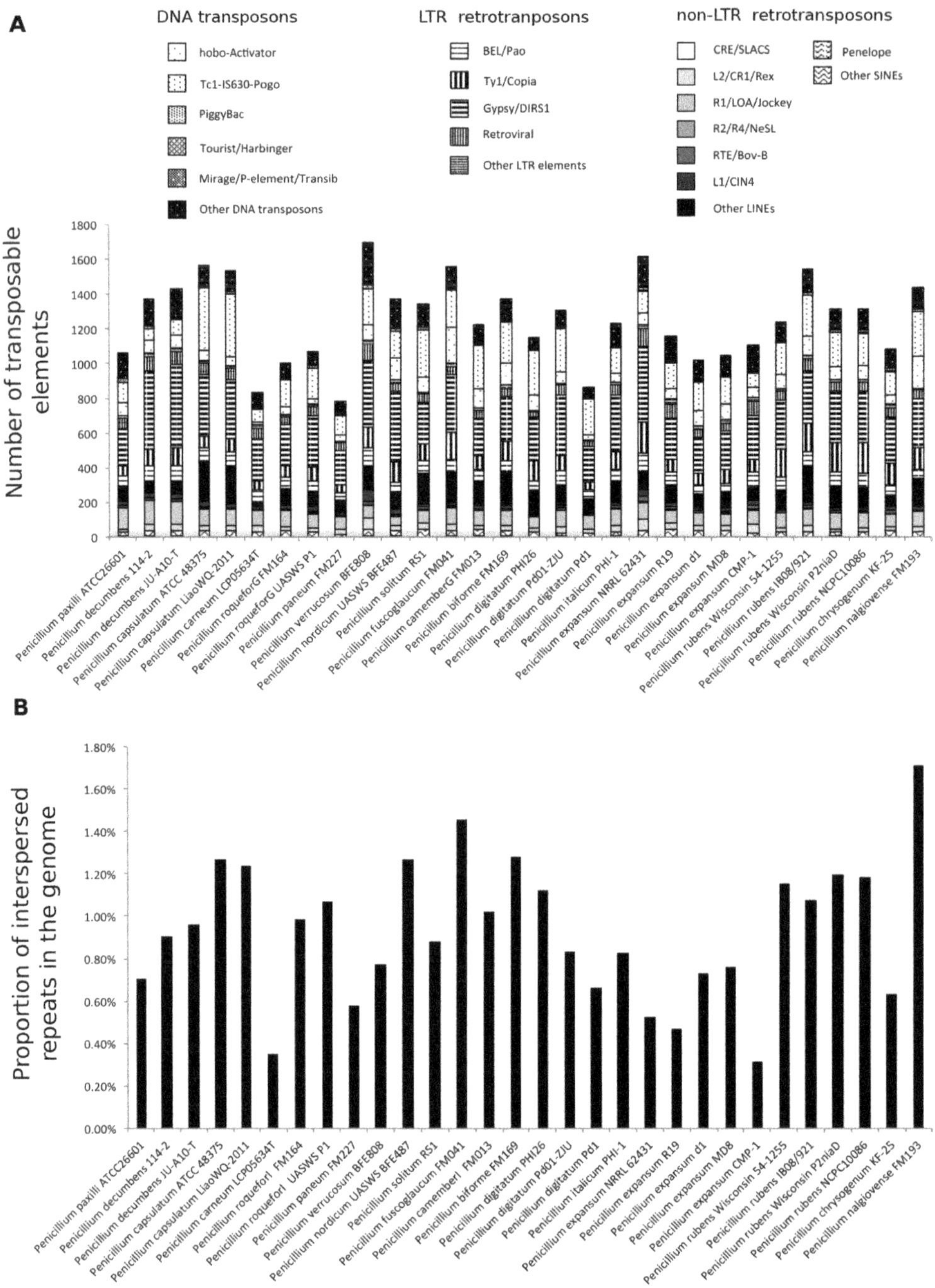

Figure 3.2 (A) Number and classification of transposable elements found in *Penicillium* genomes; (B) Percentage of the *Penicillium* genomes composed of interspersed repeats.

acquisition of new genes. In contrast, *P. digitatum* is characterized by a much lower number of genes than other *Penicillium* lineages (Marcet-Houben *et al.*, 2012; Ropars *et al.*, 2015), which probably relates to its highly specialized necrotrophic lifestyle.

Comparative genomic studies have revealed interesting patterns of gains or losses in genes involved in the production of secondary metabolites, such as penicillin, patulin or small secreted proteins acting as effectors in pathogens. *Penicillium rubens* for instance has acquired its capacity to produce high quantities of penicillin through the duplication of its penicillin biosynthetic gene cluster (Fierro *et al.*, 1995) A comparison of two industrial *P. rubens* genomes, one of a high-penicillin-producing strain and one of a low-penicillin-producing strain, revealed in the high-penicillin-producing strain an even higher number of copies of the penicillin biosynthetic gene cluster, as well as many genomic structural variations, such as translocations and gene gains/losses, likely related to an enhanced nitrogen and energy metabolism (Wang *et al.*, 2014). In the necrotrophic species *P. expansum*, a large number of secondary metabolism gene clusters were identified that were absent in other sequenced *Penicillium* genomes and may be involved in pathogenicity (Ballester *et al.*, 2015). In fact, despite a major genome contraction compared with other *Penicillium* species, *P. expansum* had the largest repertoire of secondary metabolites genes, indicating high numbers of gene gains and losses in this species. The patulin gene cluster was inferred to be present in the ancestor of *P. expansum* and *P. roqueforti*, its absence in other lineages of this clade implying gene losses (Ballester *et al.*, 2015). Overall, *Penicillium* genomes thus appear highly dynamic, with changes in gene content relating to genomic adaptations.

Horizontal gene transfers

The multiple available *Penicillium* genomes have further allowed detecting dozens of horizontal gene transfers (HGTs) that are transmissions of genetic material between species by other means than sexual reproduction. HGT events are most often detected by the existence of incongruences between gene genealogies and the species tree. Indeed, the finding of orthologues from distantly related species placed close together in a gene tree most likely indicates that this gene has recently been horizontally transferred between the two species instead of having followed vertical inheritance and divergence along the species tree. Despite long thought to be rare in eukaryotes, recent studies have shown that HGTs may play a major role in adaptation in this lineage, in particular in fungi (Gladieux *et al.*, 2014; Keeling, 2009; Keeling and Palmer, 2008; Wisecaver *et al.*, 2014).

Fungi are the eukaryotic group for which the largest number of HGT events has been described so far. Most of described HGTs have a prokaryotic origin, likely reflecting the abundance of prokaryotes in all environments and the relative ease of detecting such HGT events compared to those from a eukaryotic origin (Gladieux *et al.*, 2014). However, many HGTs between fungi have also been described, such as transfers of genes involved in secondary metabolite pathways in *Aspergillus* and *Penicillium* clades (Wisecaver *et al.*, 2014).

Among *Penicillium* species, the penicillin producer *P. rubens* may have acquired several important genes from bacteria by horizontal gene transfers, including some of the penicillin biosynthetic genes, *pcbAB* and *pcbC*, and the arsenate resistance cluster (van den Berg *et al.*, 2008). Genome analysis of the necrotrophic fungus *P. digitatum* revealed four putative genes that have been horizontally acquired from prokaryotes, including DEC1, likely playing a role in pathogenicity (Marcet-Houben *et al.*, 2012); indeed, it belongs to a gene family associated with virulence in maize infections, with homologues in plant pathogenic fungi and in bacteria, but without any homologue in non-pathogenic *Penicillium* species.

Several other horizontal gene transfers in *Penicillium* have occurred in the cheese environment, being striking by (1) the size of the transferred regions (i.e. several kilobases), (2) the eukaryotic origin of these transfers (likely among cheese-associated *Penicillium* species), (3) the number of species in which the same regions have been horizontally transferred, and (4) the very recent date of the transfers, likely associated with the human history of cheese production (Cheeseman *et al.*, 2014; Ropars *et al.*, 2015). These horizontally transferred regions (HTRs) indeed occurred between *Penicillium* species from the cheese environment, and were completely identical at the nucleotide level between distant species (otherwise having a pairwise sequence identity of 85–90%) while lacking in other closely related species. One of these HTRs, *Wallaby*, is a 575 kb region that accounts for 2% of the *P. roqueforti* genome, and it can be in a single block or in a few fragments depending on the species (Cheeseman *et al.*, 2014). Another HTR, *CheesyTer*, is 80 kb long, and is always found in a single block (Ropars *et al.*, 2015).

These two HTRs are flanked by copies of transposable elements (TEs) belonging to a specific family, the *i* non-LTR retrotransposons, that are rare elsewhere in the genomes (Ropars *et al.*, 2015). This suggests that these TEs may be involved in the horizontal gene transfers. In fungi, the transfer of genetic material is thought to occur through conjugation, natural and agrobacterial transformation, viral transduction, or anastomosis (Coelho *et al.*, 2013; Wisecaver and Rokas, 2015). In *Fusarium*, for example, the transfer of an entire chromosome can occur by simple co-incubation of mycelial of two strains (Ma *et al.*, 2010).

The gene content in *Wallaby* and *CheesyTer* suggests that these transfers may play an important role in the adaptation of these fungi to the cheese environment. Among the 248 genes that *Wallaby* was predicted to contain, two genes, *paf* and Hce2, encoded proteins that may be involved in antagonistic interactions with other microorganisms (Cheeseman *et al.*, 2014). *CheesyTer* carries 37 putative genes, including genes coding for lactose permease and beta-galactosidase, which likely provide advantages in terms of use of the cheese substrate (Ropars *et al.*, 2015). Actually, these two genes were found to be overexpressed in the first days of cheese maturation (Lessard *et al.*, 2014; Ropars *et al.*, 2015).

In *P. roqueforti*, all strains were found to carry either both or none of these two HTRs. The two HTRs were present only in strains found in the dairy environment, while lacking in some strains from cheese and in all the strains isolated from other environments, such as silage or wood (Cheeseman *et al.*, 2014; Ropars *et al.*, 2015). This further indicates an advantage conferred by these two HTRs in cheese.

Experiments of growth and competition on different media have further supported a role of the two HTRs in adaptation to cheese. Indeed, *P. roqueforti* strains carrying the two HTRs showed a significantly higher growth rate on cheese medium and a significantly lower growth rate on minimal medium (Ropars *et al.*, 2015). Furthermore, *P. roqueforti* strains carrying the two HTRs showed a significant competitive advantage, both against *P. roqueforti* strains lacking the HTRs and against other *Penicillium* species also lacking the HTRs. Interestingly, this effect was only significant when strains were grown on cheese medium and not on minimal medium (Ropars *et al.*, 2015).

Transcriptomics

The availability of *Penicillium* genomes have also facilitated transcriptomic studies, i.e. investigations of mRNA expression in different conditions, which allows studying the regulation of genes and therefore detecting important genes involved in the adaptation to particular environments. Transcriptomics in *Penicillium* has focused so far on penicillin production in *P. chrysogenum*, on secondary metabolites production in plant pathogens, such as *P. expansum*, *P. digitatum* and *P. italicum*, and on cheese-making fungi.

The transcriptome analyses of the pathogenic *P. expansum* on apple revealed the induction of several metabolic pathways during infection and thus identified putative pathogenicity factors, such as proteases, cell-wall degrading enzymes and oxidoreductases (Ballester *et al.*, 2015). Putative effectors that are able to modulate host physiology were also identified (Ballester *et al.*, 2015).

A metatranscriptome analysis of *P. camemberti* and *Geotrichum candidum* was performed in a camembert-type cheese matrix (Lessard *et al.*, 2014). The functional annotation allowed the identification of the biological processes involved in cheese ripening. Globally, similar functions appeared involved in the use of the cheese substrate in both the yeast *G. candidum* and the mould *P. camemberti* (Lessard *et al.*, 2014).

A system biology approach, including transcriptomic but also metabolome and metabolic flux analyses, was used to understand the loss of penicillin production capacity by the high-producing *P. chrysogenum* strain during long-term ethanol-limited cultivation, a phenomenon called degeneration. The findings indicated that degeneration was due to the production of a lower quantity of the first two enzymes acting during penicillin biosynthesis, which may be due to a decrease of translation efficiency (Douma *et al.*, 2011).

Population genetic diversity within *Penicillium* species and mode of reproduction

The genetic diversity has also been investigated within *Penicillium* species in some cases, although only with neutrally evolving markers so far. The population genetic variability has been found to differ drastically among *Penicillium* species. In *P. camemberti*, no genetic polymorphism could be detected using either DNA fragments or microsatellites. This is consistent with this species being a clonal lineage originating from a white mutant in the cheese-making mould formerly used for making Brie, *P. commune* (Giraud *et al.*, 2010). In contrast, the genetic diversity in *P. roqueforti* was revealed to be substantial using microsatellites (Ropars *et al.*, 2014) and DNA fragment sequences (Gillot *et al.*, 2015). A strong population structure was found, with one population containing only cheese strains, most of which carried the HTRs described above, *Wallaby* and *CheesyTer*, and a second population containing cheese and non-cheese strains, all lacking the HTRs. These two populations showed further, although weaker, subdivisions that corresponded to different morphologies and different cheese types (Gillot *et al.*, 2015).

To date, genetic diversity has not been investigated within other *Penicillium* species, with the exception of *P. chrysogenum*, which has actually led to the identification of cryptic species and in the subsequent renaming of the penicillin-producing strain in *P. rubens* (Houbraken *et al.*, 2011).

Although sexual stages have been described in many *Penicillium* species (Visagie *et al.*, 2014b), others have historically been considered to be exclusively asexual. This actually holds true for one-fifth of fungal species (Taylor *et al.*, 1999), but was mainly due to the difficulty of observing sex in this phylum in nature. Indeed, direct or indirect evidence of sex have been observed in most cases when thoroughly investigated. Indirect evidence include

(1) the presence of the complete meiotic toolbox (i.e. all the genes known to be necessary for meiosis and for mating-type determinism, with sequences apparently functional, i.e. under purifying selection), (2) footprints of recombination in populations, and (3) footprints of RIP (repeat-induced point-mutation), a defence mechanism of fungal genome inducing C/T transition mutations in repeated sequences during sexual reproduction in ascomycetes (Galagan and Selker, 2004). Recent studies have improved our knowledge of the reproduction mode and breeding system in *Penicillium*. All species studied so far were shown to be heterothallic, as haploid genomes carried a single mating-type allele, either MAT1-1 or MAT1-2 (Hoff *et al.*, 2008; Ropars *et al.*, 2012).

After discoveries of indirect evidence of sex in populations of the penicillin producer *P. rubens*, with occurrence of both mating-type alleles (Hoff *et al.*, 2008) and of RIP footprints (Braumann *et al.*, 2008), a sexual cycle could be induced in this species (Böhm *et al.*, 2013). Similarly, in the cheese species *P. roqueforti*, mating types were shown to occur in balanced ratios in populations, RIP footprints were observed and purifying selection was inferred on genes involved in mating (Ropars *et al.*, 2012). Later, population analyses showed no linkage disequilibria among markers, suggesting recurrent recombination events and fruiting bodies and recombinant sexual ascospores could be successfully produced *in vitro* (Ropars *et al.*, 2014).

Conclusions and future prospects

Genomic and transcriptomic analyses have revealed several interesting genes and mechanisms likely involved in the adaptation of *Penicillium* species to various environments. In particular, the domesticated penicillin-producing and cheese-making *Penicillium* appear ideal model eukaryotes for studying the genomic processes of adaptation, given the recent and strong selection by humans. These genomic inferences now need to be validated using functional genetics, which will be allowed by the recent development of transformation and gene silencing methods (Durand *et al.*, 1991; Gil-Durán *et al.*, 2015; Goarin *et al.*, 2014; Kosalková *et al.*, 2015; Ullán *et al.*, 2008). It will also be very interesting to explore the population genomics of adaptation within species, in particular in the *Penicillium* fungi with high genetic diversity and a variety of ecological niches, such as *P. roqueforti*. Footprints of selective sweeps for instance may reveal selection having acting recently on other genes that the horizontally transferred regions.

Acknowledgements

This work was supported by the ANR FROMA-GEN grant (ANR-12-PDOC-0030) awarded to AB, an '*attractivité*' grant from Paris-Sud University to AB, the ERC starting grant GenomeFun 309403 awarded to TG, a Marie Curie postdoctoral fellowship to RCRdlV [FP7 COFUND PRES-SUD No. 246556], and the ANR grant 'Food Microbiomes' (ANR-08-ALIA-007-02) awarded to JD.

References

Ballester, A.-R., Marcet-Houben, M., Levin, E., Sela, N., Selma-Lázaro, C., Carmona, L., Wisniewski, M., Droby, S., González-Candelas, L., and Gabaldón, T. (2015). Genome, transcriptome, and functional analyses of *Penicillium expansum* provide new insights into secondary metabolism and pathogenicity. Mol. Plant. Microbe Interact. 28, 232–248.

Barkai-Golan, R. (2001). Postharvest diseases of fruits and vegetables: development and control (Elsevier Amsterdam, The Netherlands).

van den Berg, M.A., Albang, R., Albermann, K., Badger, J.H., Daran, J.-M., Driessen, A.J.M., Garcia-Estrada, C., Fedorova, N.D., Harris, D.M., Heijne, W.H.M., *et al.* (2008). Genome sequencing and analysis of the filamentous fungus *Penicillium chrysogenum*. Nat. Biotechnol. *26*, 1161–1168.

Bernáldez, V., Córdoba, J.J., Rodríguez, M., Cordero, M., Polo, L., and Rodríguez, A. (2013). Effect of *Penicillium nalgiovense* as protective culture in processing of dry-fermented sausage "salchichón." Food Control *32*, 69–76.

Berry, D., Cox, M.P., and Scott, B. (2015). Draft genome sequence of the filamentous fungus *Penicillium paxilli* (ATCC 26601). Genome Announc. *3*, e00071–15.

Böhm, J., Hoff, B., O'Gorman, C.M., Wolfers, S., Klix, V., Binger, D., Zadra, I., Kürnsteiner, H., Pöggeler, S., Dyer, P.S., *et al.* (2013). Sexual reproduction and mating-type-mediated strain development in the penicillin-producing fungus *Penicillium chrysogenum*. Proc. Natl. Acad. Sci. U.S.A. *110*, 1476–1481.

Braumann, I., Van den Berg, M., and Kempken, F. (2008). Repeat induced point mutation in two asexual fungi, *Aspergillus niger* and *Penicillium chrysogenum*. Curr. Genet. *53*, 287–297.

Castella, G., Larsen, T.O., Cabanes, J., Schmidt, H., Alboresi, A., Niessen, L., Färber, P., and Geisen, R. (2002). Molecular characterization of ochratoxin A producing strains of the genus *Penicillium*. Syst. Appl. Microbiol. *25*, 74–83.

Cheeseman, K., Ropars, J., Renault, P., Dupont, J., Gouzy, J., Branca, A., Abraham, A.-L., Ceppi, M., Conseiller, E., Debuchy, R., *et al.* (2014). Multiple recent horizontal transfers of a large genomic region in cheese making fungi. Nat. Commun. *5*, 2876.

Coelho, M.A., Gonçalves, C., Sampaio, J.P., and Gonçalves, P. (2013). Extensive intra-kingdom horizontal gene transfer converging on a fungal fructose transporter gene. PLoS Genet. *9*, e1003587.

Darwin, C. (1868). The variation of animals and plants under domestication. John Murray (ed), London, UK.

Douma, R.D., Batista, J.M., Touw, K.M., Kiel, J.A., Krikken, A.M., Zhao, Z., Veiga, T., Klaassen, P., Bovenberg, R.A., Daran, J.-M., *et al.* (2011). Degeneration of penicillin production in ethanol-limited chemostat cultivations of *Penicillium chrysogenum*: A systems biology approach. BMC Syst. Biol. *5*, 132.

Durand, N., Reymond, P., and Fèvre, M. (1991). Transformation of *Penicillium roqueforti* to phleomycin- and to hygromycin B-resistance. Curr. Genet. *19*, 149–153.

Eckert, J., and Eaks, I. (1989). Postharvest disorders and diseases of citrus fruits. In The Citrus Industry, vol. 5, Reuther, W., Calavan, E.C., and Carman, G.E., eds. (University of California Press, Berkeley, CA, USA), pp. 179–260.

Eichler, F. (1968). Beurteilung der Auskeirnfâhigkeit von Blauschimmelsporen. Milchwissenschait *10*, 632–635.

Elmer, K.R., and Meyer, A. (2011). Adaptation in the age of ecological genomics: insights from parallelism and convergence. Trends Ecol. Evol. *26*, 298–306.

Fierro, F., Barredo, J.L., Díez, B., Gutierrez, S., Fernández, F.J., and Martín, J.F. (1995). The penicillin gene cluster is amplified in tandem repeats linked by conserved hexanucleotide sequences. Proc. Natl. Acad. Sci. U.S.A. *92*, 6200–6204.

Fleming, A. (1929). On the antibacterial action of cultures of a penicillium, with special reference to their use in the isolation of B. influenzae. Br. J. Exp. Pathol. *10*, 226.

Friesen, T.L., Stukenbrock, E.H., Liu, Z., Meinhardt, S., Ling, H., Faris, J.D., Rasmussen, J.B., Solomon, P.S., McDonald, B.A., and Oliver, R.P. (2006). Emergence of a new disease as a result of interspecific virulence gene transfer. Nat. Genet. *38*, 953–956.

Frisvad, J.C. (1981). Physiological criteria and mycotoxin production as aids in identification of common asymmetric penicillia. Appl. Environ. Microbiol. *41*, 568–579.

Frisvad, J.C., and Samson, R.A. (2004). Polyphasic taxonomy of *Penicillium* subgenus *Penicillium*. A guide to identification of food and air-borne terverticillate Penicillia and their mycotoxins. Stud. Mycol. *49*, C174.

Galagan, J.E., and Selker, E.U. (2004). RIP: The evolutionary cost of genome defense. Trends Genet. *20*, 417–423.

Gil-Durán, C., Rojas-Aedo, J.F., Medina, E., Vaca, I., García-Rico, R.O., Villagrán, S., Levicán, G., and Chávez, R. (2015). The pcz1 Gene, which encodes a Zn (II) 2Cys6 protein, is involved in the control of growth, conidiation, and conidial germination in the filamentous fungus *Penicillium roqueforti*. PLoS One *10*, e0120740.

Gillot, G., Jany, J.-L., Coton, M., Le Floch, G., Debaets, S., Ropars, J., López-Villavicencio, M., Dupont, J., Branca, A., Giraud, T., *et al.* (2015). Fifty shades of blue: insights into *Penicillium roqueforti* morphological and genetic diversity. PLoS One *10*, e0129849.

Giraud, F., Giraud, T., Aguileta, G., Fournier, E., Samson, R., Cruaud, C., Lacoste, S., Ropars, J., Tellier, A., and Dupont, J. (2010). Microsatellite loci to recognize species for the cheese starter and contaminating strains associated with cheese manufacturing. Int. J. Food Microbiol. *137*, 204–213.

Gladieux, P., Ropars, J., Badouin, H., Branca, A., Aguileta, G., Vienne, D.M., Rodríguez de la Vega, R.C., Branco, S., and Giraud, T. (2014). Fungal evolutionary genomics provides insight into the mechanisms of adaptive divergence in eukaryotes. Mol. Ecol. 23, 753–773.

Goarin, A., Silar, P., and Malagnac, F. (2014). Gene replacement in *Penicillium roqueforti*. Curr. Genet. 1–8.

Hall, C., Brachat, S., and Dietrich, F.S. (2005). Contribution of horizontal gene transfer to the evolution of *Saccharomyces cerevisiae*. Eukaryot. Cell 4, 1102–1115.

Hersbach, G.J.M., Van der Beek, C.P., and Van Dijck, P.W.M. (1984). The penicillins: properties, biosynthesis, and fermentation. In Biotechnology of Industrial Antibiotics. Drugs and the Pharmaceutical Sciences, Vandamme, E.J., ed. (Marcel Dekker, New York) vol. 22, pp. 45–140.

Hoff, B., Pöggeler, S., and Kück, U. (2008). Eighty years after its discovery, Fleming's Penicillium strain discloses the secret of its sex. Eukaryot. Cell 7, 465–470.

Houbraken, J., and Samson, R.A. (2011). Phylogeny of *Penicillium* and the segregation of Trichocomaceae into three families. Stud. Mycol. *70*, 1–51.

Houbraken, J., Frisvad, J.C., and Samson, R.A. (2011). Fleming's penicillin producing strain is not *Penicillium chrysogenum* but *P. rubens*. IMA Fungus 2, 87–95.

Jurick, W.M., Janisiewicz, W.J., Saftner, R.A., Vico, I., Gaskins, V.L., Park, E., Forsline, P.L., Fazio, G., and Conway, W.S. (2011). Identification of wild apple germplasm (*Malus* spp.) accessions with resistance to the postharvest decay pathogens *Penicillium expansum* and *Colletotrichum acutatum*. Plant Breed. *130*, 481–486.

Keeling, P.J. (2009). Functional and ecological impacts of horizontal gene transfer in eukaryotes. Curr. Opin. Genet. Dev. *19*, 613–619.

Keeling, P.J., and Palmer, J.D. (2008). Horizontal gene transfer in eukaryotic evolution. Nat. Rev. Genet. *9*, 605–618.

Khaldi, N., and Wolfe, K.H. (2008). Elusive origins of the extra genes in *Aspergillus oryzae*. PLoS One 3, e3036.

Khaldi, N., Collemare, J., Lebrun, M.H., and Wolfe, K.H. (2008). Evidence for horizontal transfer of a secondary metabolite gene cluster between fungi. Genome Biol. *9*, R18.

Kosalková, K., Domínguez-Santos, R., Coton, M., Coton, E., García-Estrada, C., Liras, P., and Martín, J.F. (2015). A natural short pathway synthesizes roquefortine C but not meleagrin in three different *Penicillium roqueforti* strains. Appl. Microbiol. Biotechnol. *99*, 7601–7612.

Lessard, M.-H., Viel, C., Boyle, B., St-Gelais, D., and Labrie, S. (2014). Metatranscriptome analysis of fungal strains *Penicillium camemberti* and *Geotrichum candidum* reveal cheese matrix breakdown and potential development of sensory properties of ripened Camembert-type cheese. BMC Genomics *15*, 235.

Li, Z.-H., Du, C.-M., Zhong, Y.-H., and Wang, T.-H. (2010). Development of a highly efficient gene targeting system allowing rapid genetic manipulations in *Penicillium decumbens*. Appl. Microbiol. Biotechnol. *87*, 1065–1076.

Liti, G., and Louis, E.J. (2005). Yeast evolution and comparative genomics. Annu. Rev. Microbiol. *59*, 135–153.

Liti, G., Barton, D.B., and Louis, E.J. (2006). Sequence diversity, reproductive isolation and species concepts in *Saccharomyces*. Genetics *174*, 839–850.

Lücke, F.-K. (1997). Fermented sausages. In Microbiology of Fermented Foods, Wood, B.J.B., ed., (Blackie Academic and Professional, London, UK), pp. 441–483.

Lücke, F.K., and Hechelmann, H. (1987). Starter cultures for dry sausages and raw ham composition and effect. Fleischwirtschaft *67*, 307–314.

Ma, L.-J., van der Does, H.C., Borkovich, K.A., Coleman, J.J., Daboussi, M.-J., Di Pietro, A., Dufresne, M., Freitag, M., Grabherr, M., Henrissat, B., *et al.* (2010). Comparative genomics reveals mobile pathogenicity chromosomes in Fusarium. Nature *464*, 367–373.

Machida, M., Asai, K., Sano, M., Tanaka, T., Kumagai, T., Terai, G., Kusumoto, K.-I., Arima, T., Akita, O., Kashiwagi, Y., *et al.* (2005). Genome sequencing and analysis of *Aspergillus oryzae*. Nature *438*, 1157–1161.

Marcet-Houben, M., Ballester, A.-R., de la Fuente, B., Harries, E., Marcos, J.F., González-Candelas, L., and Gabaldón, T. (2012). Genome sequence of the necrotrophic fungus *Penicillium digitatum*, the main postharvest pathogen of citrus. BMC Genomics *13*, 646.

McCallum, J., Tsao, R., and Zhou, T. (2002). Factors affecting patulin production by *Penicillium expansum*. J. Food Prot. *65*, 1937–1942.

Moreau, C. (1979). Nomenclature des Penicillium utiles à la préparation du Camembert. Le Lait *59*, 219–233.

Novo, M., Bigey, F., Beyne, E., Galeote, V., Gavory, F., Mallet, S., Cambon, B., Legras, J.-L., Wincker, P., and Casaregola, S. (2009). Eukaryote-to-eukaryote gene transfer events revealed by the genome sequence of the wine yeast Saccharomyces cerevisiae EC1118. Proc. Natl. Acad. Sci. U.S.A. *106*, 16333–16338.

O'Brien, M., Nielsen, K.F., O'Kiely, P., Forristal, P.D., Fuller, H.T., and Frisvad, J.C. (2006). Mycotoxins and other secondary metabolites produced *in vitro* by *Penicillium paneum* Frisvad and *Penicillium roqueforti* Thom isolated from baled grass silage in Ireland. J. Agric. Food Chem. *54*, 9268–9276.

Palou, L. (2014). Penicillium digitatum, Penicillium italicum (Green Mold, Blue Mold). Postharvest Decay Control Strategies (Elsevier, Amsterdam, The Netherlands), pp. 45–102.

Papagianni, M., Ambrosiadis, I., and Filiousis, G. (2007). Mould growth on traditional Greek sausages and penicillin production by *Penicillium* isolates. Meat Sci. *76*, 653–657.

Petersson, S., and Schnürer, J. (1999). Growth of *Penicillium roqueforti, P. carneum, and P. paneum* during malfunctioning airtight storage of high-moisture grain cultivars. Postharvest Biol. Technol. *17*, 47–54.

Pitt, J., and Leistner, L. (1991). Toxigenic *Penicillium* species. Mycotoxins Anim. Foods 81–99.

Pitt, J.I., Cruickshank, R.H., and Leistner, L. (1986). *Penicillium commune, P. camemberti*, the origin of white cheese moulds, and the production of cyclopiazonic acid. Food Microbiol. *3*, 363–371.

Raper, K.B., Alexander, D.F., and Coghill, R.D. (1944). Penicillin: II. Natural variation and penicillin production in *Penicillium notatum* and allied species. J. Bacteriol. *48*, 639.

Ropars, J., Dupont, J., Fontanillas, E., De La Vega, R.C.R., Malagnac, F., Coton, M., Giraud, T., and López-Villavicencio, M. (2012). Sex in cheese: evidence for sexuality in the fungus *Penicillium roqueforti*. PLoS One *7*, e49665.

Ropars, J., López-Villavicencio, M., Dupont, J., Snirc, A., Gillot, G., Coton, M., Jany, J.-L., Coton, E., and Giraud, T. (2014). Induction of sexual reproduction and genetic diversity in the cheese fungus *Penicillium roqueforti*. Evol. Appl. *7*, 433–441.

Ropars, J., Rodríguez de la Vega, R.C., López-Villavicencio, M., Gouzy, J., Sallet, E., Dumas, É., Lacoste, S., Debuchy, R., Dupont, J., Branca, A., *et al.* (2015). Adaptive horizontal gene transfers between cheese-associated fungi. Curr. Biol. *25*, 2562–2569.

Rosewich, U.L., and Kistler, H.C. (2000). Role of horizontal gene transfer in the evolution of fungi. Annu. Rev. Phytopathol. *38*, 325–363.

Samson, R.A., Seifert, K.A., Kuijpers, A.F.A., Houbraken, J.A.M.P., and Frisvad, J.C. (2004). Phylogenetic analysis of *Penicillium* subgenus *Penicillium* using partial β-tubulin sequences. Stud. Mycol. *49*, 175–200.

Sanderson, P., and Spotts, R. (1995). Postharvest decay of winter pear and apple fruit caused by species of *Penicillium*. Phytopathology *85*, 103–110.

Seifert, K.A., Samson, R.A., deWaard, J.R., Houbraken, J., Lévesque, C.A., Moncalvo, J.-M., Louis-Seize, G., and Hebert, P.D.N. (2007). Prospects for fungus identification using CO1 DNA barcodes, with *Penicillium* as a test case. Proc. Natl. Acad. Sci. U.S.A. *104*, 3901–3906.

Smit, A., Hubley, R., and Green, P. (2013). RepeatMasker Open-4.0, http://www.repeatmasker.org. Accessed 22 January 2016.

Stamatakis, A. (2014). RAxML version 8: a tool for phylogenetic analysis and post-analysis of large phylogenies. Bioinformatics *30*, 1312–1313.

Taylor, J., Jacobson, D., and Fisher, M. (1999). The evolution of asexual fungi: reproduction, specification and classification. Annu. Rev. Phytopathol. *37*, 197–246.

Ullán, R.V., Godio, R.P., Teijeira, F., Vaca, I., García-Estrada, C., Feltrer, R., Kosalkova, K., and Martín, J.F. (2008). RNA-silencing in *Penicillium chrysogenum* and *Acremonium chrysogenum*: validation studies using β-lactam genes expression. J. Microbiol. Methods *75*, 209–218.

Visagie, C., Hirooka, Y., Tanney, J., Whitfield, E., Mwange, K., Meijer, M., Amend, A., Seifert, K., and Samson, R. (2014a). *Aspergillus, Penicillium* and *Talaromyces* isolated from house dust samples collected around the world. Stud. Mycol. *78*, 63–139.

Visagie, C.M., Houbraken, J., Frisvad, J.C., Hong, S.-B., Klaassen, C.H.W., Perrone, G., Seifert, K.A., Varga, J., Yaguchi, T., and Samson, R.A. (2014b). Identification and nomenclature of the genus *Penicillium*. Stud. Mycol. *78*, 343–371.

Wang, F.-Q., Zhong, J., Zhao, Y., Xiao, J., Liu, J., Dai, M., Zheng, G., Zhang, L., Yu, J., Wu, J., *et al.* (2014). Genome sequencing of high-penicillin producing industrial strain of *Penicillium chrysogenum*. BMC Genomics *15*, S11.

Wisecaver, J.H., and Rokas, A. (2015). Fungal metabolic gene clusters – caravans traveling across genomes and environments. Front. Microbiol. *6*, 1–11.

Wisecaver, J.H., Slot, J.C., and Rokas, A. (2014). The evolution of fungal metabolic pathways. PLoS Genet. *10*, e1004816.

Yang, Y., Zhao, H., Barrero, R.A., Zhang, B., Sun, G., Wilson, I.W., Xie, F., Walker, K.D., Parks, J.W., Bruce, R., *et al.* (2014). Genome sequencing and analysis of the paclitaxel-producing endophytic fungus *Penicillium aurantiogriseum* NRRL 62431. BMC Genomics *15*, 69.

Approaches for Comparative Genomics in *Aspergillus* and *Penicillium*

4

Jane L. Nybo,* Sebastian Theobald,* Julian Brandl, Tammi C. Vesth and Mikael R. Andersen

Abstract

The number of available genomes in the closely related fungal genera *Aspergillus* and *Penicillium* is rapidly increasing. At the time of writing, the genomes of 62 species are available, and an even higher number is being prepared. Fungal comparative genomics is thus becoming steadily more powerful and applicable for many types of studies.

In this chapter, we provide an overview of the state-of-the-art of comparative genomics in these fungi, along with recommended methods. The chapter describes databases for fungal comparative genomics. Based on experience, we suggest strategies for multiple types of comparative genomics, ranging from analysis of single genes, over gene clusters and CaZymes to genome-scale comparative genomics. Furthermore, we have examined published comparative genomics papers to summarize the preferred bioinformatic methods and parameters for a given type of analysis, highly useful for new fungal geneticists. Moreover, the chapter contains a detailed overview of comparative genomics studies of key fungal traits such as primary metabolism, secondary metabolism, and secretome analysis.

Finally, we gaze into a possible future of the field by comparing the current state of fungal comparative genomics to the development in bacterial genomics, where the comparison of hundreds of genomes has been performed for a while.

Introduction

The advent of the genomic era for the *Aspergillus/Penicillium* group started in 2005 with a set of three publications in Nature covering the most studied organisms in the genus, *Aspergillus nidulans, A. oryzae* and *A. fumigatus* (Galagan *et al.*, 2005; Machida *et al.*, 2005; Nierman *et al.*, 2005). These analyses gave us the first genome-scale overview of the three individual species, but notably, a large part of all three manuscripts was dedicated to the comparison of the genomes to each other or other fungal genomes (e.g. *Saccharomyces cerevisiae*). It was already clear from these first studies that a large part of the understanding of the individual species and genetic features comes from *comparative genomics* in addition to the initial genome analysis. The comparative analysis gave new information about a variety of features such as mating factor variations; primary and secondary metabolism; adaptation to different environments; genome dynamics; conserved non-coding sequence; and pathogenicity. Similarly, the later publication of the genome sequence for *Penicillium chrysogenum* (van den

*These authors contributed equally.

Berg *et al.*, 2008) based the analyses of phylogeny and chromosome dynamics on comparisons with the previously published aspergilli.

Since then, the interest in comparative genomics has only increased. A notable application is in the strength of comparative genomics to infer annotations from well-studied model organisms, such as the ones mentioned above, to less studied species, which are being sequenced as part of current large genome sequencing efforts in both penicillia and aspergilli (see 'Current status of genomics', below). Furthermore, genome mining of fungi using tailored algorithms for comparative genomics has long been interesting due to the wealth of natural products found in these species (Bok *et al.*, 2006).

In this chapter, we define *comparative genomics* as *the comparison of genetic features for large parts or whole genomes* and will focus on applications and studies of this type. The comparison of single genes across organisms is clearly a highly valuable effect of the availability of genome sequences, but will not be addressed here.

It is clear from the examples presented above, and what will be shown in the following sections, that there is a high analytical power in comparing features across genomes. In this context, we wish to provide overviews of the following:

- the current status of fungal genomics in *Aspergillus* and *Penicillium*;
- the available tools and databases for fungal comparative genomics, both general and specialized applications;
- results and potentials of comparative genomics studies within the field.

Furthermore, we will present the main application areas of comparative genomics, as well as the current best practice for comparative genomics within the individual research areas. Our goal is to enable the reader to identify the current best practices for fungal comparative genomics based on an overview of algorithms and preferred parameters in the field.

Current status of genomics

The foundation of comparative genomics is the availability of relevant genome sequences. Currently (May 2015), 62 genome sequences from the *Aspergillus* (32 species) and *Penicillium* (30 species) genera are available from public repositories (Table 4.1). While the majority of the sequences are from individual species, an increasing amount of sequences are addressing multiple strains of the same species. In those cases, the primary application of the genome has been comparative genomics with the objective to identify strain-specific traits. Examples include high pencillin production in *P. chrysogenum* NCPC10086 (Wang *et al.*, 2014), studies of pathogenicity and secondary metabolism in *A. fumigatus* Af293 versus A1163 (Fedorova *et al.*, 2008; Sanchez *et al.*, 2012), and enzyme-producing phenotypes in *A. niger* CBS 513.88 (Andersen *et al.*, 2011). Details regarding these cases, as well as cross-species comparative genomics will be discussed in the following sections.

It is quite clear given the current number of multiple independent genome sequencing efforts, as well as the JGI Community Sequencing Programs, that the number of available *Aspergillus* and *Penicillium* genome sequences will increase rapidly in the near future. Current ongoing efforts include both re-sequencing of multiple isolates of the same species (for instance clinical isolates, basic research as for the *S. cerevisiae* 100 genomes project (Strope *et al.*, 2015), or results of industrial strain improvement programs) as well as a large number

Table 4.1 List of *Aspergillus* and *Penicillium* species and strains with publicly available genome sequences

Species	Strain	Reference
A. acidus	CBS 106.47	[1]
A. aculeatus	ATCC16872	
A. brasiliensis	CBS 101740	[1]
A. campestris	CBS 348.81/IBT28561	[3]
A. carbonarius	ITEM 5010	
A. clavatus	NRRL1	Fedorova *et al*., 2008
A. flavus	NRRL3357	Nierman *et al*., 2015a
A. fischerianus/N. fischeri		Fedorova *et al*., 2008
A. fumigatus	A1163	Fedorova *et al*., 2008
A. fumigatus	Af293	Nierman *et al*., 2005
A. glaucus	CBS 516.65	[1]
A. kawachii	IFO 4308	Futagami *et al*., 2011
A. nidulans	FGSC A4	Galagan *et al*., 2005
A. niger	ATCC 1015	Andersen *et al*., 2011
A. niger	CBS 513.88	Pel *et al*., 2007
A. niger	NRRL3	
A. niger van Tieghem	ATCC 13496	
A. novofumigatus	CBS117520/IBT16806	[3]
A. ochraceoroseus	CBS 550.77/IBT245754	[3]
A. oryzae	RIB40	Machida *et al*., 2005
A. parasiticus	SU-1	Linz *et al*., 2014
A. phoenicis	ATCC 13157	
A. rambelli	SRRC1468	
A. ruber/Eurotium rubrum	CBS 135680	Kis-Papo *et al*., 2014
A. steynii	CBS112812/IBT 23906	[3]
A. sydowii	CBS 593.65	[1]
A. terreus	NIH 2624	
A. tubingensis	CBS 134.48	[1]
A. ustus	3.3904	Pi *et al*., 2015
A. versicolor	CBS 583.65	[1]
A. wentii	DTO 134E9	[1]
A. zonatus		[1]
P. aethiopicum	IBT 5753	
P. aurantiogriseum	NRRL 62431	Yang *et al*., 2014
P. bilaiae	ATCC 10455	[2]
P. brevicompactum	AgRF18	[2]
P. camemberti	FM 013	Cheeseman *et al*., 2014
P. canescens	ATCC 10419	[2]
P. capsulatum	ATCC 48735	

Table 4.1 Continued

Species	Strain	Reference
P. chrysogenum	NCPC10086	Wang *et al.*, 2014
P. digitatum	Pd1	Marcet-Houben *et al.*, 2012
P. digitatum	PHI26	Marcet-Houben *et al.*, 2012
P. expansum	ATCC 24692	[2]
P. expansum	R19	Yu *et al.*, 2014
P. expansum		Li *et al.*, 2015
P. expansum	PEXP	Ballester *et al.*, 2015
P. expansum	PEX1	Ballester *et al.*, 2015
P. expansum	PEX2	Ballester *et al.*, 2015
P. fellatanum	ATCC 48694	[2]
P. glabrum	DAOM 239074	[2]
P. italicum		Li *et al.*, 2015
P. italicum	PITC	Ballester *et al.*, 2015
P. janthinellum	ATCC 10455	[2]
P. lanosocoeruleum	ATCC 48919	[2]
P. marneffei	ATCC18224	Nierman *et al.*, 2015b
P. nordicum	BFE487	
P. oxalicium (*decumbens*)	114-2/CGMCC 5302	Liu *et al.*, 2013a
P. paxilli	ATCC 26601	Berry *et al.*, 2015
P. raistrickii	ATCC 10490	[2]
P. roqueforti	FM164	Cheeseman *et al.*, 2014
P. rubens[4]	Wisconsin 54-1255	van den Berg *et al.*, 2008
P. stipitatum/Talaromyces stipitatus	ATCC 10500	Nierman *et al.*, 2015b

[1]Genomes sequenced as a part of a JGI community sequencing proposal (CSP) led by Ronald P. de Vries (CBS-KNAW, NL). Publication pending.
[2]Genomes sequenced as a part of a JGI CSP led by Dave Greenshields (Novozymes).
[3]Genomes sequenced by JGI as a part of a JBEI/DTU sequencing proposal.
[4]Species previously thought to be *P. chrysogenum*.

of *de novo* sequencing projects. Currently, it seems likely that the number of *Penicillium* and *Aspergillus* genome sequences combined will exceed 100 in 2016. Such a development renders traditional storage and analysis methods inefficient and increases the need for databases, which can make these sequences available to the research community, as well as provide tools for comparing the wealth of genomes.

Available genome databases with comparative genomics capabilities

The fungal research community has been fortunate in the availability of several data repositories dedicated to analysis of fungal genomes or specific features thereof. The original source of inspiration for many of these efforts has been the immensely successful Saccharomyces

Genome Database (SGD) (Cherry *et al.*, 2012), which formed the basis for the framework behind the Aspergillus Genome Database (AspGD) (Arnaud *et al.*, 2010; Cerqueira *et al.*, 2014). Since the launch of AspGD, several new databases and repositories have been made available, each adding specific analysis features and/or additional genomes relevant for comparisons. Table 4.2 gives an overview of fungal and general databases currently available which give access to fungal genomes and offer various types of comparative genomics tools.

Table 4.2 denotes which databases offer which types of services, but of course all tools are not identical even with similar scopes, and have different strengths. All databases furthermore have individual design philosophies and are tailored towards different communities. For the following types of comparative genomics studies based on information available in databases, we recommend the following approaches:

- *Investigation of orthologues of single genes.* An efficient approach would be to look up the gene in FungiDB, AspGD or the JGI webpages, if the gene/species of interest is present in any of these, and from there navigate to the gene orthology feature of each webpage. Alternatively, one can employ the BLAST interface present at many of the sites (Camacho *et al.*, 2009). To get a comprehensive overview including as many organisms as possible, use both the JGI portal or NCBI GenBank as these contain genomes not found elsewhere.
- *Analysis of gene clusters across organisms.* Currently, the simplest approach would be to navigate to the JGI secondary metabolism feature, identify the gene cluster of interest, and from there navigate to the JGI genome browser. Here, one can examine gene conservation in a few related species. For a more sophisticated analysis, where the gene cluster is known, one can use FungiDB or AspGD. Both databases offer detailed synteny mapping based on the SYBIL software (Crabtree *et al.*, 2007) to a larger number of organisms. This allows the user to examine whether gene order has been rearranged.
- *Analysis of CAZymes.* For both *Aspergillus* and *Penicillium* species, this is a highly interesting analysis of potential biomass-degrading enzymes. Such data are currently best examined by downloading it from the CAZy-database (Cantarel *et al.*, 2009), if it is available for the organisms of choice, and performing an analysis offline using custom comparative tools. Alternatively, the staff behind CAZy are involved in many community efforts of this type.
- *Untargeted analysis.* In some cases, it is of interest not to look at defined/known genes, but instead compare genes of a specific type, structure or function across organisms. For this type of analysis, FungiDB/EupathDB have highly sophisticated gene search methods, based on DNA and protein motifs, domains, signal peptides, association with specific metabolites, etc.
- *Genome-scale comparative genomics.* At the moment, this type of analysis is not available through any of the platforms. This is currently best undertaken by downloading genome data of the organisms of choice and performing these computationally heavy tasks on dedicated systems.
- *Download of genomics data sets.* Depending on the genomes of interest, NCBI Genomes or JGI are the best sources of data. Some genomes are only found in one database but not the other, so for a comprehensive analysis, both databases must be queried. In our hands, we find that for most of our applications, the data format downloadable from JGI is the most flexible, as the sequence data is preformatted in several different ways. Furthermore,

Table 4.2 Overview of often-used resources for fungal comparative genomics

		Comparative genomics tools											
Database	Reference(s)	BLAST	Synteny analysis	Custom analysis flows	Orthologue identification	Secondary Metabolite Cluster comparison	Visual comparison of genomes	Comp. Genome Browser	Comparison to other omics	Comp. Gene Locus information	Cross-genome searches for annotation	Metabolic Pathway Analysis	Bulk Data Download
AspGD	Arnaud *et al.*, 2010; Cerqueira *et al.*, 2014	•	•		•			•	•	•	•		•
CADRE	Gilsenan *et al.*, 2012												•
CAZy	Cantarel *et al.*, 2009										•		•
EuPathDB	Aurrecoechea *et al.*, 2013	•	•	•	•			•	•	•	•		•
FungiDB	Stajich *et al.*, 2012	•	•	•	•			•	•	•	•	•	•
JGI genome portal	Grigoriev *et al.*, 2012	•			•	•		•	•[1]	•			•
MycoCosm	Grigoriev *et al.*, 2014	•			•	•		•	•[1]	•			•
NCBI	Wheeler *et al.*, 2013	•											•

[1]EST/SRN-sequencing data are integrated as a genome browser track on many genomes.

the JGI has a specialized interface (GLOBUS) for downloading large datasets, which is very useful for studies involving the comparison of more than a few species.

For larger or more advanced, or specialized studies, it is typically advantageous to use some of the many specialized algorithms and methods developed. The next section gives an overview of these.

General methods for comparative genomics

Methods in comparative genomics are in general based on tools used for single genome analysis, and in many cases need to be tailored accordingly to be able to manage cross-species differences. Comparative genomics has been used to shed light on everything from identifying species-specific genes, transcription factor (TF) binding sites, repeat elements, and single nucleotide polymorphisms (SNPs) to predicting phylogenetic trees, genome scale models, secretomes, and horizontal gene transfers. These approaches can generally be sorted into two broad categories: Studies identifying orthologues/paralogues (see 'Identification of orthologues across species', below) and studies identifying synteny across genomes (see 'Whole genome comparisons and synteny analysis', below).

Selecting the right method for a specific analysis depends on multiple factors, the most important ones being the research topic or question and the type of data involved. The data type and quality can have a large impact on the strength of the analysis and a variety of methods have been developed to deal with items such as variations in annotation and sequencing quality. Because of these large variations and biological questions, the right tools and methods should be used in the right context. The following section describes different cases and the methods used, and presents an overview table of studies and types of analysis (Table 4.3).

Identification of orthologues across species

As is evident from Table 4.3, there is a wide range of applications, methods, and parameters for identifying orthologues. A key component of all these methods is to compare either DNA or protein sequences. Sequence alignment is a method to compare the DNA, RNA or proteins to identify similar regions. There are a large number of alignment algorithms addressing scalability, speed, and parameter optimization, where some are mentioned in Table 4.3.

The predominant alignment algorithm for comparative genomics is the Basic Local Alignment Search Tool (BLAST), which combines database searches with a fast heuristic Smith–Waterman-based local alignment approach (Altschul *et al.*, 1990). It is perhaps the most widely used tool, but does come with some pitfalls, which warrants a detailed discussion. A common orthologue selection approach is to use bidirectional (reciprocal) best hits between genomes. It is a very rigorous approach that almost certainly will find true orthologues (Wolf and Koonin, 2012), but will miss as much as 60% of potential orthologues (Dalquen and Dessimoz, 2013), in that only one hit is identified for each gene or protein. Another, less stringent approach is to use reciprocal hits in combination with additional selection criteria, among other *expectancy value* (*e-value*), *alignment identity* (%) and *alignment coverage* (%), which increases the likelihood of identifying orthologous relations and allows the identification of multiple orthologues for a given gene or protein. In general, one

Table 4.3 Overview of methods and parameters applied in comparative genomics studies of penicillia and aspergilli

	Agren *et al.*, 2013	Andersen *et al.*, 2008	Andersen *et al.*, 2011	Arnaud *et al.*, 2010	Ballester *et al.*, 2015	Braaksma *et al.*, 2010	Coutinho *et al.*, 2009	David *et al.*, 2008	Fedorova *et al.*, 2008	Flipphi *et al.*, 2009	Galagan *et al.* 2005	Gibbons and Rokas, 2009	Inglis *et al.*, 2013	Kis-Papo *et al.*, 2014	G. Liu *et al.*, 2013	J. Liu *et al.*, 2013	Machida *et al.*, 2005	Marcet-Houben *et al.*, 2012	Pel *et al.*, 2007	Pi *et al.*, 2015	Rokas *et al.*, 2007	van den Berg *et al.*, 2008	Vongsangnak *et al.*, 2008	Wang *et al.*, 2014	Toolbox	Visualization	Online	Local installation	Database
Genus																													
Penicillium	•				•									•	•			•				•		•					
Aspergillus		•	•	•		•	•	•	•	•	•	•	•			•	•		•	•	•		•						
Type of study																													
Genome-scale metabolic model	•	•	•					•								•			•				•						
Database				•																									
Genome sequenced			•		•		•		•		•			•	•		•	•	•	•		•		•					
Specialized comp. genomics										•		•																	
Focus																													
Strain	•	•	•	•	•	•	•	•	•	•	•	•	•	•	•	•	•	•	•	•	•	•	•	•					
Clade/group										•	•												•						
Core					•				•	•	•			•	•			•	•	•	•		•						
Homology/orthology																													
BLAST alignment																													
Bidirectional[f] best hits	•	•				•			•	•				•	•	•	•	•	•	•	•								
Modified mutual best hit				•																									
BLASTp	•	•		•	•	•	•	•	•	•	•			•	•	•	•	•	•	•	•		•			•	•	•	
E-value	$\geq 1e^{-30}$				$\geq 1e^{-05}$		$\geq 1e^{-30}$	$\geq 1e^{-50}$	$\geq 1e^{-05}$	$\geq 1e^{-30}$					$\geq 1e^{-10}$	$\geq 1e^{-30}$		$\geq 1e^{-05}$	$\geq 1e^{-10}$	$\geq 1e^{-10}$	$\geq 1e^{-05}$								
Bit score																	≥100												
Alignment identity	≥40%					≥40%					≥80%					≥40%				≥50%			≥40%						

Alignment coverage	≥50%			≥50%								≥60%		≥50%		≥50%									
Alignment coverage of shortest hit					≥70%																				
Alignment length	≥200aa														≥100aa				≥200aa						
Sequence length similarity					≥80%																				
tBLAST *Bit score*													≥100									•	•	•	
BLASTx *AL*[II] *≥ 25%* *AI*[III] *≥ 50 aa*																			•			•	•	•	
BLASTn *E–value*		$1e^{-75}$	•																			•	•	•	
Alignment-based clustering																									
InParanoid			•							•												•	•		•
TRIBE-MCL											•													•	
OrthoMCL												•										•	•	•	•
E-value								≥$1e^{-10}$				≥$1e^{-05}$													
Match cutoff								40%				≥50%													
CAFE[IV]											•													•	
Strain/clade/core-specific feaure identification																									
Synteny																									
Sybil			•				•		•								•				•	•	•		
BLAT																				•			•	•	
MUMmer							•									•	•	•				•		•	
PASA pipeline							•										•					•		•	
Single nucleotide polymorphism																									
SAMtools[V]														•							•		•	•	
BWA[V, VI]				•																				•	
MUSCLE[V, VII]														•								•	•		
GATK[VIII]				•																	•			•	
SOAP[IX]																				•	•			•	

Table 4.3 Continued

	Agren *et al.*, 2013	Andersen *et al.*, 2008	Andersen *et al.*, 2011	Arnaud *et al.*, 2010	Ballester *et al.*, 2015	Braaksma *et al.*, 2010	Coutinho *et al.*, 2009	David *et al.*, 2008	Fedorova *et al.*, 2008	Flipphi *et al.*, 2009	Galagan *et al.* 2005	Gibbons and Rokas, 2009
Non-coding or repeats												
Repeat-Masker									•			
RepeatScout									•			
Tandem Repeats Finder									•			
Transposon-PSI									•			
tBLASTn									•			
Emboss etandem[X]												•
Multi-lagan											•	
RSAT[XI]										•		
Cosmo[XII]		•								•		
TRANSFAC[XIII]										•		
Secondary metabolism												
SMURF[XIV]					•				•			
BLASTp					•							
E-value					$\geq 1e^{-05}$							
Alignment identity												
Alignment coverage					≥50%							
Alignment coverage (query)												

	Inglis *et al.*, 2013	Kis-Papo *et al.*, 2014	G. Liu *et al.*, 2013	J. Liu *et al.*, 2013	Machida *et al.*, 2005	Marcet-Houben *et al.*, 2012	Pel *et al.*, 2007	Pi *et al.*, 2015	Rokas *et al.*, 2007	van den Berg *et al.*, 2008	Vongsangnak *et al.*, 2008	Wang *et al.*, 2014	Toolbox	Visualization	Online	Local installation	Database
Non-coding or repeats																	
Repeat-Masker										•			•		•	•	
RepeatScout										•						•	•
Tandem Repeats Finder										•					•	•	•
Transposon-PSI										•						•	
tBLASTn															•	•	
Emboss etandem[X]														•	•		
Multi-lagan													•		•	•	
RSAT[XI]													•	•	•	•	
Cosmo[XII]													•			•	
TRANSFAC[XIII]													•	•	•	•	•
Secondary metabolism																	
SMURF[XIV]	•		•					•		•					•	•	
BLASTp								•						•	•	•	
E-value								$\geq 1e^{-10}$									
Alignment identity								≥50%									
Alignment coverage																	
Alignment coverage (query)								≥50%									

Present when % genes in cluster was found					≥50%																							
Secretome																												
SignalP(XV)						≥0.43						•		•				•	•							•	•	
PSORT(XVI)	•															•		•	•							•	•	
TMHMM(XVII)												•		•				•								•	•	
Phylogeny																												
Idenification of conserved genes																												
BLASTp					•				•						•			•		•					•	•	•	
E-value					$\geq 1e^{-05}$				$\geq 1e^{-04}$						$\geq 1e^{-10}$			$\geq 1e^{-05}$		$\geq 1e^{-10}$								
Alignment identity																				≥80%								
Alignment coverage					≥50%										≥60%			≥50%		≥80%								
Identical number of introns									•													•						
Similar length									•													≥95%						
Alignment																												
GeneDoc										•											•				•		•	
ClustalW [aa]			•						•						•				•	•					•	•	•	
ClustalX [aa]							•								•		•		•						•	•	•	
MAFFT					•									•				•							•	•	•	
MUSCLE [DNA/aa]					•				•									•				•			•	•	•	
Kalign					•																				•	•	•	
DIALIGN-TX																		•								•	•	
M-Coffee(XVIII)					•													•							•	•	•	
Evolutionary clustering																												
Gblocks														•	•										•	•	•	
RaxML					•									•				•								•	•	
PhyML/BioNJ(XIX)					•													•						•		•	•	

Table 4.3 Continued

	Agren *et al.*, 2013	Andersen *et al.*, 2008	Andersen *et al.*, 2011	Arnaud *et al.*, 2010	Ballester *et al.*, 2015	Braaksma *et al.*, 2010	Coutinho *et al.*, 2009	David *et al.*, 2008	Fedorova *et al.*, 2008	Flipphi *et al.*, 2009	Galagan *et al.* 2005	Gibbons and Rokas, 2009	Inglis *et al.*, 2013	Kis-Papo *et al.*, 2014	G. Liu *et al.*, 2013	J. Liu *et al.*, 2013	Machida *et al.*, 2005	Marcet-Houben *et al.*, 2012	Pel *et al.*, 2007	Pi *et al.*, 2015	Rokas *et al.*, 2007	van den Berg *et al.*, 2008	Vongsangnak *et al.*, 2008	Wang *et al.*, 2014	Toolbox	Visualization	Online	Local installation	Database
MEGA[XIX]										•					•				•	•					•	•		•	
Phylip[XX]									•		•											•			•	•		•	
TreeCon[XIX]			•																							•		•	
Bootstrap ML[XXI]			1000		100		1000			1000	1000	•			1000			100	1000	1000					-				
BLASTp for HGT[XXII]																													
Cluster absent in closely related species																		•		•						•	•	•	
Cluster found in distant related species																		•		•						•	•	•	
Cluster sequence identity																				≥50%						•	•	•	
Visualization																													
ETE toolkit[XXIII]					•													•							•	•	•		
FigTree[XXIII]															•											•		•	

Functional annotation																													
GO(XXIV)			•	•	•				•	•			•	•	•				•	•				•	•	•	•	•	•
Pfam(XXV)	•							•	•		•	•		•	•			•	•				•				•	•	•
AntiSMASH(XXVI)													•							•						•	•	•	•
InterPro(XXVII)	•		•		•								•	•	•			•	•					•			•	•	•
KEGG(XXVIII)	•	•	•		•					•				•	•	•		•	•				•		•	•	•		•
CAZy annotation(XXIX)					•		•								•			•							•	•	•*		•
CELLO(XXX)	•															•							•				•		•

*Only per request. (I) reciprocal; (II) alignment identity; (III) alignment length; (IV) Computational Analysis of gene Family Evolution; (V) for alignment variation detection and mapping; (VI) Burrows–Wheeler Alignment tool; (VII) Multiple Sequence Alignment tool; (VIII) Genome Analysis Toolkit; (IX) Short Oligonucleotide Analysis Package; (X) for repeats in a protein sequence; (XI) Regulatory Sequence Analysis Tool; (XII) An R package; (XIII) transcription factor binding site predictor; (XIV) Secondary Metabolite Unknown Regions Finder; (XV) signal sequence predictor; (XVI) subcellular location prediction tool; (XVII) transmembrane domain prediction tool; (XVIII) for consensus alignment; (XIX) neighbour-joining tree construction; (XX) maximum likelihood tree construction; (XXI) maximum likelihood; (XXII) horizontal gene transfer; (XXIII) tree visualization; (XXIV) Gene Ontology database tool; (XXV) Protein family prediction database; (XXVI) antibiotics & Secondary Metabolite Analysis Shell; (XXVII) Protein sequence analysis & classification database; (XXVIII) Kyoto Encyclopedia of Genes and Genomes; (XXIX) carbohydrate-active enzyme annotation; (XXX) Subcellular Localization Predictive System. aa, amino acid.

Methods included in this table are referenced as follows: BLAST (Camacho *et al*., 2009); BLAT (Kent, 2002); BWA (Li and Durbin, 2009); CAFE (De Bie *et al*., 2006); ClustalW/X (Larkin *et al*., 2007); Cosmo (Bembom *et al*., 2007); DIALIGN-TX (Subramanian *et al*., 2010); Emboss (Rice *et al*., 2000); ETE Toolkit (Huerta-Cepas *et al*., 2010); FigTree (Rambaut, 2009); GATK (McKenna *et al*., 2010); Gblocks (Castresana, 2000); InParanoid (O'Brien *et al*., 2005); Kalign (Lassmann and Sonnhammer, 2005); MAFFT (Katoh *et al*., 2002); M-Coffee (Wallace *et al*., 2006); MEGA (Kumar *et al*., 1994); Multi-lagan (Brudno *et al*., 2003); MUMmer (Delcher *et al*., 1999a); MUSCLE (Edgar, 2004); PASA (Haas, 2003); Phylip (Retief, 2000); PhyML (Criscuolo, 2011); PSORT (Horton *et al*., 2007); RaxML (Stamatakis, 2014); Repeat-Masker (http://www.repeatmasker.org/); RepeatScout (Price *et al*., 2005); RSAT (Thomas-Chollier *et al*., 2008); SAMtools (Li *et al*., 2009); SignalP (Emanuelsson *et al*., 2007); SMURF (Khaldi *et al*., 2010); SOAP (Li *et al*., 2008); Sybil (Van De Peer and De Wachter, 1994) (Crabtree *et al*., 2007; Riley *et al*., 2012); Tandem Repeats Finder (Benson, 1999); TMHMM (Emanuelsson *et al*., 2007); TRANSFAC (Wingender *et al*., 2000); Transposon-PSI (Brian Haas, http://transposonpsi.sourceforge.net); TreeCon; TRIBE-MCL (Enright *et al*., 2002); OrthoMCL (Li *et al*., 2003).

should be cautious with using e-values as sole selection criterion, especially when comparing across studies, as it is directly derived from database size and is biased towards long sequence matches and sensitive to data bias. This is particularly challenging in fungi, as there are very large proteins (e.g. polyketide synthases), which may have partial matches to each other, which are longer than some full-length proteins. For this reason, the main motivation for e-value cutoffs should be to reduce the number of false/random hits, and *not* to select true orthologues. This is best achieved by bidirectional hits in tandem with alignment coverage and identity. This also comes with a note of caution. In particular alignment identity is dependent on phylogenetic distance of the species investigated. For this reason, it is not possible to recommend a single set of parameters for BLAST for orthologue detection. Instead we suggest examining the data set and species first, and choose parameters based on the values generated by known orthologues. Typically used parameter settings are described in Table 4.3.

Orthology case studies

A primary application of the identification of putative orthologues is the transfer of annotation information between genomes. In a recent example, the genome of *A. ustus* was sequenced by Pi *et al.* (2015) and the potential protein-coding sequences were found using AUGUSTUS software. These sequences were primarily annotated by homology search against the NCBI non-redundant (nr) protein database (www.ncbi.nlm.nih.gov) using BLASTp with the selection criteria (e-value $< 1e^{-03}$, alignment identity > 25%, query alignment coverage > 50%). Their gene ontology annotations were transferred to *A. ustus* when bidirectional best hits between *A. ustus* and *A. nidulans* were fitting the strict parameters (e-value $< 1e^{-10}$, alignment identity > 50%, query alignment coverage > 50%) (Pi *et al.*, 2015).

Orthologues have also been used in evolutionary studies. Phylogeny has often been predicted from a single or few conserved genes, but depending on the taxonomic span, the resolution might need to include numerous genes. An approach could be to find all the orthologues shared among the compared species (the core genome) and use these to generate the phylogenetic relation between the species. Investigating shared genes can also give insight into evolutionary development such as horizontal gene transfer and selective pressure.

In a study of pathogenic filamentous fungi, the evolutionary relations between seven *Aspergillus* strains were generated by pairwise comparing genomes using bidirectional best BLASTp hits with an e-value $< 1e^{-05}$. From the core genome, 90 genes were selected on the basis of similar lengths and identical number of intron/exon structures. From this analysis Fedorova *et al.* (2008) additionally found genes that potentially were horizontally transferred between *Chaetomium globosum* and *A. fischerianus* (*Neosartorya fischeri*).

Whole-genome comparisons and synteny analysis

Whole genome-to-genome comparison is a highly desired analysis, which includes the study of genetic variations in genes and non-coding regions, chromosomal rearrangements and genetic co-localization (synteny).

Although, the algorithms used today are not ideal as the number of genomes increases (comparison of a large number of genomes is in essence a multidimensional problem) studies with 3–5 whole genome comparisons have not been uncommon (e.g. Fedorova *et al.*, 2008; Galagan *et al.*, 2005; Rokas *et al.*, 2007). The main issue with increasing the number of genomes is computational time, which increases exponentially with the number of genomes in the comparison, and poses challenges with visualizing the output. Should new algorithms solve this issue, it will allow us to efficiently study more complex dynamics of fungal genomes.

Synteny is commonly addressed using sequence alignments, comparing the localization of genes in one genome with the positions of similar genes in another genome. The alignments are often performed using BLASTp, but other more specialized algorithms as SYBIL (Crabtree *et al.*, 2007; Riley *et al.*, 2012) and MUMmer (Delcher *et al.*, 1999a) have been used (Table 4.3). MUSCLE (Edgar, 2004), MAFFT (Katoh *et al.*, 2002), CLUSTAL X/W (Larkin *et al.*, 2007), and Kalign (Lassmann and Sonnhammer, 2005) are all multiple sequence alignment tools optimized for either long DNA or protein sequences which have been applied in fungal phylogeny research. Additionally, M-Coffee has an extra unique feature where it combines all alignments from other tools generating one consensus alignment (Wallace *et al.*, 2006).

Case examples

Synteny case studies

When the genome of *P. chrysogenum* was released, van den Berg *et al.* (2008) aligned *P. chrysogenum* to *A. nidulans, A. niger, A. fumigatus, and A. oryzae* showing that reshuffling of the chromosomes has occurred after divergence of the *Aspergillus* and *Penicillium* lineages. In particular, they pairwise aligned the *Aspergillus* contigs larger than 100 kb to the 14 largest supercontigs of *P. chrysogenum* using the Promer program of the MUMmer package (van den Berg *et al.*, 2008).

These different cases provide insight to what comparative genomics can be used for. Even if the methods are based on the tools used for single genome analysis, comparing genomes with each other provides the means to study species evolution, gene and chromosomal diversity, physiological traits and novel as well as horizontally transferred genes. Furthermore, comparative genomics provides the opportunity of combining evidence of single genome studies, adding strength to knowledge deduced from the analysis.

Primary metabolism

Primary metabolism can be defined as the set of metabolic reactions directly supporting the propagation of a species by providing precursors for growth and reproduction. Primary metabolism of filamentous fungi is of great interest due to the intriguing phenotype of overproducing large quantities of di- or tri-carboxylic acids derived from intermediates of the citric acid cycle in different species of *Aspergillus,* as well as providing flavour compounds for food production and precursor molecules for production of bioactive compounds from *Penicillium* species. Examples of industrially interesting processes include the production of citric acid by *A. niger* (Karaffa and Kubicek, 2003) or production of itaconic acid by *A. terreus* (Okabe *et al.*, 2009). In order to understand these inter- and intraspecific differences, comparative genomics has been applied as a tool for examining the genomic basis of different phenotypes, e.g. the citric acid-producing strain ATCC 1015 and protein overproducing strain CBS 513.88 of *A. niger* (Andersen *et al.*, 2011). Differences in copy number and predicted functions can be correlated to productivity and therefore represent an invaluable tool for genetic engineering and target identification. Primary metabolism has been examined on an individual pathway basis in the past, identifying and characterizing the function of individual genes involved in the different pathways.

General comparisons of primary metabolism

Flipphi *et al.* (2009) used the genomes of eight *Aspergillus* species for the comparison of the primary metabolism genes in these organisms. Using this dataset, the authors could identify gene duplications at several steps of primary metabolism indicating the potential of these species to adapt to their individual niches. By comparing multiple genomes the evolution of individual genes can be reconstructed and gene duplication events narrowed down in time.

Genome-scale metabolic models as sources for comparative genomics

Another way of using the wealth of information provided by fungal sequencing efforts is the construction of genome-scale metabolic models. Similar to genome sequencing, where the complete list of genes of a species is enumerated, genome-scale models aim at assembling the complete set of metabolic reactions for a selected species. Due to the unsolved challenge of characterizing metabolic enzymes in a high-throughput manner, understanding differences in primary metabolic phenotype between individual strains or species, relies on external information. In the classical bioinformatics approach, domain and functional predictions are based on sequence motifs and structural properties of the gene products. In addition to this approach, comparative genomics also aims at inferring information of uncharacterized genes from similar characterized genes in a close neighbouring species. Comparative genomics is therefore frequently used during the construction of genome-scale models or in order to fill gaps in known pathways as well as for the construction of less well-characterized organism. There are multiple genome-scale metabolic reconstructions of filamentous fungi available (Table 4.3) and their usage has been reviewed recently (Brandl and Andersen, 2015). Many of these models have been reconstructed using previously published models as templates or source of metabolic reactions (Fig. 4.1). Here, the relatively well-annotated fungal genome of *S. cerevisiae* has served as an initial source of information to generate the first genome-scale models of *Aspergillus* species, which has then propagated to other aspergilli, and a recent model for *P. chrysogenum.*

This use of comparative genomics propagates biological evidence across species thereby increasing the level of experimental support of the individual models. However, attention should be paid to the possibility of propagating errors between different reconstructions as well as to a potential 'dilution' of the experimental backing. This dilution arises from the fact that gene–function associations are inferred from similarities to a gene in another organism. A cutoff has to be chosen for these comparisons in order to classify gene comparisons as meaningful. By chaining these comparisons the experimental evidence is diluted, as new sequences are no longer compared to the original source but to an inferred gene function itself.

Comparative genomics is not the only source for model generation, as manual reconstructions rely on experimental evidence and bibliomic data, but it is heavily used by algorithms for the automatic generation of genome-scale models (e.g. as described by Pitkänen *et al.*,

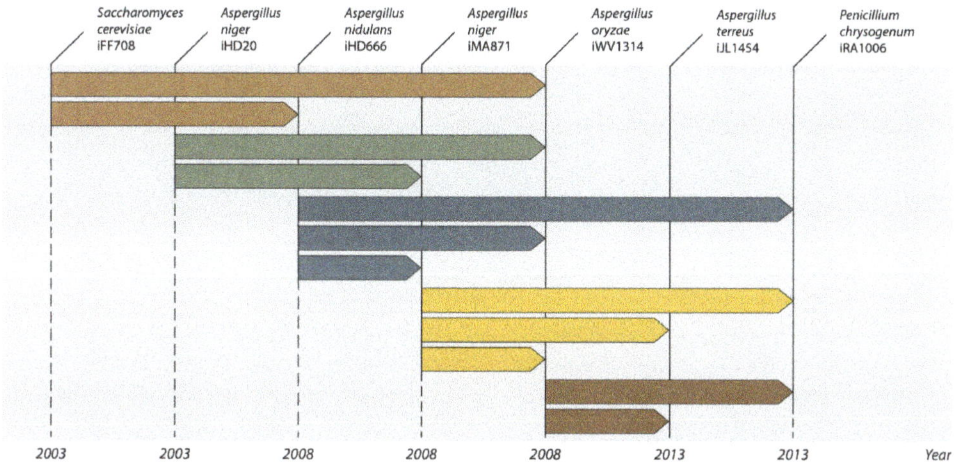

Figure 4.1 Information transfer between different fungal genome-scale models. Propagation of information between the genome-scale models of *S. cerevisiae* (Förster *et al.*, 2003), *A. niger* (Andersen *et al.*, 2008a; David *et al.*, 2003), *A. nidulans* (David *et al.*, 2008), *A. oryzae* (Vongsangnak *et al.*, 2008), *A. terreus* (Liu *et al.*, 2013b) and *P. chrysogenum* (Agren *et al.*, 2013).

2014). The combined network from multiple organisms is a strong tool for rapid tentative annotation of primary metabolism in new sequenced organisms.

Secondary metabolism

Comparative genomics is a powerful tool to analyse secondary metabolite producing gene clusters across fungal genomes. In several cases horizontal gene transfer of polyketide synthases (PKS) from bacteria to filamentous fungi has been observed, as well as transfer of gene clusters between fungal species, enabling the fungus to create new compounds by implementation of other genes in the cluster or exchanging cluster members, e.g. in the case of aflatoxin and sterigmatocystin (Osbourn, 2010).

Comparison of gene sequences also led to the identification of new gene clusters, e.g. the aflatrem genes in *A. flavus* have been characterized due to their homology to paxilline from *P. paxilli* (Nicholson *et al.*, 2009), which is an elegant example of how slight modifications in sequence can yield another compound. There are also examples of plant genes, which have been characterized based on fungal sequence information. The cluster coding for helvolic acid synthesis was first identified in *A. fumigatus*, and then cloned into *S. cerevisiae* verifying helvolic acid production. The sequence of the helvolic acid gene cluster was used to determine the biosynthetic genes in plants, shedding light on biosynthetic pathways (Mitsuguchi *et al.*, 2009). Several examples of similar types have been reviewed recently (Sanchez *et al.*, 2012).

Using comparative genomics, one can either search manually (e.g. using sequence alignments) for conserved backbone proteins or their domains in combination with their accessory tailoring genes. Alternatively, one can use predictor tools at the genome level in combination with sequence databases to compare gene clusters throughout organisms.

Table 4.4 presents an overview of such databases and predictors, which are relevant for fungi. A more comprehensive overview also including bacterial tools can be found in Weber *et al.* (2014).

In general, Table 4.4 demonstrates that there are multiple algorithms to choose from, if one wants to comparatively investigate secondary metabolism in fungi. In particular, the predictors can be divided into identifying gene clusters and predicting the specificity of synthases and synthetases in the gene clusters.

Prediction of secondary metabolite gene clusters

In general, there are two main methods applied within the field of predicting secondary metabolite clusters: the Antibiotics and Secondary Metabolite Analysis SHell (AntiSMASH) and the Secondary Metabolite Unique Regions Finder (SMURF) (Khaldi *et al.*, 2010; Weber *et al.*, 2015).

AntiSMASH has recently been upgraded to version 3.0 and optimized to perform faster on large scale projects. The algorithm was built using a profile Hidden Markov Model

Table 4.4 Overview of databases of secondary metabolites, their genes and predictors for fungal secondary metabolite gene clusters

Name	URL	Reference	Organism	Focus[1]
Databases				
Clustermine 360	http://www.clustermine360.ca/	Conway and Boddy, 2013	Fungi, bacteria	Clusters
Norine	http://bioinfo.lifl.fr/norine/	Caboche *et al.*, 2008	Fungi, bacteria	Compound
Novel Antibiotics Database	http://www0.nih.go.jp/~jun/NADB/search.html		Fungi, bacteria	Compound
PubChem	https://pubchem.ncbi.nlm.nih.gov/		Fungi, OTHER	Compound
Prediction of secondary metabolite gene clusters				
antiSMASH3	http://antismash.secondarymetabolites.org/	Weber *et al.*, 2015	Fungi, bacteria	Clusters
SMURF	http://jcvi.org/smurf/index.php	Khaldi *et al.*, 2010	Fungi, bacteria	Clusters
Prediction of substrate specificity				
antiSMASH3	http://antismash.secondarymetabolites.org/	Weber *et al.*, 2015	Fungi, bacteria	PKS, NRPS
MAPSI/ASMPKS	http://gate.smallsoft.co.kr:8008/pks/	Park, 2009	Fungi, bacteria	PKS
NRPSpredictor2	http://nrps.informatik.uni-tuebingen.de/Controller?cmd=SubmitJob	Röttig *et al.*, 2011	Fungi, bacteria	NRPS
LSI-based A-domain function predictor	http://bioserv7.bioinfo.pbf.hr/LSIpredictor/AdomainPrediction.jsp	Baranašić *et al.*, 2014	Bacteria, Fungi	NRPS

[1]PKS, polyketide synthase; NRPS, non-ribosomal peptide synthetase.

(pHMM) on mostly bacterial but also fungal clusters from the NCBI protein database. Once a cluster protein domain (including PKS, nonribosomal peptide synthases (NRPS), terpenes, aminoglycosides, aminocoumarins, indolocarbazoles, lantibiotics, bacteriocins, nucleosides, siderophores or melanin) is found, the program determines accessory genes, by scanning the genome in a 10 kb range and then searches outward by 5, 10 or 20 kb depending on the cluster type. This is highly efficient, but an observed problem with this method is that clusters are predicted to have too many genes included (as commented by Inglis *et al.*, 2013). It rarely underpredicts. A new feature of version 3 is the implementation of ClusterFinder (Cimermancic *et al.*, 2014), which uses a HMM algorithm based on 732 bacterial gene clusters to determine gene clusters of unknown type, filling a gap in the AntiSMASH functionality. However, the training set only consisted of bacterial gene clusters and the performance of this algorithm on fungal sequences still has to be evaluated.

SMURF was launched prior to AntiSMASH, and works in a similar fashion, but only fungal conserved domains were explored to build a HMM algorithm, thus giving it a more specific focus. It also examines a narrower set of cluster domains compared to AntiSMASH, namely dimethyl-allyltransferases including prenyltransferases, hybrids (combinations of PKS and NRPS domains), PKS, PKS-like, NRPS, NRPS-like, terpenecyclases. Once all backbones are found, the algorithm searches for members of 27 secondary metabolite-defining domains within a frame of 20 genes. The user can set intergenic distance between genes in the cluster and the maximum number of secondary metabolite domain negative genes.

In general, AntiSMASH has become sort of a *de facto* standard for secondary metabolite gene cluster predictions, and has developed a user-friendly web interface. This should be applied if one is interested in comparing gene clusters in a given genome. However, given that SMURF is trained on a fungal-specific set, and has a runtime that is significantly below AntiSMASH, we recommend that both algorithms are applied and compared if possible.

Prediction of substrate specificity

With gene clusters identified, one can use a variety of tools to predict the chemicals generated by such clusters (Table 4.4). AntiSMASH (see section on 'Prediction of secondary metabolite gene clusters') includes functions to predict the specificity of both PKSs and NRPSs and is possibly the go-to method at the moment. As an alternative for NRPSs, NRPSpredictor2 (Röttig *et al.*, 2011) uses support vector machines to predict NRPS adenylation domain specificity. The majority of the training set consisted of bacterial gene clusters, while only 16% were of fungal origin. Fortunately, a predictor for fungal gene clusters was added to make investigation of fungal sequences possible.

Supplementing methods for cluster identification

A few methods have been published, which use alternative methods for identifying gene clusters. In particular, (Inglis *et al.*, 2013) used comparative genomics between *A. nidulans*, *A. fumigatus*, *A. niger*, *A. oryzae* to refine the borders of gene clusters returned by antiSMASH and SMURF. Andersen *et al.* (2013) designed an algorithm for the use of transcriptomics data in tandem with genome sequences to determine gene cluster boundaries. A benefit of this method is that it can identify superclusters, i.e. clusters on different chromosomal loci, such as austinol, dehydroaustinol and prenyl xanthones. Additionally, it is not limited by the annotation of the genes, which causes some gene members to be left out in other algorithms. The MIDDAS-M algorithm (Umemura *et al.*, 2013) offers the implementation of a transcriptomics-driven method using the same principles as the one of (Andersen *et al.*, 2013).

Examples of comparative genomics in secondary metabolism

Until now, comparative studies have shown that only few clusters are conserved in filamentous fungi. Khaldi *et al.* (2010) showed that only five secondary metabolite gene clusters involved in protection against oxidative stress are conserved between the closely related species *A. fumigatus* Af293, *A. clavatus* and *A. fischerianus* (*Neosartorya fischeri*), implying that the other clusters are species-specific. This study showed the power of comparative genomics to reveal common traits of species. However, in this analysis combined with the work by (Lind *et al.*, 2015), one should note that the relatively large diversity in the compared species may underestimate the number of shared clusters. In the future, including new genomes to the analysis could redefine the 'core' clusters and reveal additional common traits.

Chiang *et al.* (2010) employed comparative genomics and protein domain predictions to compare multiple types of PKSs to infer shared traits and modular functions of the proteins, thus identifying shared traits across species.

The study by Inglis *et al.* (2013), mentioned above, has been a first approach to process gene clusters in fungi at larger scale. They identified a total of 261 non-redundant clusters in multiple species using their predicted gene clusters refined by alignment. This is an excellent example of how comparative genomics can complement the quality of automated prediction and can accelerate the research on secondary metabolites in the future.

Secretome analysis and plant polysaccharide degradation

Given the role of filamentous fungi as prolific producers of extracellular enzymes, it is not surprising that a number of studies have employed comparative genomics to investigate the degradation potential of different species. The applications span both the proteases and polysaccharide degrading enzymes.

A general investigation has been performed by (Braaksma *et al.*, 2010). Here, the authors attempted to study the secretome of *A. niger*, and did this by comparing the genome sequences of two *A. niger* strains to *A. oryzae*, *A. fumigatus* and *A. nidulans* to identify all theoretically secreted proteins. This has been combined with a proteomics analysis and allowed a refinement of the identification of secreted proteins from *A. niger*.

Examining proteases has been in a study spanning seven *Aspergillus* species (Ozturkoglu Budak *et al.*, 2014). The authors performed an extensive study of proteases using comparative genomics, comparative proteomics, and protease assays. The study confirmed that the aspergilli examined have protease capacities of similar magnitude, and that the different species have divergent profiles of proteases dependent on the growth medium, thus suggesting specialized life styles.

Polysaccharide degradation potential is a trait of interest in most species, and has been uncovered in many of the recent genome publications (Ballester *et al.*, 2015; Marcet-Houben *et al.*, 2012; Pel *et al.*, 2007). For a more specialized investigation, Coutinho *et al.* (2009) examined the polysaccharide degradation potential of three *Aspergillus* species in detail, using a combination of CAZy predictions (Cantarel *et al.*, 2009; Lombard *et al.*, 2014), comparative genomics, and growth assays on polysaccharides. Furthermore, the authors included a comparison to *Podospora anserina*. The main discovery in this study was the finding that the fungi had relatively similar degradation potential, but the actual degradation profiles of the fungi were different, possibly due to differential regulation and natural ecological niches.

Other comparative 'omics'

In the context of comparative genetics, it can be quite informative to also examine other types of comparative omics and examine what can be learned from these approaches. In this case, we see comparative omics as performing the same type of 'omics' analysis in multiple species and comparing the results using an identification of orthologues between the species. As the genome is used for orthologue determination (see 'Identification of orthologues across species', above), it can be seen as an additional dimension added to the comparative genomics data by an experimental analysis. Some studies of this type have been performed, in particular comparative transcriptomics and proteomics.

Comparative transcriptomics

Comparative transcriptomics is particularly interesting, when one wishes to examine and identify conserved transcriptional regulation across species. To our knowledge the first example of this in *Aspergillus* and/or *Penicillium* was by Andersen *et al.* (2008), where the regulatory response to growing on L-xylose relative to D-glucose was examined. The authors carried out DNA microarray experiments for *A. niger*, *A. nidulans* and *A. oryzae*, identified a set of genes with the same transcriptional response in all three examined species, and a common regulatory motif for binding of the xylanolytic transcriptional regulator (XlnR).

Salazar *et al.* (2009) performed a similar study using glucose and glycerol as the carbon sources, which gave interesting findings on the fungal response to osmotic stress. It was seen that the *Aspergillus* response to glycerol-induced osmotic stress has conserved elements and regulatory motifs to similar systems in *S. cerevisiae* and humans.

The study by Coutinho *et al.* (2009), which we discussed in the section on 'Secretome analysis and plant polysaccharide degradation', also employs comparative transcriptomics based on the data set of Andersen *et al.* (2008b) to study the secretory response and regulation, concluding a differential regulation of secreted polysaccharide-degrading enzymes.

The responses to hypoxia in *A. oryzae* and *A. nidulans* have been studied by Terabayashi *et al.* (2012) by comparing DNA microarray experiments of *A. oryzae* with available data for *A. nidulans*. Cultures were performed under hypoxic conditions and identified both conserved and species-specific responses to hypoxia.

Comparative proteomics

In this context, it is important to define that we here see 'comparative proteomics' as the comparison of transcriptomic results from multiple species or diverse strains, and not, as is also seen in the literature, the comparison of proteomics from multiple conditions [e.g. Stoll *et al.* (2014) or other notable papers as also reviewed by Kniemeyer (2011)].

Within our definition of comparative proteomics, an interesting aspect is, that this approach allows identification of conserved responses as well as unique traits across multiple species.

One such example in the area of immunogenicity, is the work of Benndorf *et al.* (2008), who studied allergens from spores of *A. versicolor, A. fumigatus,* and *P. expansum* by performing 2D gels in combination with immunoblotting to identify allergenic proteins in individual and multiple species.

A very comprehensive study of secreted proteases in seven different *Aspergillus* species has been conducted (Ozturkoglu Budak *et al.*, 2014). Here, the authors mine the genomes for putative proteases, employ comparative genomics to identify shared and unique proteases, and followed up by conducting proteome profiling on the secreted proteins from all seven species. Results were further validated with enzymatic protease assays. The analysis identified highly conserved proteins across the species, but with very different expression profiles even on the same media. An important hypothesis was that the species may have different physiologies at the same time point although cultivated in the same manner; similar to what was concluded for the polysaccharide-degrading enzymes by Coutinho *et al.* (2009).

In general, both comparative transcriptomics and proteomics, while expensive to conduct, have the advantage of letting us extrapolate what is known in one species across multiple species, thus convincingly allowing the researchers to identify whether a specific trait is unique for a species or a general biological phenomenon. Such studies have interesting potentials for understanding biology at a more general level.

Future perspectives – what can we learn from bacteria?

If we are to guess at what the future of fungal comparative genetics will bring, it is tempting to look at the kingdom of bacteria. Genome sequencing started in bacteria 10 years before aspergilli and penicillia with the first sequencing of the prokaryote *Haemophilus influenzae* in 1995. As a consequence of this head start, and the advantage of the lower price of sequencing the smaller bacterial genomes, the number of bacterial genomes now number in the tens of thousands (Land *et al.*, 2014), and some individual species have more than 2000 isolates sequenced (Cook and Ussery, 2013). Furthermore, by the time the first fungal genome was published in 2005, sequence databases, online analysis tools, gene calling pipelines and functional annotation models had been created and optimized for bacteria and were applied to fungal genomes. In some aspects, the technology of analysing bacterial genomes is both the future and the past of fungal comparative genomics.

Data sharing, maintenance and standards

An area, where the experience from bacterial genomics has been a great advantage, is in the construction and use of sequence databases. Through resources like GenBank, SWISS-PROT and RefSeq, experiences and ideas have been gained and used to create systems with more optimized downloads, searches, data submission and community involvement. Optimal design of data sharing resources has continually led to the introduction of new setups in resources like GenBank (Wheeler *et al.*, 2013). The fungal community has the opportunity to take advantage of these experiences and create databases that suit the needs of the research community. Some of this work has already been addressed in the structure and integration of FungiDB (Stajich *et al.*, 2012) and EuPathDB (Aurrecoechea *et al.*, 2013) as discussed above (section on 'Available genome databases with comparative genomics capabilities').

A key component in setting up data resources is data standards, for downloads. Although different data formats are widely used (FASTA, TAB, XML) the specifications on what information to include in these formats are minimal. Problems include naming of organisms, identification numbers of organisms and genes, relations between gene and protein sequences as well as metadata like literature references and growth conditions. All of these problems make it difficult to relate data resources to one another and utilize the full potential of these efforts. Comparative genomics relies heavily on the availability and quality of data without which comparisons become biased and less informative (Schnoes *et al.*, 2013).

The genetic data from sequencing of fungal genomes will increase rapidly in the coming years and efforts should be made to establish community standards for data, metadata, data submission and download to ensure the future use and reproducibility of *in silico* analyses. Efforts should also be made to encourage incorporation of experimental knowledge with sequence data to gain maximum information and understanding from years of experimental knowledge and future sequencing efforts.

The increased number of data and the pace at which it is being generated, presents a technical problem for the field of bioinformatics. In recent years the rate at which sequencing data has grown exceeds the pace of Moore's law (a doubling every 2 years). This puts a new pressure on the development of methods that can deal with large amounts of data without relying on larger computers. The issue of storage must also be addressed, with funding being granted to maintain online resources and publishing houses insisting on data submissions for research that does *in silico* analyses.

Comparative genomics-driven insights

Comparative genomics has been used in a wide range of bacterial research areas and some applications have yet to be widely studied in fungal research. One such topic is the genetic variation of species, genus and family. Some attention has been given to the core genome concept, identifying genes found in all strains of a specific set (Fedorova *et al.*, 2008; Galagan *et al.*, 2005). Genes found in specific strain subsets can also be used to connect possible phenotypes to genotypes, aiding experimental design (Cheeseman *et al.*, 2014; Coutinho *et al.*, 2009). However, these studies have been limited by the number of available genomes and general genetic variance of fungi is yet to be investigated. As more data becomes available along with well-documented metadata, it will be possible to use comparative genomics to identify genes involved in specific traits such as pathogenicity or identify sets of strain-specific genes. Comparative genomics can also aid in construction of evolutionary models by including whole genome patterns, adding deeper resolution to the known taxonomy (Rokas *et al.*, 2007; Snipen and Ussery, 2010), especially in the light of the expansion in number of fungal genome sequences discussed in the section on 'Current status of genomics'.

The role of fungi in biotechnology has long made them target for optimization of production. One comparative genomics-driven method of optimizing this is modelling of primary metabolism, another feature which has primarily been successful in bacteria and yeast (Heavner *et al.*, 2013; Monk *et al.*, 2013; Orth *et al.*, 2011). Genome-scale metabolic models will facilitate metabolic engineering by offering *in silico* predicted targets before experimental testing (Brandl and Andersen, 2015). Comparative genomics can greatly increase the evidence for these targets by looking for patterns across multiple genome sequences.

The general concept of a model, a mathematical description of a pattern in data, can in theory be used to describe almost any kind of pattern assuming that enough example data is provided. Creating models requires training data that represents enough variation to generalized the pattern (Ansari *et al.*, 2004). Sequence models can for example be used to predict secreted proteins (Braaksma *et al.*, 2010) or identify carbohydrate-active enzymes (CAZy) (Cantarel *et al.*, 2009).

Genome sequencing and annotation

A fungal research area that has extensively used comparative genomics is the field of secondary metabolism. High-throughput *in silico* screening of fungal genome sequences has revealed a large number of possible gene clusters involved in secondary metabolism, opening up for targeted experimental studies and discovery of new compounds for industrial production. The topics addressed including identification of novel clusters (Chiang *et al.*, 2009; Hansen *et al.*, 2011; Inglis *et al.*, 2013), locating known clusters in new genomes, assigning functions to genes in clusters (Szewczyk *et al.*, 2008) and locating silent gene clusters (Bergmann *et al.*, 2007). The use of *in silico* methods within this field has developed rapidly, including machine learning approaches as well as manually curated data collections (Weber, 2014).

Secondary metabolism has also been studied extensively in bacteria, exploiting the relative ease of sequencing many of these species for genome mining (Duncan *et al.*, 2015). However, despite this wealth of data and studies, there are still problems in the field, in particular with gene finding. Problems include inconsistencies due to different sequencing technologies or quality as well as use of different gene callers. Some of the most common problems involve over-annotation – prediction of genes that are not actually expressed. There is currently no clear way of determining *in silico* if a gene is truly a gene, experiments can reveal if a gene is expressed under some conditions but testing that a gene is not expressed under any conditions is not feasible on any scale. Comparative genomics has been used to address some of the annotation problems including assembly of genomes based on other genomes and gap filling (Delcher *et al.*, 1999a) and annotation, gene finding and functions (Inglis *et al.*, 2013).

The scale of over- or under-annotation becomes increasingly obvious as more genomes are sequenced. In one study, 4.94 million genes were identified *in silico* in 1474 prokaryotic organisms and 7% of these were not found in the published annotations (Wood *et al.*, 2012). The study showed that 60% of these genes were genes of length between 110 and 300 amino acids, thus suggesting that short genes in general are underpredicted. Another aspect of gene finding is the setting of parameters used for different prediction programs. Although some settings are widely used, actual benchmarking of different settings and the effect of these should be performed and published for the general community as these can have a major effect on the analysis outcome (Mancheron *et al.*, 2011).

All of these topics offer types of information and data ideal for modelling and machine learning. Data driven models have been extensively used in bacterial research to describe and identify functional protein domains such as PFAM or PANTHER (Finn *et al.*, 2014; Mi *et al.*, 2005) or creating gene calling models such as GLIMMER, SnowyOwl or Prodigal (Delcher *et al.*, 1999b; Hyatt *et al.*, 2010; Reid *et al.*, 2014). The success of these methods relies heavily on available training data, connections to experimental data and manual curation. Although these tools can be used for fungal genomes and gene sequences, the training data is currently biased towards bacteria due to the larger number of available data (e.g. AntiSMASH). In order for fungal annotation to improve, more fungal data should be included in the model training as well as manual curation of literature and sequence models. Creating new tools specifically for fungi is also an option, as in the SnowyOwl gene calling algorithm (Reid *et al.*, 2014).

We strongly believe that the post-genome future of fungal genomes will be highly driven by the application of comparative genomics approaches. The future of fungal comparative genomics promises to produce more and better data, addressing broader research fields and integrating into more projects. Establishing good practices on how this integration is created, how it should be published, documented and shared will be vital for its success and contribution to research. Making use of the experiences from other fields should shape the way the fungal community deals with these issues and affect the way grants are funded, publications are written and projects are planned.

References

Agren, R., Liu, L., Shoaie, S., Vongsangnak, W., Nookaew, I., and Nielsen, J. (2013). The RAVEN Toolbox and its use for generating a genome-scale metabolic model for *Penicillium chrysogenum*. PLoS Comput. Biol. *9*.

Altschul, S.F., Gish, W., Miller, W., Myers, E.W., and Lipman, D.J. (1990). Basic local alignment search tool. J. Mol. Biol. *215*, 403–410.

Andersen, M., Nielsen, M., and Nielsen, J. (2008a). Metabolic model integration of the bibliome, genome, metabolome and reactome of Aspergillus niger. Mol. Syst. Biol. *4*, 178.

Andersen, M.R., Vongsangnak, V., Panagiotou, G., Salazar, M.P., Lehmann, L., and Nielsen, J. (2008b). A trispecies *Aspergillus* microarray: comparative transcriptomics of three Aspergillus species. Proc. Natl. Acad. Sci. U.S.A. *105*, 4387–4392.

Andersen, M.R., Salazar, M.P., Schaap, P.J., van de Vondervoort, P.J.I., Culley, D., Thykaer, J., Frisvad, J.C., Nielsen, K.F., Albang, R., Albermann, K., *et al.* (2011). Comparative genomics of citric-acid-producing *Aspergillus niger* ATCC 1015 versus enzyme-producing CBS 513.88. Genome Res. *21*, 885–897.

Andersen, M.R., Nielsen, J.B., Klitgaard, A., Petersen, L.M., Zachariasen, M., Hansen, T.J., Blicher, L.H., Gotfredsen, C.H., Larsen, T.O., Nielsen, K.F., *et al.* (2013). Accurate prediction of secondary metabolite gene clusters in filamentous fungi. Proc. Natl. Acad. Sci. U.S.A. *110*, E99–E107.

Ansari, M.Z., Yadav, G., Gokhale, R.S., and Mohanty, D. (2004). NRPS-PKS: A knowledge-based resource for analysis of NRPS-PKS megasynthases. Nucleic Acids Res. *32*.

Arnaud, M.B., Chibucos, M.C., Costanzo, M.C., Crabtree, J., Inglis, D.O., Lotia, A., Orvis, J., Shah, P., Skrzypek, M.S., Binkley, G., *et al.* (2010). The Aspergillus Genome Database, a curated comparative genomics resource for gene, protein and sequence information for the Aspergillus research community. Nucleic Acids Res. *38*, D420–D427.

Aurrecoechea, C., Barreto, A., Brestelli, J., Brunk, B.P., Cade, S., Doherty, R., Fischer, S., Gajria, B., Gao, X., Gingle, A., *et al.* (2013). EuPathDB: The eukaryotic pathogen database. Nucleic Acids Res. *41*, 684–691.

Ballester, A., Marcet-houben, M., Levin, E., Sela, N., Selma-lázaro, C., Carmona, L., Wisniewski, M., Droby, S., González-candelas, L., and Gabaldón, T. (2015). Genome, transcriptome, and functional analyses of *Penicillium expansum* provide new insights into secondary metabolism and pathogenicity. Mol. Plant-Microbe Interact. *28*, 232–248.

Baranašić, D., Zucko, J., Diminic, J., Gacesa, R., Long, P.F., Cullum, J., Hranueli, D., and Starcevic, A. (2014). Predicting substrate specificity of adenylation domains of nonribosomal peptide synthetases and other protein properties by latent semantic indexing. J. Ind. Microbiol. Biotechnol. *41*, 461–467.

Bembom, O., Keles, S., and van der Laan, M.J. (2007). Supervised detection of conserved motifs in DNA sequences with cosmo. Stat. Appl. Genet. Mol. Biol. *6*, Article 8.

Benndorf, D., Müller, a., Bock, K., Manuwald, O., Herbarth, O., and Von Bergen, M. (2008). Identification of spore allergens from the indoor mould Aspergillus versicolor. Allergy Eur. J. Allergy Clin. Immunol. *63*, 454–460.

Benson, G. (1999). Tandem repeats finder: A program to analyze DNA sequences. Nucleic Acids Res. *27*, 573–580.

van den Berg, M.A., Albang, R., Albermann, K., Badger, J.H., Daran, J.-M., Driessen, A.J.M., Garcia-Estrada, C., Fedorova, N.D., Harris, D.M., Heijne, W.H.M., *et al.* (2008). Genome sequencing and analysis of the filamentous fungus *Penicillium chrysogenum*. Nat. Biotechnol. *26*, 1161–1168.

Bergmann, S., Schümann, J., Scherlach, K., Lange, C., Brakhage, A. a, and Hertweck, C. (2007). Genomics-driven discovery of PKS–NRPS hybrid metabolites from *Aspergillus nidulans*. Nat. Chem. Biol. *3*, 213–217.

Berry, D., Cox, M.P., and Scott, B. (2015). Draft genome sequence of the filamentous fungus *Penicillium paxilli* (ATCC 26601). Genome Announc. *3*, e00071–15.

De Bie, T., Cristianini, N., Demuth, J.P., and Hahn, M.W. (2006). CAFE: a computational tool for the study of gene family evolution. Bioinformatics *22*, 1269–1271.

Bok, J.W., Hoffmeister, D., Maggio-Hall, L.A., Murillo, R., Glasner, J.D., and Keller, N.P. (2006). Genomic mining for Aspergillus natural products. Chem. Biol. *13*, 31–37.

Braaksma, M., Martens-Uzunova, E.S., Punt, P.J., and Schaap, P.J. (2010). An inventory of the *Aspergillus niger* secretome by combining *in silico* predictions with shotgun proteomics data. BMC Genomics *11*, 584.

Brandl, J., and Andersen, M.R. (2015). Current state of genome-scale modeling in filamentous fungi. Biotechnol. Lett. *37*, 1131–1139.

Brudno, M., Do, C.B., Cooper, G.M., Kim, M.F., Davydov, E., Green, E.D., Sidow, A., and Batzoglou, S. (2003). LAGAN and Multi-LAGAN: Efficient tools for large-scale multiple alignment of genomic DNA. Genome Res. *13*, 721–731.

Caboche, S., Pupin, M., Leclère, V., Fontaine, A., Jacques, P., and Kucherov, G. (2008). NORINE: a database of nonribosomal peptides. Nucleic Acids Res. *36*, D326–D331.

Camacho, C., Coulouris, G., Avagyan, V., Ma, N., Papadopoulos, J., Bealer, K., and Madden, T.L. (2009). BLAST+: architecture and applications. BMC Bioinformatics *10*, 421.

Cantarel, B.L., Coutinho, P.M., Rancurel, C., Bernard, T., Lombard, V., and Henrissat, B. (2009). The Carbohydrate-Active EnZymes database (CAZy): an expert resource for glycogenomics. Nucleic Acids Res. *37*, D233–D238.

Castresana, J. (2000). Selection of conserved blocks from multiple alignments for their use in phylogenetic analysis. Mol. Biol. Evol. *17*, 540–552.

Cerqueira, G.C., Arnaud, M.B., Inglis, D.O., Skrzypek, M.S., Binkley, G., Simison, M., Miyasato, S.R., Binkley, J., Orvis, J., Shah, P., *et al.* (2014). The Aspergillus Genome Database: Multispecies curation and incorporation of RNA-Seq data to improve structural gene annotations. Nucleic Acids Res. *42*, 705–710.

Cheeseman, K., Ropars, J., Renault, P., Dupont, J., Gouzy, J., Branca, A., Abraham, A.-L., Ceppi, M., Conseiller, E., Debuchy, R., *et al.* (2014). Multiple recent horizontal transfers of a large genomic region in cheese making fungi. Nat. Commun. *5*, 2876.

Cherry, J.M., Hong, E.L., Amundsen, C., Balakrishnan, R., Binkley, G., Chan, E.T., Christie, K.R., Costanzo, M.C., Dwight, S.S., Engel, S.R., *et al.* (2012). Saccharomyces Genome Database: The genomics resource of budding yeast. Nucleic Acids Res. *40*.

Chiang, Y.-M., Szewczyk, E., Davidson, A.D., Keller, N., Oakley, B.R., and Wang, C.C.C. (2009). A gene cluster containing two fungal polyketide synthases encodes the biosynthetic pathway for a polyketide, asperfuranone, in *Aspergillus nidulans*. J. Am. Chem. Soc. *131*, 2965–2970.

Chiang, Y.-M., Oakley, B.R., Keller, N.P., and Wang, C.C.C. (2010). Unraveling polyketide synthesis in members of the genus Aspergillus. Appl. Microbiol. Biotechnol. *86*, 1719–1736.

Cimermancic, P., Medema, M.H., Claesen, J., Kurita, K., Wieland Brown, L.C., Mavrommatis, K., Pati, A., Godfrey, P.A., Koehrsen, M., Clardy, J., *et al.* (2014). Insights into secondary metabolism from a global analysis of prokaryotic biosynthetic gene clusters. Cell *158*, 412–421.

Conway, K.R., and Boddy, C.N. (2013). ClusterMine360: A database of microbial PKS/NRPS biosynthesis. Nucleic Acids Res. *41*, 402–407.

Cook, H., and Ussery, D.W. (2013). Sigma factors in a thousand *E. coli* genomes. Environ. Microbiol. *15*, 3121–3129.

Coutinho, P.M., Andersen, M.R., Kolenova, K., VanKuyk, P.A., Benoit, I., Gruben, B.S., Trejo-Aguilar, B., Visser, H., van Solingen, P., Pakula, T., *et al.* (2009). Post-genomic insights into the plant polysaccharide degradation potential of *Aspergillus nidulans* and comparison to *Aspergillus niger* and *Aspergillus oryzae*. Fungal Genet. Biol. *46* (Suppl. 1), S161–S169.

Crabtree, J., Angiuoli, S.V., Wortman, J.R., and White, O.R. (2007). Sybil: methods and software for multiple genome comparison and visualization. Methods Mol. Biol. *408*, 93–108.

Criscuolo, A. (2011). MorePhyML: Improving the phylogenetic tree space exploration with PhyML 3. Mol. Phylogenet. Evol. *61*, 944–948.
Dalquen, D.A., and Dessimoz, C. (2013). Bidirectional best hits miss many orthologs in duplication-rich clades such as plants and animals. Genome Biol. Evol. *5*, 1800–1806.
David, H., Akesson, M., and Nielsen, J. (2003). Reconstruction of the central carbon metabolism of *Aspergillus niger*. Eur. J. Biochem. *270*, 4243–4253.
David, H., Ozçelik, I.S., Hofmann, G., and Nielsen, J. (2008). Analysis of *Aspergillus nidulans* metabolism at the genome-scale. BMC Genomics *9*, 163.
Delcher, A.L., Kasif, S., Fleischmann, R.D., Peterson, J., White, O., and Salzberg, S.L. (1999a). Alignment of whole genomes. Nucleic Acids Res. *27*, 2369–2376.
Delcher, A.L., Harmon, D., Kasif, S., White, O., and Salzberg, S.L. (1999b). Improved microbial gene identification with GLIMMER. Nucleic Acids Res. *27*, 4636–4641.
Duncan, K.R., Crüsemann, M., Lechner, A., Sarkar, A., Li, J., Ziemert, N., Wang, M., Bandeira, N., Moore, B.S., Dorrestein, P.C., *et al.* (2015). Molecular networking and pattern-based genome mining improves discovery of biosynthetic gene clusters and their products from Salinispora species. Chem. Biol.
Edgar, R.C. (2004). MUSCLE: multiple sequence alignment with high accuracy and high throughput. Nucleic Acids Res. *32*, 1792–1797.
Emanuelsson, O., Brunak, S., von Heijne, G., and Nielsen, H. (2007). Locating proteins in the cell using TargetP, SignalP and related tools. Nat. Protoc. *2*, 953–971.
Enright, A.J., Van Dongen, S., and Ouzounis, C.A. (2002). An efficient algorithm for large-scale detection of protein families. Nucleic Acids Res. *30*, 1575–1584.
Fedorova, N.D., Khaldi, N., Joardar, V.S., Maiti, R., Amedeo, P., Anderson, M.J., Crabtree, J., Silva, J.C., Badger, J.H., Albarraq, A., *et al.* (2008). Genomic islands in the pathogenic filamentous fungus *Aspergillus fumigatus*. PLoS Genet. *4*, e1000046.
Finn, R.D., Bateman, A., Clements, J., Coggill, P., Eberhardt, R.Y., Eddy, S.R., Heger, A., Hetherington, K., Holm, L., Mistry, J., *et al.* (2014). Pfam: The protein families database. Nucleic Acids Res. *42*, D290–D301.
Flipphi, M., Sun, J., Robellet, X., Karaffa, L., Fekete, E., Zeng, A.-P., and Kubiecek, C.P. (2009). Biodiversity and evolution of primary carbon metabolism in *Aspergillus nidulans* and other *Aspergillus* spp. Fungal Genet. Biol. 46(Suppl. 1), S19–S44.
Förster, J., Famili, I., Fu, P., Palsson, B., and Nielsen, J. (2003). Genome-scale reconstruction of the *Saccharomyces cerevisiae* metabolic network. Genome Res. *13*, 244–253.
Futagami, T., Mori, K., Yamashita, A., Wada, S., Kajiwara, Y., Takashita, H., Omori, T., Takegawa, K., Tashiro, K., Kuhara, S., *et al.* (2011). Genome sequence of the white koji mold Aspergillus kawachii IFO 4308, used for brewing the Japanese distilled spirit shochu. Eukaryot. Cell *10*, 1586–1587.
Galagan, J.E., Calvo, S.E., Cuomo, C., Ma, L.-J., Wortman, J.R., Batzoglou, S., Lee, S.-I., Baştürkmen, M., Spevak, C.C., Clutterbuck, J., *et al.* (2005). Sequencing of Aspergillus nidulans and comparative analysis with A. fumigatus and A. oryzae. Nature *438*, 1105–1115.
Gilsenan, J.M., Cooley, J., and Bowyer, P. (2012). CADRE: The Central Aspergillus Data REpository 2012. Nucleic Acids Res. *40*, 660–666.
Grigoriev, I.V., Nordberg, H., Shabalov, I., Aerts, A., Cantor, M., Goodstein, D., Kuo, A., Minovitsky, S., Nikitin, R., Ohm, R.A., *et al.* (2012). The genome portal of the Department of Energy Joint Genome Institute. Nucleic Acids Res *40*, D26–D32.
Grigoriev, I.V., Nikitin, R., Haridas, S., Kuo, A., Ohm, R., Otillar, R., Riley, R., Salamov, A., Zhao, X., Korzeniewski, F., *et al.* (2014). MycoCosm portal: Gearing up for 1000 fungal genomes. Nucleic Acids Res. *42*, 699–704.
Haas, B.J. (2003). Improving the *Arabidopsis* genome annotation using maximal transcript alignment assemblies. Nucleic Acids Res. *31*, 5654–5666.
Hansen, B.G., Genee, H.J., Kaas, C.S., Nielsen, J.B., Regueira, T.B., Mortensen, U.H., Frisvad, J.C., and Patil, K.R. (2011). A new class of IMP dehydrogenase with a role in self-resistance of mycophenolic acid producing fungi. BMC Microbiol. *11*, 202.
Heavner, B.D., Smallbone, K., Price, N.D., and Walker, L.P. (2013). Version 6 of the consensus yeast metabolic network refines biochemical coverage and improves model performance. Database (Oxford) *2013*, bat059.
Horton, P., Park, K.J., Obayashi, T., Fujita, N., Harada, H., Adams-Collier, C.J., and Nakai, K. (2007). WoLF PSORT: Protein localization predictor. Nucleic Acids Res. *35*, W585–W587.
Huerta-Cepas, J., Dopazo, J., and Gabaldón, T. (2010). ETE: a python Environment for Tree Exploration. BMC Bioinformatics *11*, 24.

Hyatt, D., Chen, G.-L., Locascio, P.F., Land, M.L., Larimer, F.W., and Hauser, L.J. (2010). Prodigal: prokaryotic gene recognition and translation initiation site identification. BMC Bioinformatics *11*, 119.

Inglis, D.O., Binkley, J., Skrzypek, M.S., Arnaud, M.B., Cerqueira, G.C., Shah, P., Wymore, F., Wortman, J.R., and Sherlock, G. (2013). Comprehensive annotation of secondary metabolite biosynthetic genes and gene clusters of *Aspergillus nidulans*, *A. fumigatus*, *A. niger* and *A. oryzae*. BMC Microbiol *13*, 91.

Karaffa, L., and Kubicek, C.P. (2003). Aspergillus niger citric acid accumulation: do we understand this well working black box? Appl. Microbiol. Biotechnol. *61*, 189–196.

Katoh, K., Misawa, K., Kuma, K., and Miyata, T. (2002). MAFFT: a novel method for rapid multiple sequence alignment based on fast Fourier transform. Nucleic Acids Res. *30*, 3059–3066.

Kent, W.J. (2002). BLAT – The BLAST-Like Alignment Tool. Genome Res. *12*, 656–664.

Khaldi, N., Seifuddin, F.T., Turner, G., Haft, D., Nierman, W.C., Wolfe, K.H., and Fedorova, N.D. (2010). SMURF: Genomic mapping of fungal secondary metabolite clusters. Fungal Genet. Biol. *47*, 736–741.

Kis-Papo, T., Weig, A.R., Riley, R., Peršoh, D., Salamov, A., Sun, H., Lipzen, A., Wasser, S.P., Rambold, G., Grigoriev, I.V., *et al.* (2014). Genomic adaptations of the halophilic Dead Sea filamentous fungus Eurotium rubrum. Nat. Commun. *5*, 3745.

Kniemeyer, O. (2011). Proteomics of eukaryotic microorganisms: The medically and biotechnologically important fungal genus *Aspergillus*. Proteomics *11*, 3232–3243.

Kumar, S., Tamura, K., and Nei, M. (1994). MEGA: Molecular Evolutionary Genetics Analysis software for microcomputers. Comput. Appl. Biosci. *10*, 189–191.

Land, M.L., Hyatt, D., Jun, S.-R., Kora, G.H., Hauser, L.J., Lukjancenko, O., and Ussery, D.W. (2014). Quality scores for 32,000 genomes. Stand. Genomic Sci. *9*, 20.

Larkin, M.A., Blackshields, G., Brown, N.P., Chenna, R., McGettigan, P.A., McWilliam, H., Valentin, F., Wallace, I.M., Wilm, A., Lopez, R., *et al.* (2007). Clustal W and Clustal X version 2.0. Bioinformatics *23*, 2947–2948.

Lassmann, T., and Sonnhammer, E.L.L. (2005). Kalign – an accurate and fast multiple sequence alignment algorithm. BMC Bioinformatics *6*, 298.

Li, H., and Durbin, R. (2009). Fast and accurate short read alignment with Burrows-Wheeler transform. Bioinformatics *25*, 1754–1760.

Li, B., Zong, Y., Du, Z., Chen, Y., Zhang, Z., Qin, G., Zhao, W., and Tian, S. (2015). Genomic characterization reveals insights into patulin biosynthesis and pathogenicity in *Penicillium* species. Mol. Plant Microbe Interact. *28*, 635–647.

Li, H., Handsaker, B., Wysoker, A., Fennell, T., Ruan, J., Homer, N., Marth, G., Abecasis, G., and Durbin, R. (2009). The Sequence Alignment/Map format and SAMtools. Bioinformatics *25*, 2078–2079.

Li, L., Stoeckert, C.J., and Roos, D.S. (2003). OrthoMCL: identification of ortholog groups for eukaryotic genomes. Genome Res. *13*, 2178–2189.

Li, R., Li, Y., Kristiansen, K., and Wang, J. (2008). SOAP: Short oligonucleotide alignment program. Bioinformatics *24*, 713–714.

Lind, A.L., Wisecaver, J.H., Smith, T.D., Feng, X., Calvo, A.M., and Rokas, A. (2015). Examining the evolution of the regulatory circuit controlling secondary metabolism and development in the fungal genus *Aspergillus*. PLOS Genet. *11*, e1005096.

Linz, J.E., Wee, J., and Roze, L.V. (2014). *Aspergillus parasiticus* SU-1 genome sequence, predicted chromosome structure, and comparative gene expression under aflatoxin-inducing conditions: evidence that differential expression contributes to species phenotype. Eukaryot. Cell *13*, 1113–1123.

Liu, G., Zhang, L., Wei, X., Zou, G., Qin, Y., Ma, L., Li, J., Zheng, H., Wang, S., Wang, C., *et al.* (2013a). Genomic and secretomic analyses reveal unique features of the lignocellulolytic enzyme system of *Penicillium decumbens*. PLoS One *8*, e55185.

Liu, J., Gao, Q., Xu, N., and Liu, L. (2013b). Genome-scale reconstruction and *in silico* analysis of *Aspergillus terreus* metabolism. Mol. Biosyst. *9*, 1939–1948.

Lombard, V., Golaconda Ramulu, H., Drula, E., Coutinho, P.M., and Henrissat, B. (2014). The carbohydrate-active enzymes database (CAZy) in 2013. Nucleic Acids Res. *42*, 490–495.

Machida, M., Asai, K., Sano, M., Tanaka, T., Kumagai, T., Terai, G., Kusumoto, K.-I., Arima, T., Akita, O., Kashiwagi, Y., *et al.* (2005). Genome sequencing and analysis of *Aspergillus oryzae*. Nature *438*, 1157–1161.

Mancheron, A., Uricaru, R., and Rivals, E. (2011). An alternative approach to multiple genome comparison. Nucleic Acids Res. 1–11.

Marcet-Houben, M., Ballester, A.-R., de la Fuente, B., Harries, E., Marcos, J.F., González-Candelas, L., and Gabaldón, T. (2012). Genome sequence of the necrotrophic fungus Penicillium digitatum, the main postharvest pathogen of citrus. BMC Genomics *13*, 646.

McKenna, A., Hanna, M., Banks, E., Sivachenko, A., Cibulskis, K., Kernytsky, A., Garimella, K., Altshuler, D., Gabriel, S., Daly, M., *et al.* (2010). The genome analysis toolkit: A MapReduce framework for analyzing next-generation DNA sequencing data. Genome Res. *20*, 1297–1303.

Mi, H., Lazareva-Ulitsky, B., Loo, R., Kejariwal, A., Vandergriff, J., Rabkin, S., Guo, N., Muruganujan, A., Doremieux, O., Campbell, M.J., *et al.* (2005). The PANTHER database of protein families, subfamilies, functions and pathways. Nucleic Acids Res. *33*, D284–D288.

Mitsuguchi, H., Seshime, Y., Fujii, I., Shibuya, M., Ebizuka, Y., and Kushiro, T. (2009). Biosynthesis of steroidal antibiotic fusidanes: Functional analysis of oxidosqualene cyclase and subsequent tailoring enzymes from *Aspergillus fumigatus*. J. Am. Chem. Soc. *131*, 6402–6411.

Monk, J.M., Charusanti, P., Aziz, R.K., Lerman, J.A., Premyodhin, N., Orth, J.D., Feist, A.M., and Palsson, B.Ø. (2013). Genome-scale metabolic reconstructions of multiple *Escherichia coli* strains highlight strain-specific adaptations to nutritional environments. Proc. Natl. Acad. Sci. U.S.A. *110*, 20338–20343.

Nicholson, M.J., Koulman, A., Monahan, B.J., Pritchard, B.L., Payne, G.A., and Scott, B. (2009). Identification of two aflatrem biosynthesis gene loci in *Aspergillus flavus* and metabolic engineering of *Penicillium paxilli* to elucidate their function. Appl. Environ. Microbiol. *75*, 7469–7481.

Nierman, W.C., Pain, A., Anderson, M.J., Wortman, J.R., Kim, H.S., Arroyo, J., Berriman, M., Abe, K., Archer, D.B., Bermejo, C., *et al.* (2005). Genomic sequence of the pathogenic and allergenic filamentous fungus *Aspergillus fumigatus*. Nature *438*, 1151–1156.

Nierman, W.C., Yu, J., Fedorova-Abrams, N.D., Losada, L., Cleveland, T.E., Bhatnagar, D., Bennett, J.W., Dean, R., and Payne, G.A. (2015a). Genome sequence of *Aspergillus flavus* NRRL 3357, a strain that causes aflatoxin contamination of food and feed. Genome Announc. *3*, e00168–15.

Nierman, W.C., Fedorova-Abrams, N.D., and Andrianopoulos, A. (2015b). Genome sequence of the AIDS-associated pathogen *Penicillium marneffei* (ATCC18224) and its near taxonomic relative *Talaromyces stipitatus* (ATCC10500). Genome Announc. *3*, e01559–14.

O'Brien, K.P., Remm, M., and Sonnhammer, E.L.L. (2005). Inparanoid: a comprehensive database of eukaryotic orthologs. Nucleic Acids Res. *33*, D476–D480.

Okabe, M., Lies, D., Kanamasa, S., and Park, E.Y. (2009). Biotechnological production of itaconic acid and its biosynthesis in Aspergillus terreus. Appl. Microbiol. Biotechnol. *84*, 597–606.

Orth, J.D., Conrad, T.M., Na, J., Lerman, J.A., Nam, H., Feist, A.M., and Palsson, B.Ø. (2011). A comprehensive genome-scale reconstruction of *Escherichia coli* Metabolism. Mol. Syst. Biol. *7*, 535.

Osbourn, A. (2010). Secondary metabolic gene clusters: evolutionary toolkits for chemical innovation. Trends Genet. *26*, 449–457.

Ozturkoglu Budak, S., Zhou, M., Brouwer, C., Wiebenga, A., Benoit, I., Di Falco, M., Tsang, A., and de Vries, R.P. (2014). A genomic survey of proteases in aspergilli. BMC Genomics *15*, 523.

Park, K. (2009). Development of an analysis program of type I polyketide synthase gene clusters using homology search and profile hidden Markov model. J. Microbiol. Biotechnol. *19*, 140–146.

Van De Peer, Y., and De Wachter, R. (1994). Treecon for windows: A software package for the construction and drawing of evolutionary trees for the microsoft windows environment. Bioinformatics *10*, 569–570.

Pel, H.J., de Winde, J.H., Archer, D.B., Dyer, P.S., Hofmann, G., Schaap, P.J., Turner, G., de Vries, R.P., Albang, R., Albermann, K., *et al.* (2007). Genome sequencing and analysis of the versatile cell factory *Aspergillus niger* CBS 513.88. Nat. Biotechnol. *25*, 221–231.

Pi, B., Yu, D., Dai, F., Song, X., Zhu, C., Li, H., and Yu, Y. (2015). A genomics based discovery of secondary metabolite biosynthetic gene clusters in *Aspergillus ustus*. PLoS One *10*, e0116089.

Pitkänen, E., Jouhten, P., Hou, J., Syed, M.F., Blomberg, P., Kludas, J., Oja, M., Holm, L., Penttilä, M., Rousu, J., *et al.* (2014). Comparative genome-scale reconstruction of gapless metabolic networks for present and ancestral species. PLoS Comput. Biol. *10*, e1003465.

Price, A.L., Jones, N.C., and Pevzner, P.A. (2005). *De novo* identification of repeat families in large genomes. Bioinformatics *21*, i351–i358.

Rambaut, A. (2009). FigTree, a graphical viewer of phylogenetic trees. Inst. Evol. Biol. Univ. Edinburgh. Available online: http://tree.bio.ed.ac.uk/software/figtree/. Accessed 21 March 2016.

Reid, I., O'Toole, N., Zabaneh, O., Nourzadeh, R., Dahdouli, M., Abdellateef, M., Gordon, P.M.K., Soh, J., Butler, G., Sensen, C.W., *et al.* (2014). SnowyOwl: accurate prediction of fungal genes by using RNA-Seq and homology information to select among *ab initio* models. BMC Bioinformatics *15*, 229.

Retief, J.D. (2000). Phylogenetic analysis using PHYLIP. Methods Mol. Biol. *132*, 243–258.

Rice, P., Longden, I., and Bleasby, A. (2000). EMBOSS: the European Molecular Biology Open Software Suite. Trends Genet. *16*, 276–277.

Riley, D.R., Angiuoli, S.V., Crabtree, J., Dunning Hotopp, J.C., and Tettelin, H. (2012). Using Sybil for interactive comparative genomics of microbes on the web. Bioinformatics *28*, 160–166.

Rokas, A., Payne, G., Fedorova, N.D., Baker, S.E., Machida, M., Yu, J., Georgianna, D.R., Dean, R.A., Bhatnagar, D., Cleveland, T.E., *et al.* (2007). What can comparative genomics tell us about species concepts in the genus *Aspergillus*? Stud. Mycol. *59*, 11–17.

Röttig, M., Medema, M.H., Blin, K., Weber, T., Rausch, C., and Kohlbacher, O. (2011). NRPSpredictor2 – A web server for predicting NRPS adenylation domain specificity. Nucleic Acids Res. *39*.

Salazar, M., Vongsangnak, W., Panagiotou, G., Andersen, M., and Nielsen, J. (2009). Uncovering transcriptional regulation of glycerol metabolism in Aspergilli through genome-wide gene expression data analysis. Mol. Genet. Genomics *282*, 571–586.

Sanchez, J.F., Somoza, A.D., Keller, N.P., and Wang, C.C.C. (2012). Advances in *Aspergillus* secondary metabolite research in the post-genomic era. Nat. Prod. Rep. *29*, 351–371.

Schnoes, A.M., Ream, D.C., Thorman, A.W., Babbitt, P.C., and Friedberg, I. (2013). Biases in the experimental annotations of protein function and their effect on our understanding of protein function space. PLoS Comput. Biol. *9*, e1003063.

Snipen, L., and Ussery, D.W. (2010). Standard operating procedure for computing pangenome trees. Stand. Genomic Sci. *2*, 135–141.

Stajich, J.E., Harris, T., Brunk, B.P., Brestelli, J., Fischer, S., Harb, O.S., Kissinger, J.C., Li, W., Nayak, V., Pinney, D.F., *et al.* (2012). FungiDB: An integrated functional genomics database for fungi. Nucleic Acids Res. *40*, 675–681.

Stamatakis, A. (2014). RAxML version 8: A tool for phylogenetic analysis and post-analysis of large phylogenies. Bioinformatics *30*, 1312–1313.

Stoll, D. a., Link, S., Kulling, S., Geisen, R., and Schmidt-Heydt, M. (2014). Comparative proteome analysis of *Penicillium verrucosum* grown under light of short wavelength shows an induction of stress-related proteins associated with modified mycotoxin biosynthesis. Int. J. Food Microbiol. *175*, 20–29.

Strope, P.K., Skelly, D.A., Kozmin, S.G., Mahadevan, G., Stone, E.A., Magwene, P.M., Dietrich, F.S., and McCusker, J.H. (2015). The 100-genomes strains, an S. cerevisiae resource that illuminates its natural phenotypic and genotypic variation and emergence as an opportunistic pathogen. Genome Res. *25*, 762–774.

Subramanian, A.R., Hiran, S., Steinkamp, R., Meinicke, P., Corel, E., and Morgenstern, B. (2010). DIALIGN-TX and multiple protein alignment using secondary structure information at GOBICS. Nucleic Acids Res. *38*, W19–W22.

Szewczyk, E., Chiang, Y.-M., Oakley, C.E., Davidson, A.D., Wang, C.C.C., and Oakley, B.R. (2008). Identification and characterization of the asperthecin gene cluster of *Aspergillus nidulans*. Appl. Environ. Microbiol. *74*, 7607–7612.

Terabayashi, Y., Shimizu, M., Kitazume, T., Masuo, S., Fujii, T., and Takaya, N. (2012). Conserved and specific responses to hypoxia in Aspergillus oryzae and Aspergillus nidulans determined by comparative transcriptomics. Appl. Microbiol. Biotechnol. *93*, 305–317.

Thomas-Chollier, M., Sand, O., Turatsinze, J.V., Janky, R., Defrance, M., Vervisch, E., Brohée, S., and van Helden, J. (2008). RSAT: regulatory sequence analysis tools. Nucleic Acids Res. *36*, W119–W127.

Umemura, M., Koike, H., Nagano, N., Ishii, T., Kawano, J., Yamane, N., Kozone, I., Horimoto, K., Shin-ya, K., Asai, K., *et al.* (2013). MIDDAS-M: Motif-independent *de novo* detection of secondary metabolite gene clusters through the integration of genome sequencing and transcriptome data. PLoS One *8*, e84028.

Vongsangnak, W., Olsen, P., Hansen, K., Krogsgaard, S., and Nielsen, J. (2008). Improved annotation through genome-scale metabolic modeling of *Aspergillus oryzae*. BMC Genomics *9*, 245.

Wallace, I.M., O'Sullivan, O., Higgins, D.G., and Notredame, C. (2006). M-Coffee: combining multiple sequence alignment methods with T-Coffee. Nucleic Acids Res. *34*, 1692–1699.

Wang, F.-Q., Zhong, J., Zhao, Y., Xiao, J., Liu, J., Dai, M., Zheng, G., Zhang, L., Yu, J., Wu, J., *et al.* (2014). Genome sequencing of high-penicillin producing industrial strain of Penicillium chrysogenum. BMC Genomics 15 (Suppl. 1), S11.

Weber, T. (2014). In silico tools for the analysis of antibiotic biosynthetic pathways. Int. J. Med. Microbiol. *304*, 230–235.

Weber, T., Blin, K., Duddela, S., Krug, D., Kim, H.U., Bruccoleri, R., Lee, S.Y., Fischbach, M. a., Muller, R., Wohlleben, W., *et al.* (2015). antiSMASH 3.0 – a comprehensive resource for the genome mining of biosynthetic gene clusters. Nucleic Acids Res. 1–7.

Wheeler, D.L., Chappey, C., Lash, A.E., Leipe, D.D., Madden, T.L., Schuler, G.D., Tatusova, T.A., and Rapp, B.A. (2013). Database resources of the National Center for Biotechnology Information. Nucleic Acids Res. *28*, 10–14.

Wingender, E., Chen, X., Hehl, R., Karas, H., Liebich, I., Matys, V., Meinhardt, T., Prüss, M., Reuter, I., and Schacherer, F. (2000). TRANSFAC: an integrated system for gene expression regulation. Nucleic Acids Res. *28*, 316–319.

Wolf, Y.I., and Koonin, E.V. (2012). A tight link between orthologs and bidirectional best hits in bacterial and archaeal genomes. Genome Biol. Evol. *4*, 1286–1294.

Wood, D.E., Lin, H., Levy-Moonshine, A., Swaminathan, R., Chang, Y.-C., Anton, B.P., Osmani, L., Steffen, M., Kasif, S., and Salzberg, S.L. (2012). Thousands of missed genes found in bacterial genomes and their analysis with COMBREX. Biol. Direct *7*, 37.

Yang, Y., Zhao, H., Barrero, R. a, Zhang, B., Sun, G., Wilson, I.W., Xie, F., Walker, K.D., Parks, J.W., Bruce, R., *et al.* (2014). Genome sequencing and analysis of the paclitaxel-producing endophytic fungus Penicillium aurantiogriseum NRRL 62431. BMC Genomics *15*, 69.

Yu, J., Jurick, W.M., Cao, H., Yin, Y., Gaskins, V.L., Losada, L., Zafar, N., Kim, M., Bennett, J.W., and Nierman, W.C. (2014). Draft genome sequence of *Penicillium expansum* strain R19, which causes postharvest decay of apple fruit. Genome Announc. *2*, 2013–2014.

Blue Mould to Genomics and Beyond: Insights into the Biology and Virulence of Phytopathogenic *Penicillium* Species

5

Wayne M. Jurick II, Jiujiang Yu and Joan W. Bennett

Abstract

Apples and pears are economically important pome fruits grown and consumed worldwide. The United States is the second largest producer of apples and pears in the world behind China. Fruit decay caused by blue moulds, *Penicillium expansum* and other *Penicillium* spp., results in significant economic losses to the apple and pear fruit packing and processing industries. Since blue mould fungi produce several mycotoxins, they also are a food safety concern. In recent years, various '-omics' technologies have been used to learn more about fungal virulence and toxigenic potential, and to find possible news ways to control *Penicillium* species that infect pome fruits. Several *P. expansum* genomes have been sequenced and compared with the genomes of other sequenced *Penicillium* species. In this chapter, the current status of blue mould biology, control, epidemiology, genomics and functional genomics are discussed.

Introduction

Commercial pome fruit production and storage

Pome is a botanical term that refers to fruits having several seed chambers and an outer fleshy part largely derived from a greatly swollen receptacle. The major cultivated pome fruits are apples, pears, and quince, grown domestically and commercially in temperate climates around the world. The World Apple and Pear Association reports that global production of apples and pears reached 76.3 and 23.5 million tons, respectively. Behind China, the USA is the second largest producer of pome fruits in the world. Washington, New York, Michigan and Pennsylvania are the top apple-producing states while Washington, Oregon and California are the major pear producers in the USA. Approximately two-thirds of these pome fruits are used for fresh consumption while the remaining third are processed. US exports of fresh pome fruit in 2012 were valued at \$1.1 billion for apples and \$188.7 million for pears. Blue mould fungi are major postharvest pathogens that can cause spoilage and mycotoxin contamination of pome fruit (Spotts *et al.*, 1999). The control of spoilage and mycotoxin production by blue mould fungi is the focus of international scientific efforts.

After harvesting, pome fruits typically are stored for extended periods of time: usually 6 months in cold storage and up to 1 year in controlled atmosphere cold storage. While these long storage times provide consumers with high quality fresh fruit year round, the prolonged storage also makes stored fruits susceptible to many different fungal rots and physiological disorders (i.e. superficial scald, cracking, water core). Postharvest rots, caused mostly by fungal storage pathogens, impact tight profit margins by necessitating additional expenditures in the form of labour for repacking, multiple fungicide applications, rejection from processors, and overall product losses. Currently, blue moulds are controlled by using a combination of packinghouse sanitation, cultural methods, biocontrol and fungicide applications

Economic impact of postharvest decays

Postharvest diseases caused by filamentous fungal spoilage pathogens are a globally significant problem for the apple and pear fruit packing and processing industries. Specifically, blue mould diseases caused by *P. expansum* and other *Penicillium* spp., are the most common and economically important postharvest fruit rots (Rosenberger, 1990; Xiao and Boal, 2009). For example, in Washington State, blue mould accounted for 28% of fruit decays in storage (Kim and Xiao, 2008), while in Pennsylvania, a two year study conducted in commercial storage facilities by Jurick found that *Penicillium* spp. caused over 50% of the postharvest decay on apples. In addition to the direct loss due to spoilage, blue moulds produce a wide variety of mycotoxins that contaminate processed fruit products like juices, sauces, baby foods and butters. Patulin is the most important toxin produced by blue moulds, although other mycotoxins are also found. Because patulin is heat stable and survives pasteurization, it can contaminate processed fruit products like juices, sauces and butters in addition to fresh fruits (Bandoh *et al.*, 2009)

Causative species, symptoms, detection, and epidemiology

Penicillium expansum is the major cause of blue mould of pome fruits, but several other *Penicillium* species also have been implicated. In the USA, a survey of apples and pears with blue mould symptoms from Oregon and Washington found *P. aurantiogriseum, P. commune, P. glabrum, P. roqueforti, P. rugulosum, P. solitum,* and *P. verrucosum* in addition to *P. expansum* (Sanderson, 1995). A similar investigation in Canada from British Columbia found *P. brevicompactum, P. crustosum,* and *P. expansum* (Sholberg, 1996). *Penicillium carneum,* generally associated with meat and cheese spoilage, was found on apples at a commercial storage facility in Pennsylvania, and was comparably virulent to *P. expansum* under simulated storage conditions (Peter *et al.*, 2012).

Fruits infected with blue mould develop mushy, tan-to-brown decayed areas with a sharp margin between diseased and healthy tissues. During advanced stages of decay, bluish coloured conidia develop on the fruit surface, giving the disease its name. Blue mould infection occurs primarily through wounds caused by the stem punctures, bruises that occur during harvesting and handling, or via natural openings such as lenticels and the calyx.

Penicillium moulds are successful environmental saprophytes so they are ubiquitous in orchard soil and leaf litter (Lennox *et al.*, 2003). Bins of fruit contaminated with soil and leaf litter introduce spores into the packinghouse environment thereby imitating blue mould epidemics. Epidemiological studies have shown the importance of reducing blue mould inoculum levels in order to abate postharvest decay (Spotts, 1994; Spotts and Cervantes,

2001; Lennox *et al.*, 2003). Curiously, inoculum levels in the orchard, packinghouse air, or orchard soil/litter cannot be used as predictors of postharvest decay whereas the amount of inoculum present on the fruit surface is a strong predictive factor (Lennox *et al.*, 2003). The controlled environment of the packinghouse provides a unique opportunity for applying various sanitation methods in order to minimize the presence of fungal spores (Lennox *et al.*, 2003).

The generic name *Penicillium* was introduced in 1809 by Link for three species: *P. candidum*, *P. expansum* and *P. glaucum*. Since Thom's pioneering work on growing filamentous fungi in pure culture, *P. expansum* has been considered the lectotype for the genus (Thom, 1910; Pitt, 1979). *Penicillium* taxonomy remains challenging with more than 150 recognized species (Pitt, 1979; Pitt and Hocking, 2009). Traditionally, species have been identified by a combination of morphological and cultural techniques, followed by microscopic examination of the conidiophores by a trained scientist. Cultural methods rely on the use of specific media and growth a variety of temperatures, are labour intensive, and often yield variable results. (Frisvad and Samson, 2004). Fortunately, several molecular methods have been developed that enable rapid species-specific fungal identification and complement traditional morphological and cultural methods. These include restriction length fragment polymorphism, (RFLP), large subunit ribosomal RNA sequences, and internal transcribed spacer (ITS) bar codes (DuPont *et al.*, 1999). The most widely implemented molecular method is conventional PCR amplification of the β-tubulin locus for *Penicillium*; by targeting the β -tubulin locus this method can concomitantly indicate resistance to the postharvest fungicide Mertect® (active ingredient = thiabendazole) at codon #198 (Sholberg *et al.*, 2005).

Chemical, biological, and cultural control

A postharvest drench with the fungicides Penbotec™ (pyrimethanil), Scholar® (fludioxonil), or Mertec® (thiabendazole) is commonly applied to the fruits on the day of harvest to control *Penicillium* blue mould decay on pome fruits during storage. Thiabendazole, which inhibits β-tubulin synthesis, was labelled for blue mould control in 1968. Fludioxonil is a contact fungicide that works by inhibiting the osmosensing signalling pathway thereby preventing spore germination and halting mycelial growth. Both Scholar® and Penbotec™ were labelled for postharvest use in 2004 and are effective in controlling decay as they penetrate the fruit skin and travel through the cortical tissue (Xiao and Boal, 2009). Pyrimethanil has both preventative and curative effects as it inhibits spore germination and germ tube elongation via secretion of cell wall degradation enzymes and methionine biosynthesis (Kanetis *et al.*, 2008; Li, 2008).

In addition to chemical treatments, several commercial biological control products have been developed to control postharvest decay. BioSave™, which contains a saprophytic strain of *Pseudomonas syringae*, was introduced in 1996 and is labelled in the USA to control postharvest diseases of fruits and vegetables including blue and grey mould on pome and stone fruits (Janisiewicz and Jeffers, 1997). The product has had its registration expanded for use on cherries, potatoes, and sweet potatoes (Janisiewicz, 2010). The commercial implementation of BioSave™ has been steadily increasing since its release and during the 2000–2001 storage season it was applied to over 34 million bushel boxes of produce (Janisiewicz and Korsten 2002). Aspire™, containing the yeast *Candida oleophila*, was registered in the United States for postharvest application to control blue mould and green mould on citrus in 1995 (Droby *et al.*, 1998). However, this product was taken off the market 3 years after its large-scale commercial introduction. Outside of the USA other commercial formulations include YieldPlus™ (*Cryptococcus albidus*), Avogreen™ (*Bacillus subtilis*), and Shemer™ (*Metschnikowia fructicola*) although they have not had widespread use.

Decay can be reduced by careful harvest and handling practices that decrease inoculum level. Good cultural practices in the orchard include removal of organic debris and decayed fruits. After harvest, bin sterilization, stringent packinghouse sanitation, care to pack dry fruit, the culling of bruised and infected fruit, and maintenance of proper storage temperature, are essential for managing postharvest decay (Hansen *et al.*, 2010; Rosenberger, 2012). Specific sanitation methods include: treating bins with steam, cleaning packinghouse walls and floors with quaternary amines or peroxyacetic acids, and maintaining proper chlorine levels in sizing flumes (Spotts, 1994; Rosenberger, 2012). In addition to being effective on their own merits, cultural practices function synergistically when integrated with either chemical and or biological control methods. Thus, they should be practised as part of an integrated control strategy to achieve the highest level of postharvest decay management.

Current status of blue mould research

Cell wall-degrading enzymes as funal virulence factors

An abundance of physiological and biochemical research implicate fungal cell wall degrading enzymes in virulence; in fact, they were the first class of virulence factors studied in detail. Polygalacturonases (PG), the initial class of lytic enzyme deployed by fungal plant pathogens, are secreted pectic enzymes that aid in the acquisition of nutrients and facilitate host invasion and colonization. PG enzyme activity is highly correlated with decay characterized by tissue maceration and soft rot (Bateman and Basham, 1976). Purified PGs from both *P. expansum* and *P. solitum*-decayed pome fruit can macerate host tissue; moreover, they function at various temperatures and pHs *in vitro* (Jurick *et al.*, 2009, 2010). Both of these species can secrete host specific PGs in pear and apple fruit. The PG isozymes produced by *P. expansum* during postharvest decay of apple fruit differs from the isozyme produced by this pathogen during pear fruit decay (Yao *et al.*, 1996; Jurick *et al.*, 2010). Similarly, *P. solitum* produces a PG isozyme during postharvest decay on apples that is distinct from the PG isozyme produced during postharvest decay on pears (Jurick *et al.*, 2009). Nevertheless, despite the significant amount of biochemical knowledge concerning PGs, the genes and signalling pathways that regulate their production in *P. expansum* and *P. solitum* are poorly understood and will require further detailed investigations. Genomics approaches may provide information to target production, secretion, and activity of PGs and other important cell wall-degrading enzymes, thus interfering with the pathogens' virulence systems and guiding commodity-specific strategies to block decay.

Resistance to postharvest fungicides

Only a limited number of chemicals are labelled to control postharvest decay on apple and pear fruit. An industry consortium, the Fungicide Resistance Action Committee (FRAC) works to develop policies to extend the effectiveness of fungicides. It is critical for producers to rotate chemicals with different modes of action following the FRAC Code List in order to minimize the development of fungicide resistance. Routine monitoring can detect shifts in baseline sensitivities which may indicate the development of fungicide resistance. In general, baseline information is derived from a representative pathogen population before an active ingredient is introduced to market (Russell, 2004). This allows researchers to establish a minimum inhibitory concentration (MIC) for a specific chemical. A shift from baseline is often indicative of resistance that manifests in control failures in the field.

ARS scientists, led by Dr Wayne M. Jurick II, have been monitoring blue mould populations from commercial packing facilities in Pennsylvania since 2010. During this time, many *Penicillium* isolates were found able to grow *in vitro* on thiabendazole well above the labelled application rate (Jurick, unpublished results). This widespread resistance is not surprising because this compound has been used for more than 50 years to control blue mould. Moreover, resistance in *Penicillium* spp. to thiabendazole in other parts of the US, Canada and Europe has been well documented (Rosenberger *et al.*, 1991; Baraldi *et al.*, 2003; Sholberg *et al.*, 2005). Penbotec™ and Scholar® have been on the market since 2004. Findings from independent research groups on the east and west coast of the US have found pyrimethanil-resistant isolates capable growing on fungicide concentrations many folds higher than baseline. These resistant P. *expansum* isolates caused decay on Penbotec™ treated apple fruit (Yan *et al.*, 2014). Gaskins *et al.* (2015) sampled water sizing flumes from a commercial pome fruit packinghouse in Pennsylvania which yielded 'reduced sensitive' (= tolerant) strains of *P. expansum* capable of growing on 0.5 ppm of fludioxonil, the active ingredient in Scholar® fungicide. While still controlled by labelled rates of Scholar®, the isolation of these reduced-sensitive strains indicate that resistance to fludioxonil is developing in *P. expansum*. In summary, fungicide resistant and fungicide tolerant strains have manifested in the USA. It is hoped that application of functional genomic technologies can provide insights that will minimize the emergence of widespread fungicide resistance and perhaps generate additional strategies for developing sustainable control for postharvest decay (Ma and Michailides, 2005).

Mycotoxins and their impact on food safety

It is estimated that approximately one hundred different *Penicillium* species are mycotoxin producers. Published reports of mycotoxins made by members of the genus *Penicillium* include brevianamide A, chaetoglobosins, communesins, citrinin, cyclopenin, cyclopiazonic acid, expansolides, ochratoxin, patulin, penicillic acid, penitrem A, and rubratoxin B, roquefortine A, viomellein, viridicatin, and xanthomegnin (Frisvad, 1988; Pitt, 1997). However, only a few of these mycotoxins are found regularly in the food commodities that are of primary concern to humans and animals (Frisvad *et al.*, 2004).

P. expansum is not only the main species that causes postharvest decay of apple and pears (Frisvad and Samson, 2004; Jurick *et al.*, 2010) but it is the main producer of patulin (McKinley and Carlton, 1991). Furthermore, *P. expansum* isolates also make citrinin, chaetoglobosins, communesins, roquefortine C, and expansolides A and B, and there are reports in the literature that *P. expansum* produces other mycotoxins such as ochratoxin A, penitrem A, and rubratoxin B (Andersen *et al.*, 2004).

Patulin is of historical interest because it was first isolated in 1943 by Birkinshaw *et al.* (1943) from *Penicillium griseofulvum* and *Penicillium expansum* during early screening efforts to find new fungal produced antibiotics following the discovery of penicillin. Patulin was originally classified as an antibiotic because of its antibacterial, antiprotozoal and antiviral activity. However, it is also toxic to vertebrates and was eventually reclassified as a mycotoxin (Bennett and Klich, 2003). In addition to *P. griseofulvum* and *P. expansum*, patulin is produced by a number of other *Penicillium* species that occupy different biological niches (Pitt, 1997). In fact, patulin producers have been found among 30 different genera (Moake *et al.*, 2005). These include species of *Aspergillus* like *A. clavatus, A. giganteus* and *A. longivesica* (Varga *et al.*, 2007) and *Penicillium* species such as *P. carneum, P. clavigerum, P. concentricum, P. coprobium, P. dipodomyicola, P. expansum, P. glandicola, P. gladioli, P. griseofulvum, P. marinum, P. paneum, P. sclerotigenum, P. vulpinum* (Frisvad *et al.*, 2004).

Patulin has a strong affinity for sulfhydryl groups and inhibits many mammalian enzymes. Its acute toxicity was demonstrated in rodents with oral LD_{50} ranging between 29 and 55 mg/kg body weight. Toxic signs include agitation, dyspnoea, pulmonary congestion, oedema, ulceration, hyperaemia and distension of the gastro intestinal tract (WHO, 1998). Signs of sub-acute toxicity include neurotoxicity (tremors, convulsions) as well as inhibition of several intestinal enzymes (Devaraj *et al.*, 1986). Patulin affects lipid metabolism (Devaraj and Devaraj, 1987). Genotoxicity, teratogenicity, mutagenicity, and immunotoxicity are implied, but inconclusive (Puel *et al.*, 2010). Due to its toxic and potentially carcinogenic effects, the European Union (EU) has set limits on the maximum amount of patulin to 50 μg/l for fruit juices and derived products, 25 μg/l for solid apple products, and 10 μg/l for juices and foods destined for babies and young infants (European-Union 2003). The U.S. Food and Drug Administration (FDA), as well as other countries, have imposed limits on permissible amounts of patulin in fruit juices and processed pome fruit products for human consumption at 50 μg/l (Puel *et al.* 2010).

Citrinin was originally isolated in 1931 by Hetherington and Raistrick from a culture of *Penicillium citrinum* (Hetherington and Raistrick, 1931). *P. citrinum* is the main producer of citrinin but it is also made by a variety of other fungi. *Aspergillus ochraceus, P. verrucosum* and *P. nodicum* are known to produce ochratoxin A (OTA), while *P. verrucosum* produces both ochratoxin and citrinin. Citrinin acts as a nephrotoxin in all species studied and causes mycotoxic nephropathy in livestock (Bennett and Klich, 2003). The reported acute toxicity of citrinin varies with species. For example, the LD_{50} at 50 mg/kg for oral administration for rat, 35–58 mg/kg for intraperitoneal (ip) for mouse and 19 mg/kg for ip for rabbit. Citrinin causes kidney damage and mild liver damage in the form of fatty tissue formation. Other effects include vasodilation, constriction of the bronchi and increased muscular tone. There is only limited evidence for the carcinogenicity of citrinin in experimental animals. Currently, there is no regulated limit set for citrinin and other mycotoxins produced from *Penicillium* species in food and feed (Abou-Zeid, 2012).

Current status of *Penicillium* genomics

Whole genome sequencing and comparative genomics

Whole genome sequencing efforts for members of the genus *Penicillium* have lagged behind other economically important fungal genera such as *Saccharomyces* and *Aspergillus* (Galagan *et al.*, 2005; Machida *et al.*, 2005; Nierman *et al.*, 2005). Not surprisingly, the first *Penicillium* genome sequenced was a penicillin-producer, namely the laboratory strain *P. chrysogenum*, Wisconsin 54-1255 (van den Berg *et al.*, 2008). This strain was originally isolated from an infected cantaloupe in Peoria, Illinois during World War II and is the progenitor of all current industrial production strains. More recently, the genomes of several other penicillin producers have been sequenced and reported. The high penicillin-producing strain, *P. chrysogenum* NCPC10086, was sequenced and found be 32.3 Mbp in size (Wang *et al.*, 2014). Comparative analysis with the Wisconsin 54-1255 strain identified 69 unique genes in NCPC 10086 that were not present in the Wisconsin 54-1255 strain. Some of the unique genes are involved in energy, nitrogen, and glutathione metabolism. A 53.7 kb 'new shift fragment' in seven copies was discovered that could be related to high penicillin biosynthesis in NCPC10086. The industrial progenitor strain *P. chrysogenum* P2niaD18 in a chromosome-scale genome assembly also was reported (Specht *et al.*, 2014). The 32.5 Mbp P2niaD18 strain is distinguished from the Wisconsin 54-1255 strain by major chromosomal rearrangements leading to a modified genomic architecture (Specht *et al.*, 2014). Additional *P. chrysogenum* strains such as IB 08/921 (32.24 Mbp, Accession#: JPDR00000000) (Böhm *et al.*, 2015) and KF-25 (29.92 Mbp, accession#: AZRB00000000.1) were published and/or submitted to the publicly available GenBank database.

Other sequenced genomes of *Penicillium* species with industrial and agricultural importance include the main postharvest pathogen of citrus *P. digitatum* (Marcet-Houben *et al.*, 2012), the lignocellulose-degrading *P. oxalicum* (*P. decumbens*) (Liu *et al.*, 2013), two cheese-related *Penicillium* species, *P. roqueforti* and *P. camemberti* (Cheeseman *et al.*, 2014), and the endophytic fungal species, *P. aurantiogriseum* (Yang *et al.*, 2014). A highly virulent strain of *P. expansum* (R19) that causes postharvest pome fruit decay was sequenced (Accession #: JHUC 00000000) and published in Genome Announcements (Yu *et al.*, 2014) along with the less aggressive pome fruit pathogen, *P. solitum* (RS1) (Accession #: JYNM 00000000). The calculated genome size of *P. expansum* (R19) contained 31,415,732 bp (Yu *et al.*, 2014). This is consistent with the genome size reported for *P. chrysogenum* (van den Berg *et al.*, 2008). Preliminary annotation demonstrated that the *P. expansum* R19 genome harbours 10,554 predicted genes with an average gene length of 1,599 bp (Table 5.1).

Ballester *et al.* (2015) sequenced three *P. expansum* strains that cause pome fruit decay and one *P. italicum* strain that causes citrus green mould, and then compared them with the published genomes of 24 other fungal species. The whole genome sequences of *P. expansum* strain T01 (33.52 Mb) and *P. italicum* B3 (28.99 Mb) in comparison with *P. digitatum* (Marcet-Houben *et al.*, 2012) were reported recently (Li *et al.*, 2015). The *P. expansum* strain T01 was assembled into 108 scaffolds. The eight largest scaffolds represent 98% DNA sequence (32.9 Mbp) which could more than likely constitute the eight arms of the four chromosomes in the *Penicillium* genome. The data of the three genome comparison are summarized in Table 5.1. The *P. expansum* T01 sequence data are also listed.

Except for patulin, genetic and genomic studies focused on the biosynthesis of mycotoxins from *Penicillium* species are not as well understood as *Aspergillus* toxins like aflatoxins or *Fusarium* toxins like fumonisins and trichothecenes (Yu *et al.*, 2015). Nevertheless, progress has been made on understanding both ochratoxin and patulin biosynthesis. In *P. nordicum*, three genes involved with ochratoxin biosynthesis were found in a 10 kb DNA region (Geisen *et al.*, 2006). The patulin biosynthetic pathway is chemically well-characterized. The patulin gene cluster was first identified in *Aspergillus clavatus* by whole genome sequencing and predicted by SMURF to consist of 15 genes in the following order: *patH*, *patG*, *patF*, *patE*, *patD*, *patC*, *patB*, *patA*, *patM*, *patN*, *patO*, *patL*, *patI*, *patJ*, and *patK* (Fedorova

Table 5.1 Comparison of three *Penicillium* species genomes

Species	*P. expansum*	*P. solitum*	*P. chrysogenum*	*P. expansum*
Strain	R19	RS1	Wisconsin 54-1255	T01
Accession	JHUC00000000	JYNM 00000000	AM920416-AM920464.	AYHP00000000
Reference	Yu *et al.*, 2014		van den Berg *et al.*, 2008	Li *et al.*, 2015
Scaffolds	1231	454	49	108
Genome size (Mbp)	31.41	32.36	32.22	33.52
G+C content (%)	48.24	48.68	48.90	47.10
Coding seq (%)	53.70	52.69	56.60	
Proteins (≥100 aa)	10,554	10,661	12,943	11,770
Putative pseudogenes	47	74	592	
Mean gene length	1599	1599	1515	1891
tRNA genes	120	127	145	
5S rRNA genes	48	50	28	
Unique genes	314	225	2146	
Secondary metabolism (SM) gene clusters predicted by SMURF	59	58		55
SM gene clusters predicted by AntiSMASH	57	65	51	

et al., 2008). The functions of two of the patulin pathway genes encoding for cytochrome P450 type enzymes in *A. clavatus* were characterized (Artigot *et al.*, 2009). The patulin gene cluster in *P. expansum* also was reported (Tannous *et al.*, 2014). These genes share 60–70% sequence identity to those in *A. clavatus* (Fedorova *et al.*, 2008; Artigot *et al.*, 2009) although their gene order was different.

Using SMURF (Khaldi *et al.*, 2010), 59 gene clusters putatively involved in the biosynthesis of secondary metabolism (SM) were identified in the *P. expansum* (R19) genome (Table 5.1). This is similar to the result predicted by the Antibiotics and Secondary Metabolite Analysis Shell program or AntiSMASH program (Medema *et al.*, 2011) which predicts 57 SM clusters. Using SMURF, 55 SM gene clusters were predicted in P. *expansum* T01 genome including a cluster of 15 genes (*pePatA* to *pePatO*) inferred to be involved in patulin biosynthesis. (Li *et al.*, 2015). The differences in number of predicted SM clusters using SMURF (59 for *P. expansum* strain R19 vs. 55 for P. *expansum* strain T01) may merely reflect the difference in the number of scaffolds. The 1231 scaffolds used for *P. expansum* R19 might have a higher chance of separating a single SM gene cluster in a way that gets predicted as two. Only 108 scaffolds were used in *P. expansum* T01 thereby reducing the chances of separating and over-estimating the number of SM gene clusters.

Functional genomics via transcriptome sequencing and gene deletion

With the advent of next-generation sequencing, especially RNA-Seq technologies via the Illumina platform, functional genomics studies through transcriptomic analysis are now widely adopted. SM genes and clusters of genes are normally expressed in concert. The low cost, quick turn around and accurate gene expression data of the entire genome make it more feasible to profile genes involved in a specific pathway. More experiments using RNA-Seq will be performed in the future in our lab and in those of collaborators for studying *Penicillium* biology, pathogenicity, mycotoxin biosynthesis, and regulation of gene expression.

Analysis of three *P. expansum* strains from Israel, Spain, and the United States, and a *P. italicum* strain from Spain, indicated that there are about 10,000 to 11,000 protein-coding genes (Ballester *et al.*, 2015). Comparisons of the *P. expansum* strains with the other fungal genomes indicated that only *A. clavatus* and *P. expansum* contain a complete patulin gene cluster consisting of 15 genes. Interestingly, infection studies using patulin-producing and non-producing *P. expansum* strains demonstrated that patulin and citrinin are not directly involved in *P. expansum* pathogenesis (Ballester *et al.*, 2015). Our studies of over 100 *P. expansum* isolates also indicated that the inability or ability of strains to produce patulin and/or citrinin has no direct correlation to their ability to cause postharvest decay on apple (Yu and Jurick, unpublished data).

Expression analysis of genes potentially involved in patulin biosynthesis under patulin-permissive and non-permissive conditions identified 1642 up-regulated genes, including 15 predicted patulin genes (Li *et al.*, 2015). Gene deletion experiments confirmed that *PePatL* and *PePatK* play crucial roles in the biosynthesis of patulin (Li *et al.*, 2015). Virulence studies, using *PePatL* and *PePatK* knock-out mutants, demonstrated that there was no direct correlation between patulin production and virulence during *P. expansum* infection on apple (Li *et al.*, 2015). It was shown that the *P. expansum* T01 contains more pathogenicity genes and SM gene clusters than *P. italicum* B3 or *P. digitatum* (Li *et al.*, 2015). The secondary metabolite global regulatory gene, *laeA*, initially identified in *Aspergillus nidulans* (Bok and Keller, 2004) by chromatin modification mediated by the velvet complex also might be an important regulatory factor in *Penicillium* species.

Future trends

With the availability of NGS technologies, sequencing a fungal genome is no longer a million dollar project requiring many years to complete. The low cost, efficiency, and rapid determination of genome sequences using NGS technologies enables scientists quickly to acquire a detailed genetic blueprint of a given fungal organism. Nevertheless, acquisition of the genome sequence information is just the beginning of a long quest to understand genetics and physiology. With each passing year, biologists become increasingly aware that gene expression and regulation are much more complex than we once expected. The real challenge in the post-genomic era will be to understand the function and regulation of the genes annotated through genome sequencing. Non-automated bioinformatics analysis of the data and interpretation of gene function lag behind simple acquisition of raw sequence data and computer annotation.

Genomics and functional genomics approaches provide a varied tool set for better understanding and controlling postharvest decay caused by *Penicillium* species.

Comparative genomic information will refine species-level delineations advancing phylogenetic studies and aid in the development of innovative nucleic acid based methods for rapid and efficient pathogen detection. Data from experimental proteomics can be used to refine genomic annotations and are particularly well suited to non-model species such as *P. expansum* (Armengaud *et al.*, 2014). It is hoped that functional genomic insights into *Penicillium* virulence will unlock control points for fungal infections that can be exploited to develop targeted control strategies to block decay during storage. Uncovering the genetic mechanisms of fungicide resistance will help tailor specific classes of chemicals that provide durable control with lower likelihood for developing resistance. Furthermore, the power of -omics-enabled studies should also be brought to bear on biocontrol organisms such as yeasts and *Pseudomonads*. If we could explicate the genetic mechanisms and interspecific signalling involved in growth perturbation of *Penicillium* species when in the presence of these biocontrol species, it might give us hints as to find new ways for minimizing decay. Furthermore, in order to achieve the goal of long-lasting control, we must build a functional genomics arsenal that incorporates knowledge derived from traditional studies in biology, epidemiology, and agriculture. New tools such as CRISP/Cas9-mediated genome editing platforms, RNAi, and sophisticated image analysis should be integrated into experimental frameworks. If we can expand and accelerate our knowledge of the gene networks that *Penicillium* species have evolved in order to become such effective contaminants, we will be more likely to find ways to sabotage their virulence. Synergistic use of genomics, functional genomics, wet laboratory research and controlled trials in commercial facilities will create a powerful amalgamation that will impact the control of blue mould diseases and also provide a model for similar scientific tactics for other fungal phytopathogens.

References

Abou-Zeid, A.M. (2012). Review on citrinin: production, effect of some plant extracts and gene involved in its biosynthesis. J. Civil Environment. Eng. 2, 113.

Andersen, B., Smedsgaard, J., and Frisvad, J.C. (2004). *Penicillium expansum*: Consistent production of patulin, chaetoglobosins, and other secondary metabolites in culture and their natural occurrence in fruit products. J. Agric. Food Chem. 52, 2421–2428.

Armengaud, J., Trapp, J., Pible, O., Geffard, O., Chaumot, A., and Hartmann, E.M. (2014). Non-model organisms, a species endangered by proteogenomics. J. Proteomics *105*, 5–18.

Artigot, M.P., Loiseau, N., Laffitte, J., Mas-Reguieg, L., Tadrist, S., Oswarld, I., and Puel, O. (2009). Molecular cloning and functional characterization of two CYP619 cytochrome P450s involved in biosynthesis of patulin in *Aspergillus clavatus*. Microbiol. Immunol. *155*, 1738–1747.

Ballester, A.R., Marcet-Houben, M., Levin, E., Sela, N., Selma-Lázaro, C., Carmona, L., Wisniewski, M., Droby, S., González-Candelas, L., and Gabaldón, T. (2015). Genome, transcriptome, and functional analyses of *Penicillium expansum* provide new insights into secondary metabolism and pathogenicity. Mol. Plant Microbe Interact. *28*, 232–248.

Bandoh, S., Takeuchi, M., Ohsawa, K., Higashihara, K., Kawamoto, Y., and Goto. T. (2009). Patulin distribution in decayed apple and its reduction. Internat. Biodeterior. Biodegradation *63*, 379–382.

Baraldi, E., Mari, M., Chierici, E., Ponderelli, M., Bertolini, P., and Pratella, G.C. (2003). Studies of thiabendazole resistance of *Penicillium expansum* of pears: pathogenic fitness and genetic characterization. Plant Pathol. *53*, 362–370.

Bateman, D.F., and Basham, H.G. (1976). Degradation of plant cell walls and membranes by microbial enzymes. Encycl. Plant Physiol. New. Ser. *4*, 316–355.

Bennett, J.W., and Klich, M. (2003). Mycotoxins. Clin. Microbiol. Rev. *16*, 497–516.

Birkinshaw, J.H., Michael, S.E., Bracken, A., and Raistrick, H. (1943). Patulin in the common cold collaborative research on a derivative of *Penicillium patulum* Bainier. II. Biochemistry and Chemistry. Lancet *245*, 625.

Böhm, J., Dahlmann, T.A., Gümüşer, H., and Kück, U. (2015). A MAT1-2 wild-type strain from *Penicillium chrysogenum*: functional mating-type locus characterization, genome sequencing and mating with an industrial penicillin-producing strain. Mol. Microbiol. *95*, 859–874.

Bok, J.-W., and Keller, N.P. (2004). LaeA, a regulator of secondary metabolism in *Aspergillus* spp. Eukaryot. Cell 3, 527–535.

Cheeseman, K., Ropars, J., Renault, P., Dupont, J., Gouzy, J., Branca, A., Abraham, A.L., Ceppi, M., Conseiller, E., Debuchy, R., *et al.* (2014). Multiple recent horizontal transfers of a large genomic region in cheese making fungi. Nat. Commun. *5*, 2876.

Devaraj, H., and Devaraj, N. (1987). Rat intestinal lipid changes in patulin toxicity. Indian J. Exp. Biol. *25*, 637–638.

Devaraj, H., Suseela, R.E., and Devaraj, N. (1986). Patulin toxicosis in chicks. Curr. Sci. *55*, 998–999.

Droby, S., Cohen, L., Daus, A., Weiss, B., and Horev, B. (1998). Commercial testing of Aspire: a yeast preparation for the biological control of postharvest decay of citrus. Biol. Control. *12*, 97–101.

Dupont, J., Magnin, S., Marti, A., and Brousse, M. (1999). Molecular tools for identification of *Penicillium* starter cultures used in food industry. Internat. J. Food Microbiol. *49*, 109–118.

European Union (2003). Commission Regulation (EC) No. 1425/2003 of 11 August 2003 amending Regulation (EC) N°466/2001 as regards patulin, Official J. European Union L 203/13.

Fedorova, N.D., Khaldi, N., Joardar, V.S., Maiti, R., Amedeo, P., Anderson, M.J., Crabtree, J., Silva, J.C., Badger, J.H., Albarraq, A., *et al.* (2008). Genomic islands in the pathogenic filamentous fungus *Aspergillus fumigatus*. PLoS Genet. *4*, e1000046.

Frisvad, J.C. (1988). Fungal species and their specific production of mycotoxins. In: Introduction to Food-Born Fungi, R.A. Samson and E. Van Reenen-Hoekstra, eds., pp 239–249. Baarn: Centraalbureau voor Schimmelcultures.

Frisvad, J.C., and Samson, R.A. (2004). Polyphasic taxonomy of *Penicillium* subgenus *Penicillium*: A guide to identification of food and air-borne terverticillate *Penicillia* and their mycotoxins. Stud. Mycol. *49*, 1–174.

Frisvad, J.C., Smedsgaard, J., Larsen, T.O., and Samson, R.A. (2004). Mycotoxins, drugs and other extrolites produced by species in *Penicillium* subgenus *Penicillium*. Stud. Mycol. *49*, 201–242.

Galagan, J.E., Calvo, S.E., Cuomo, C., Ma, L.J., Wortman, J.R., Batzoglou, S., Lee, S.I., Basturkmen, M., Spevak, C.C., Clutterbuck, J., *et al.* (2005). Sequencing of *Aspergillus nidulans* and comparative analysis with *A. fumigatus* and *A. oryzae*. Nature *438*, 1105–1115.

Gaskins, V.L., Vico, I., Yu, J., and Jurick II, W.M. (2015). First report of *Penicillium expansum* isolates with reduced sensitivity to fludioxonil from a commercial packinghouse in Pennsylvania. Plant Dis. *99*, 1182.

Geisen, R., Schmidt-Heydt, M., and Karolewiez, A. (2006). A gene cluster of the ochratoxin A biosynthetic genes in *Penicillium*. Mycotoxin Res. *22*, 134–141.

Hansen, J., Xiao, C.L., and Kupferman, G. (2010). Bin Sanitation: an effective was to reduce codling moth and fungal decay spores. WSU publication. *1*, 1–3.

Hetherington, A.C., and Raistrick, H. (1931). On the production and chemical constitution of a new yellow colouring matter, citrinin, produced from glucose by *Penicillium citrinum*. Philos. Trans. R. Soc. London. Ser. B. *220*, 269–295.

Janisiewicz, W.J. (2010). Quo vadis of biological control of postharvest diseases. Postharvest Pathology 2, 137–148.

Janisiewicz, W.J., and Jeffers, S.N. (1997). Efficacy of commercial formulation of two biofungicides for control of blue mold and gray mold of apples in cold storage. Crop Prot. *16*, 629–633.

Janisiewicz, W.J., and Korsten, L. (2002). Biological control of postharvest diseases of fruits. Ann. Rev. Phytopathol. *40*, 411–441.

Jurick, W.M. II, Vico, I., Gaskins, V.L., Garrett, W.M., Whitaker, B.D., Janisiewicz, W.J., and Conway, W.S. (2010). Purification and biochemical characterization of polygalacturonase produced by *Penicillium expansum* during postharvest decay of 'Anjou' pear. Phytopathology *100*, 42–48.

Jurick, W.M. II, Vico, I., McEvoy, J.L., Whitaker, B.D., Janisiewicz, W.J., and Conway, W.S. (2009). Isolation, purification, and characterization of a polygalacturonase produced in *Penicillium solitum*-decayed 'Golden Delicious' apple fruit. Phytopathology *99*, 636–641.

Kanetis, L., Förster, H., Jones, C.A., Borkovich, K.A., and Adaskaveg, J.E. (2008). Characterization of genetic and biochemical mechanisms of fludioxonil and pyrimethanil resistance in field isolates of *Penicillium digitatum*. Phytopathology *98*, 205–214.

Khaldi, N., Seifuddin, F.T., Turner, G., Haft, D., Nierman, W.C., Wolfe, K., and Fedorova, N.D. (2010). SMURF: Genomic mapping of fungal secondary metabolite clusters. Fungal Genet. Biol. *47*, 736–741.

Kim, Y.K., and Xiao, C.L. (2008). Distribution and incidence of Sphaeropsis rot in apple in Washington State. Plant Dis. *92*, 940–946.
Lennox, C.L., Spotts, R.S., and Cervantes, L.A. (2003). Populations of *Botrytis cinerea* and *Penicillium* spp. on pear fruit, and in orchards and packinghouses, and their relationship to postharvest decay. Plant Dis. *87*, 639–644.
Li, B., Zong, Y., Du, Z., Chen, Y., Zhang, Z., Qin, G., Zhao, W., and Tian, S. (2015). Genomic characterization reveals insights into patulin biosynthesis and pathogenicity in Penicillium species. Mol. Plant Microbe Interact. *28*, 635–647.
Li, H.X., and Xiao, C.L. (2008). Characterization of fludioxonil-resistant and pyrimethanil-resistant phenotypes of *Penicillium expansum* from apple. Phytopathology *98*, 427–435.
Liu, G., Zhang, L., Wei, X., Zou, G., Qin, Y., Ma, L., Li, J., Zheng, H., Wang, S., Wang, C., Xun, L., Zhao, G.P., Zhou, Z., and Qu, Y. (2013). Genomic and secretomic analyses reveal unique features of the lignocellulolytic enzyme system of *Penicillium decumbens*. PLoS One *8*, ePub, e55185.
Ma, Zhonghua and Michailides, Themis, J. (2005). Advances in understanding molecular mechanisms of fungicide resistance and molecular detection of resistant genotypes in phytopathogenic fungi. Crop Prot. *24*, 853–863.
Machida, M., Asai, K., Sano, M., Tanaka, T., Kumagai, T., Terai, G., Kusumoto, K., Arima, T., Akita, O., Kashiwagi, Y., *et al.* (2005). Genome sequencing and analysis of *Aspergillus oryzae*. Nature *438*, 1157–1161.
Marcet-Houben, M., Ballester, A.R., de la Fuente, B., Harries, E., Marcos, J.F., Gonzalez-Candelas, L., and Gabaldon, T. (2012). Genome sequence of the necrotrophic fungus *Penicillium digitatum*, the main postharvest pathogen of citrus. BMC Genomics *13*, 646.
McKinley, E.R., and Carlton, W.W. (1991). Patulin. In Mycotoxins and Phytoalexins. R.P. Sharma, and D.K. Salunkhe, eds. (Boca Raton, FL: CRC Press), pp.191–236.
Medema, M.H., Blin, K., Cimermancic, P., de Jager, V., Zakrzewski, P., Fischbach, M.A., Weber, T., Breitling, R., and Takano, E. (2011). antiSMASH: Rapid identification, annotation and analysis of secondary metabolite biosynthesis gene clusters. Nucleic Acids Res. *39*, 339–346.
Moake, M.M., Padilla-Zakour, O.I., and Worobo, R.W. (2005). Comprehensive review of patulin control methods in foods. Compr. Rev. Food Sci. Food Saf. *1*, 8–21.
Nierman, W.C., Pain, A., Anderson, M.J., Wortman, J.R., Kim, H.S., Arroyo, J., Berriman, M., Abe, K., Archer, D.B., Bermejo, C., *et al.* (2005). Genomic sequence of the pathogenic and allergenic filamentous fungus *Aspergillus fumigatus*. Nature *438*, 1151–1156.
Peter, K.A., Vico, I., Gaskins, V.L., and Jurick II, W.M. (2012). First report of *Penicillium carneum* causing blue mold decay of apple fruit in Pennsylvania. Plant Dis. *96*, 1823.
Pitt, J.I. (1979). The Genus *Penicillium* and its Teleomorphic States *Eupenicillium* and *Talaromyces*. (London, UK: Academic Press).
Pitt, J.I. (1997). Toxigenic *Penicillium* species. In Food Microbiology, Fundamentals and Frontiers, M.P. Doyle, L.R. Beuchat, and T.J. Montville, eds. (Washington, DC: Norfolk Press), pp. 406–418.
Puel, O., Galtier, P., and Oswald, I.P. (2010). Biosynthesis and toxilogical effects of patulin. Toxins *2*, 613.
Rosenberger, D.A. (1990). Blue mold. In Compendium of Apple and Pear Diseases. A.L. Jones, and H.S. Aldwinkle, eds. (Saint Paul, Minnesota; APS Press), pp.54–55.
Rosenberger, D.A. (2012). Sanitize apple storage rooms to minimize postharvest decays. Scaffolds Fruit J. *21*, 4–5.
Rosenberger, D.A., Wicklow, D.T., Korjagin, V.A., and Rondinaro, S.M. (1991). Pathogenicity and benzimidazole resistance in *Penicillium* species recovered from flotation tanks in apple packinghouses. Plant Dis. 75, 712–715.
Russell, P.E. (2004). Sensitivity baselines in fungicide resistance research and management. FRAC Monograph 3, 1–54.
Sanderson, P.G., and Spotts, R.A. (1995). Postharvest decay of winter pear and apple fruit caused by species of *Penicillium*. Phytopathology *85*, 103–110.
Sholberg, P.L., and Haag, P.D. (1996). Incidence of postharvest pathogens of stored apples in British Columbia. Can. J. Plant Pathol. *18*, 81–85.
Sholberg, P.L., Harlton, C., Haag, P., Lévesque, C.A., Gorman, D.O., and Seifert, K. (2005). Benzimidazole and diphenylamine sensitivity and identity of *Penicillium* spp. that cause postharvest blue mold of apples using Β-tubulin gene sequences. Postharvest Biol. Technol. *36*, 41–49.
Specht, T., Dahlmann, T.A., Zadra, I., Kürnsteiner, H., and Kück, U. (2014). Complete sequencing and chromosome-scale genome assembly of the industrial progenitor strain P2niaD18 from the penicillin producer *Penicillium chrysogenum*. Genome Announc. *2*, ePub: e00577-14.

Spotts, R.A., and Cervantes, L.A. (1994). Contamination of harvest bins with pear decay fungi and evaluation of disinfestants on plastic and wood bin material. Acta Hort. *367*, 419–425.

Spotts, R.A., and Cervantes, L.A. (2001). Disease incidence-inoculum dose relationships for *Botrytis cinerea* and *Penicillium expansum* and decay of pear fruit using dry, air-borne conidia. Plant Dis. *85*, 755–759.

Spotts, R.A., Cervantes, L.A., and Mielke, E.A. (1999). Variability in postharvest decay among apple cultivars. Plant Dis. *83*, 1051.

Tannous, J., El Khoury, R., Snini, S.P., Lippi, Y., El Khoury, A., Atoui, A., Lteif, R., Oswald, I.P., and Puel, O. (2014). Sequencing, physical organization and kinetic expression of the patulin biosynthetic gene cluster from *Penicillium expansum*. Int. J. Food Microbiol. *189*, 51–60.

Thom, C. (1910). Cultural studies of species of *Penicillium*. Bur. Anim. Ind. U.S. Dept. Agriculture. *118*, 1–109.

Van den Berg, M.A., Albang, R., Albermann, K., Badger, J.H., Daran, J.M., Driessen, A.J., Garcia-Estrada, C., Fedorova, N.D., Harris, D.M., Heijne, W.H., *et al.* (2008). Genome sequencing and analysis of the filamentous fungus *Penicillium chrysogenum*. Nat. Biotechnol. *26*, 1161–1168.

Varga, J., Due, M., Frisvad, J.C., and Samson, R.A. (2007). Taxonomic revision of *Aspergillus* section *Clavati* based on molecular, morphological and physiological data. Stud. Mycol. *59*, 89–106.

Wang, F.Q., Zhong, J., Zhao, Y., Xiao, J., Liu, J., Dai, M., Zheng, G., Zhang, L., Yu, J., Wu, J., and Duan, B. (2014). Genome sequencing of high-penicillin producing industrial strain of Penicillium chrysogenum. BMC Genomics *15*, 1–11.

WHO (1998). Position paper on patulin. Joint FAO/WHO Expert Committee on Food additives (JECFA). The Hague, The Netherlands.

Xiao, C.L., and Boal, R.J. (2009). Residual activity of fludioxonil and pyrimethanil against *Penicillium expansum* on apple fruit. Plant Dis. *93*, 1003–1008.

Yan, H., Gaskins, V.L., Vico, I., Luo, Y., and Jurick II, W.M. (2014). First report of *Penicillium expansum* isolates resistant to pyrimethanil from stored apple fruit in Pennsylvania. Plant Dis. *98*, 7.

Yang, Y., Zhao, H., Barrero, R.A., Zhang, B., Sun, G., Wilson, I.W., Xie, F., Walker, K.D., Parks, J.W., Bruce, R., *et al.* (2014). Genome sequencing and analysis of the paclitaxel-producing endophytic fungus *Penicillium aurantiogriseum* NRRL 62431. BMC Genomics *15*, 69.

Yao, C., Conway, W.S., and Sams, C.E. (1996). Purification and characterization of a polygalacturonase produced by *Penicillium expansum* in apple fruit. Phytopathology *86*, 1160–1166.

Yu, J., Jurick II, W.M., and Bennett, J.W. (2015). Current status of genomics research on mycotoxigenic fungi. Int. J. Plant Biol. Res. *3*, 1035–1041.

Yu, J., Jurick, W.M., Cao, H.Y., Yin, Y., Gaskins, V.L., Losada, L., Zafar, N., Kim, M., Bennett, J.W., and Nierman, W. (2014). Draft genome sequence of *Penicillium expansum* R19, which causes postharvest decay of apple fruit. Genome Announc. *2*, e00635–14.

Post-genomic Approaches to Dissect Carbon Starvation Responses in Aspergilli

6

Jolanda M. van Munster, Anne-Marie Burggraaf, István Pócsi, Melinda Szilágyi, Tamás Emri and Arthur F.J. Ram

Abstract

Most filamentous fungi have a saprophytic lifestyle and proliferate on organic materials from plants. They contain a large arsenal of enzymes that are expressed and secreted in response to available carbon sources. Regulatory networks in which carbon-specific transcription factors and wide-domain transcription factors play essential roles tightly control the expression of these enzymes. While a suitable carbon source is present, the entire metabolism of the fungus is directed to grow as fast as possible via hyphal tip extension. At a certain point however, the available exogenous carbon source will be limited, resulting in a carbon starvation response. Filamentous fungi react to carbon starvation with some very specific responses including the induction of glycosyl hydrolases involved in fungal cell wall degradation (autolysis) and the onset of asexual spore formation. Recent advances in genome-wide transcriptomic and proteomic approaches have revealed new insights into molecular mechanisms involved in these fungal-specific carbon starvation responses. Together with the genetic accessibility of filamentous fungi, the omics technologies have made a large contribution in broadening our understanding of the carbon starvation response in recent years. In this review, we have focussed on summarizing and integrating the most important cellular responses of filamentous fungi towards carbon starvation. As most studies to date have been carried out with *Aspergillus nidulans* and *Aspergillus niger*, the focus will be further on these two species and to compare their responses to carbon starvation.

Introduction

The morphology of filamentous fungi is characterized by the presence of hyphae. Filamentous growth starts with the germination of a spore and the subsequent formation of a germling. Via branching or hyphal fusion events, a network of hyphae is formed which is referred to as a hyphal mycelium. Such a hyphal mycelium can be considered as a multicellular structure. In most filamentous fungi, except the zygomycete fungi, hyphae are divided into individual cells via the formation of internal walls named septa. Each septum contains a septal pore, which is large enough to allow transport of organelles and nutrients from one hyphal compartment to the other. Filamentous fungi can control the flow of cytosol or organelles between compartments by the controlled closure of the septal pore via Woronin bodies. Woronin bodies are peroxisome-derived, spherical organelles that are located near the septal pore. By plugging the septal pores with Woronin bodies, a multicellular structure with individual cells is formed. The opening and closing of septal pores are highly dynamic and well-controlled events (Bleichrodt *et al.*, 2015).

A fungal mycelium is typically formed on solid medium such as agar plates. To analyse growth on agar plates, spores are often inoculated in the centre of the plate from where they will gradually cover the plate through radial extension. During growth, the fungus will utilize the available nutrients until they become limiting (for most of the media this will be the carbon source). At that point, the mycelium starts a developmental program leading to sexual or asexual reproduction. Since nutrient limitation first occurs in the centre of the colony, differentiation will start here also. Whether the sexual or asexual developmental pathway is chosen depends on environmental factors such as light (Bayram and Braus, 2012). Some fungi, such as *Aspergillus niger*, are only known to proliferate asexually. The ability to reproduce sexually has either been lost through evolution or remains undiscovered.

Besides the growth on solid medium, fungi are also cultured in liquid medium, referred to as submerged cultivation. To start a submerged cultivation, spores can be used as an inoculum. Depending on the spore density and ambient pH (and possibly other factors), fungal hyphae develop as a dispersed fungal mycelium (individual hyphal filamentous cells) or as pellets (dense aggregates of filamentous cells). The tendency of a culture to form pellets is mainly caused by aggregation of the spores during the initial stages of spore swelling. Low pH values reduce the aggregation of swollen spores and are therefore used to prevent pellet formation. The macromorphology of the submerged culture (dispersed mycelia or pellets) can have important consequences on the productivity and heterogeneity of the culture and is therefore an important parameter. As on solid medium, nutrients can become limiting during batch growth and filamentous fungi respond to starvation with the activation of asexual reproduction program leading to the formation of conidiospores (Nitsche *et al.*, 2012; Sarkar *et al.*, 2012). The advantages of studying starvation responses in filamentous fungi using bioreactors are reviewed in the next paragraph.

Controlled cultivation of filamentous fungi to study starvation responses

Responses to carbon starvation have been most elaborately studied in submerged cultures. Batch cultivation in controlled bioreactors is preferred as growth can be monitored on-line and is more reproducible (Nitsche *et al.*, 2012). Batch cultures in shake flasks have also been successfully used to perform C-starvation responses (van Munster *et al.*, 2014; de Sousa Lima *et al.*, 2014; Szilágyi *et al.*, 2013; Xiong *et al.*, 2014). However, more detailed measurements regarding the physiology of the fungus cannot be measured in shake-flask cultures. On-line analysis of O_2 consumption and CO_2 production in a bioreactor is a convenient method to monitor carbon limitation as the parameters will dramatically change upon starvation (Nitsche *et al.*, 2012). Yet another advantage of bioreactor cultivation is that when sufficient large vessels are used (e.g. 5 l), multiple samples for analysis can be withdrawn without affecting the total volume of the culture to such an extent that it changes the physiology of the culture. Substrate-limited growth has also been studied in chemostat cultures.

There is an important difference between carbon starvation in batch growth and C-limited chemostat cultures, as additional carbon source is continuously added during chemostat cultivations. However, at low dilution rates (resulting in low growth rates), starvation-like responses such as sporulation have been observed (Bushell and Bull, 1999; Jørgensen, personal communication). An extreme form of chemostat cultivation which leads to a very low growth rate is retentostat cultivation. In a retentostat system, the biomass is retained in the bioreactor while keeping the dilution rate of the medium constant, leading to a severe reduction of the energy source per biomass unit. Cultivation of *A. niger* in retentostats also results in the onset of conidiation (Jørgensen *et al.*, 2010) indicating that limited metabolic fluxes might be sufficient to activate the sporulation program. Based on the phenotype of conidial mutants both on solid plates and in submerged conditions, it is likely that the same genetic regulatory programme is involved under both conditions (Nitsche and Jørgensen, personal communication). Although there are obvious benefits, maintaining the environmental parameters (e.g. pH, oxygen levels and medium composition) at a constant/controlled level may alter the response of the fungus. In nature, fungi are continuously modifying their environment by secreting enzymes and metabolites during carbon starvation and these changes (e.g. alkalizing pH, liberation of nutrients from the hydrolysing cell wall) can alter their behaviour. When controlling conditions in the bioreactor, we can gain precise data on chosen stress responses, but may lose information on the complexity of the whole stress response.

Carbon starvation responses

Carbon starvation responses in filamentous fungi have been studied both on surface cultures (mostly agar plates) and in submerged conditions. Several transcriptomic and proteomic studies have been performed in the last few years to describe the carbon starvation responses and are discussed in the next two paragraphs.

Carbon starvation responses on surface cultures

The Wösten group developed an elegant ring system to analyse differential gene expression in a fungal colony and defined five zones, representing differently aged fungal mycelium. The differential gene expression between the different zones as well as the different extracellular proteomes demonstrates the multicellular nature of a fungal colony (Krijgsheld *et al.*, 2012; Levin *et al.*, 2007). Comparison of mRNA profiles between the central zone of an *A. niger* colony and the most outer zone revealed major differences in expression of genes from various functional classes (Levin *et al.*, 2007). Only the broad functional categories 01 (metabolism) including 01.01.10 (amino acid degradation), 01.05.01 and 01.05.07 (C-compound and carbohydrate utilization and transport) and 01.06.01 (lipid, fatty acid and isoprenoid biosynthesis), 06 (protein fate) including 06.07.03 (modifications by (de)-phosphorylation), 06.07.07 (modifications by (de)-ubiquitination) and 06.13.04 (lysosomal and vacuolar degradation), were consistently enriched in the central zones independent of the carbon source used (Levin *et al.*, 2007). It should be noted that in the experiments by Levin *et al.* (2007) the *A. niger* colony was covered with a perforated polycarbonate membrane to prevent sporulation in the centre of the colony. Therefore genes related to conidiation were not induced in these experiments.

mRNA profiling to dissect colony development has also been studied in *Neurospora crassa* (Kasuga and Glass, 2008) and despite the relatively large evolutionary distance between *A. niger* and *N. crassa* a comparative study between the zonal transcriptomes between the two filamentous fungi has been performed. Even though such a global comparison has its limitations, as it is highly dependent on the annotation and the classification in GO-terms or functional categories, the cross-species examination indicated conserved patterns of gene expression during colony development. These conserved patterns were mostly found in the outer parts of the colony and included genes involved in metabolism of lipids and fatty acids, which were predicted to be involved in membrane biosynthesis, and genes involved in biogenesis of cellular components which included genes involved in actin cytoskeleton organization (Kasuga and Glass, 2008).

Carbon starvation responses under submerged conditions

Several genome-wide transcriptomic or proteomic studies have been reported in the last few years to study the response of filamentous fungi towards carbon starvation under submerged growth conditions. Initial genome-wide transcriptomic studies were performed with *A. niger* for which a genome sequence and microarrays were available (van Munster *et al.*, 2015a; Nitsche *et al.*, 2012; Pel *et al.*, 2007). Since then, other reports have been published, including carbon starvation responses in the medically relevant fungus *Paracoccidioides* spp. (de Sousa Lima *et al.*, 2014) and a transcriptome study including qRT-PCR for 99 selected genes in *Aspergillus nidulans* to analyse the response to carbon starvation (Szilágyi *et al.*, 2013). Because of the different experimental designs in each of the studies, a detailed comparison between the studies is challenging. Whereas in *A. niger* three different time points were analysed (16 h, 60 h and 140 h after carbon starvation), studies in *Paracoccidioides* and *A. nidulans* included a single time point (6 h and 4 h after starvation, respectively). Starvation leads to important morphological changes, including vacuolization, the formation of empty compartments, hyphal fragmentation, the formation of thin hyphae and conidiation. These morphological changes are accompanied by massive changes in gene expression patterns. About 50% of the genes (7292 out of 13,989) are differentially expressed during starvation compared to the exponential growth phase in *A. niger* (Nitsche *et al.*, 2012). Analysis of the differentially expressed genes during carbon starvation in submerged conditions identified several commonly induced functional categories or GO terms. As expected, many of these terms were related to dealing with the carbon limitation and the use of endogenous or reserve carbon sources. Obviously, these endogenous carbon sources include extracellular sources such as the fungal cell wall, but also intracellular sources. Therefore, in all three genome-wide studies performed (Nitsche *et al.*, 2012; de Sousa Lima *et al.*, 2014; Szilágyi *et al.*, 2013) terms related to amino acid metabolism, fatty acid oxidation, C-compound and carbohydrate metabolism were up-regulated in response to starvation. Terms related to cell biological processes such as cell cycle arrest (*A. niger*), protein synthesis and secretion (*A. nidulans*), autophagy (*A. niger* and *A. nidulans*), vacuolar transports (*A. niger* and *A. nidulans*) were enriched in the up-regulated genes in response to carbon starvation. The energy and building blocks obtained from recycling its own mycelium, allow the fungus to respond to the starvation condition whereby the energy is used for maintenance and to fuel conidiation. In the next parts, we will further focus on the morphological responses observed in carbon starved cultures of filamentous fungi.

Morphological responses during C-limitation

Carbon starvation induces hyphal fragmentation in submerged *A. nidulans* cultures (Emri *et al.*, 2004, 2005; Pócsi *et al.*, 2009; Szilágyi *et al.*, 2010a, 2013). During this process, hyphal fragments are liberated from the surface of pellets and as a consequence the pellet size is decreased until it is completely disintegrated (Emri *et al.*, 2004). The disintegration of hyphae is followed by the complete degradation of the released hyphal branches until only small numbers of 'yeast like cells' (hyphal fragments of one to two cells) are present in the fermentation broth (Emri *et al.*, 2004). These short fragments can grow out and form hyphae after glucose supplementation (Emri *et al.*, 2004). Carbon starvation in *A. niger* is also accompanied by several morphological changes which include: the formation of empty compartments (starting to emerge after 16 hours of glucose depletion), the outgrowth of thin hyphae or thin hyphae within empty hyphae (16 hours after glucose depletion), and conidiation (60 hours after glucose depletion) (Nitsche *et al.*, 2012). These morphological changes are similar to the morphological changes upon starvation which have been described before (White *et al.*, 2002). The formation of thin hyphae in response to carbon starvation was reported also for *Aspergillus oryzae* (Pollack *et al.*, 2008). An important difference in comparison to morphological changes upon starvation in *A. nidulans* (Emri *et al.*, 2004) is that in *A. niger* no mycelial fragmentation was observed. Additional unpublished studies have shown that the lack of fragmentation in *A. niger* is probably due to the low pH value (pH 3.0) which is maintained in the bioreactor during the experiments. Fragmentation of the mycelium upon starvation in *A. niger* is observed when the pH is not controlled and reaches values around 6.0 during starvation (Nitsche, unpublished results). It is likely that the low pH (3.0) conditions lower the activity of cell wall hydrolytic enzymes that are involved in hyphal fragmentation. In *A. nidulans*, fragmentation and formation of empty hyphae is also accompanied by the production of cell wall glycosyl hydrolases (Emri *et al.*, 2004, 2005; Pócsi *et al.*, 2009; Szilágyi *et al.*, 2010a, 2013). *A. nidulans* mutants with defects in cell wall glycosyl hydrolase production show reduced hyphal fragmentation but these mutations do not inhibit vacuolization and formation of empty hyphae (Emri *et al.*, 2005; Pócsi *et al.*, 2009; Szilágyi *et al.*, 2010a). It clearly demonstrates that hyphal fragmentation is an enzymatic process and not the passive consequence of mechanic forces occurring in shaken cultures.

In addition to the formation of empty compartments and hyphal fragmentation, some hyphal cells remain viable and these surviving cells differentiate and can produce conidiospores (Emri *et al.*, 2004; Jørgensen *et al.*, 2010; Nitsche *et al.*, 2012; Schrickx *et al.*, 1993; Szilágyi *et al.*, 2013). BrlA encodes a transcription factor which is essential for asexual reproduction in aspergilli both in surface cultures as well as in submerged cultures (Adams *et al.*, 1998; van Munster *et al.*, 2015a; Twumasi-Boateng *et al.*, 2009). The regulation of conidiation in submerged conditions as well as the regulation of conidiation at surface cultures is dependent on BrlA and therefore BrlA is likely to be a key regulator for conidiation under both conditions. Asexual development (conidiogenesis) is supposed to be initiated by internal signals and/or several stress conditions including carbon starvation (Adams *et al.*, 1998; Krijgsheld *et al.*, 2013). The stress-dependent activation of conidiogenesis is mandatory during submerged cultivation but thought to be secondary in surface cultures (Adams *et al.*, 1998; Krijgsheld *et al.*, 2013; Skromne *et al.*, 1995). In surface cultures, asexual development generally starts well before carbon depletion and continues during starvation (Adams *et al.*, 1998). Autolytic processes (see below) induced by carbon starvation are supposed to support asexual development with nutrients during this period (Emri *et al.*, 2008; Pócsi *et al.*, 2009).

In addition to asexual development (formation of conidiophores and conidia), developmental processes also involve sexual development (formation of cleistothecia and ascospores) in aspergilli with a sexual cycle such as *A. nidulans* (Adams *et al.*, 1998; Dyer and O'Gorman, 2012; Etxebeste *et al.*, 2010; Krijgsheld *et al.*, 2013; Park and Yu, 2012). Carbon starvation induces sexual development in several pezizomycete fungi (Dyer and O'Gorman, 2012; Dyer *et al.*, 1992). *A. nidulans* is among the exceptions where sufficient nutrition is needed for the initiation of cleistothecia formation (Dyer and O'Gorman, 2012; Han *et al.*, 2003). However, further research is needed to elucidate whether carbon starvation can block sexual development if it has already started. If not, autolytic processes could support sexual development with nutrients as was suggested in the case of asexual development.

Recycling responses to deal with carbon starvation

Under carbon starvation, fungi have two options to cope with this type of environmental stress: (i) to recycle the biopolymers of old hyphal parts to provide energy for surviving cells (autolysis) or (ii) to use other, non-easily degradable substrates as carbon source (e.g. plant materials). The recycling of cell contents is coupled tightly to the (controlled) death of old cells and proliferation of other cells, and includes the utilization of inner cell material by autophagy as well as the degradation of cell wall constituents by extracellular hydrolytic enzymes. In the next few paragraphs we will discuss in more detail the role of autophagy during starvation and the roles of glycosyl hydrolases (GH) during cell wall recycling and cell wall active enzymes involved in conidiation. We will also discuss the strategies employed by fungi to scout for non-easily degradable substrates that could be used as alternative C-sources.

Recycling responses: The role of autophagy in the carbon starvation response

To survive carbon starvation conditions and support the formation of reproductive structures and substrate exploring hyphae, endogenous recycling of nutrients has been proposed to be an important strategy (Richie *et al.*, 2007; Shoji and Craven, 2011; Shoji *et al.*, 2006; White *et al.*, 2002). As such, the autophagic pathway is one of the most dominantly induced processes during carbon limitation in filamentous fungi (Ellström *et al.*, 2015; Krohn *et al.*, 2014; Nitsche *et al.*, 2012; Szilágyi *et al.*, 2013). Autophagy is a non-selective catabolic pathway which targets cytoplasmic contents to the vacuole for degradation and recycling of the building blocks of the cell. It includes the sequestering of cytosolic components and organelles in double-membrane vesicles which subsequently fuse with the vacuole, releasing their contents.

The autophagy machinery is negatively regulated by the kinase activities of the target of rapamycin (Tor) and the cAMP-dependent protein kinase (PKA), which independently act through control of the interaction between the autophagy-related proteins Atg1 and Atg13 (Kamada *et al.*, 2000; Stephan *et al.*, 2009). Under nutrient-rich conditions, autophagy is blocked by Tor-mediated hyperphosphorylation of Atg13, which reduces its affinity for Atg1. Inhibition of Tor by nutrient starvation conditions reduces the phosphorylation state of Atg13, enabling it to form a complex with Atg1 and thereby initiating the autophagy process. Both Atg1 and Atg13 are direct substrates for PKA, but its phosphorylation sites are distinct from those targeted by Tor (Stephan *et al.*, 2009). The concerted regulation of autophagy by Tor and PKA is rather complex and currently not fully understood. Responses are likely dependent on the nature of the limiting nutrients, providing the cell with more flexibility to adapt to changing environments (Dechant and Peter, 2008).

To date, more than 30 *atg* genes have been identified in yeast and filamentous fungi that control the autophagic machinery (Feng *et al.*, 2014). Genome-wide transcriptional profiling of *A. niger* and *Paxillus involutus* has shown that the majority of these genes is up-regulated under carbon starvation conditions (Ellström *et al.*, 2015; Nitsche *et al.*, 2012). In addition, extensive transcriptome analysis demonstrated the transcriptional induction of specific *atg* genes in *A. nidulans* (Szilágyi *et al.*, 2013). Analysis of *atg* deletion mutants in *Aspergillus* and *Penicillium* showed that phenotypic effects vary among different species. While the deletion of *atg1* severely reduced conidiation in *A. fumigatus* (Richie *et al.*, 2007) and *P. chrysogenum* (Bartoszewska *et al.*, 2011), conidiospore formation was only mildly reduced in the *A. niger* Δ*atg1* mutant (Nitsche *et al.*, 2013). Similarly, the effect on radial growth and conidiation upon disruption of *atg8* was less pronounced in *A. niger* (Nitsche *et al.*, 2013) than in *A. oryzae* (Kikuma *et al.*, 2006). In *A. nidulans*, the deletion of either *atg1*, *atg4*, *atg5*, *atg7* or *atg9* resulted in a deficiency in conidiation at 42°C (Pinar *et al.*, 2013). During submerged carbon starvation, *A. niger* Δ*atg1* and Δ*atg8* mutants showed growth profiles similar to wild-type, but the emergence of empty compartments and the formation of thin hyphae was clearly accelerated (Nitsche *et al.*, 2013). This suggests that autophagy promotes survival during carbon starvation by endogenous recycling of nutrients and protecting older hyphal compartments from cell death.

Recycling responses: GH specific for cell wall degradation

During carbon starvation, the fungal cell wall is thought to undergo drastic remodelling or degradation. Fungal carbohydrate-active enzymes are key effectors in these cell wall changes. The production of such hydrolytic activities in response to carbon starvation has been observed in a number of fungi (Jaroszuk-Scisel *et al.*, 2011; Lahoz *et al.*, 1976; McIntyre *et al.*, 2000; Perez-Leblic *et al.*, 1982; White *et al.*, 2002). Recent investigations of the transcriptome and proteome of carbon starving *A. niger* (van Munster *et al.*, 2015a; Nitsche *et al.*, 2012) and *A. nidulans* (Szilágyi *et al.*, 2013) cultures have provided an overview of the response of fungal cell wall active carbohydrate hydrolases.

The cell wall is a rigid, complex structure, which maintains the shape of the cell and is responsible for counteracting turgor pressure and for protecting the plasma membrane against mechanical damage. The fungal cell wall accounts for a considerable percentage of cellular biomass (about 20%) and thus it represents a rich carbon source during carbon starving conditions. The fungal cell wall consists mainly of polysaccharides. For almost all fungi, the central core of the cell wall is a β-1,3-glucan with β-1,6 branch points. Chitin is covalently bound to the non-reducing end of β-1,3-glucan side chains via β-1.4-linkages. The *Aspergillus* cell wall also contains α-1,3-glucan, galactomannan (α-1,2/α-1,6-linked mannan with short β-1,5-galactofuranose side chains), and galactosaminogalactan (a linear heterogenous polymer containing α-1,4-galactose/galactosamine) (Latgé, 2007). Cell walls also contain several heavily glycosylated proteins including GPI-modified cell wall proteins or hydrophobins (De Groot *et al.*, 2005).

Cell wall homeostasis undergoes a dramatic change during carbon starvation. For example, processes responsible for the synthesis of cell wall components are down-regulated (Szilágyi *et al.*, 2013). In *A. nidulans*, *fksA* encoding the catalytic subunit of β-1,3-glucan synthase (Kelly *et al.*, 1996); *chsB* and *chsF* coding for III type chitin synthases (Horiuchi, 2009) and *agsB* encoding α-1,3-glucan synthase (Fujioka *et al.*, 2007) were down-regulated in carbon starving cultures (Szilágyi *et al.*, 2013). In both *A. niger* and *A. nidulans*, genes coding for GPI-anchored enzymes, which may function in the remodelling of rigid cell wall to a looser structure during growth – e.g. *chiA* in the case of chitin (Yamazaki *et al.*, 2008) or *eglC* in the case of β-1,3-glucan (Choi *et al.*, 2005) – were also down-regulated during carbon starvation (van Munster *et al.*, 2015a; Szilágyi *et al.*, 2013).

In contrast to the down-regulation of cell wall synthesis, cell wall-degrading processes are up-regulated in carbon starving cultures (Szilágyi *et al.*, 2013). In *A. nidulans* and *A. niger*, elevated levels of extracellular protease, β-1,3-glucanase, β-glucosidase, α-1,3-glucanase and chitinase were detected during carbon starvation (Emri *et al.*, 2004; Nitsche *et al.*, 2012; Szilágyi *et al.*, 2010a; Zonneveld, 1974) and genes encoding such autolysis-related hydrolytic enzymes were up-regulated (van Munster *et al.*, 2015a; Nitsche *et al.*, 2012; Szilágyi *et al.*, 2013). However, only a number of these enzymes have been investigated in detail.

The *A. nidulans* EngA plays a role in biomass decrease during autolysis (Szilágyi *et al.*, 2010a). Its orthologue in *A. fumigatus* has been characterized as an endo-acting β-1,3-glucanase, and was found to be associated with the fungal cell wall (Fontaine *et al.*, 1997; Mouyna *et al.*, 2002). *A. nidulans* EngA also has β-1,3-endoglucanase activity, and its activity and stability at a broad pH range means that EngA preserves its activity even at alkaline pH values typical of late carbon starving cultures. EngA is the main secreted β-1,3-glucanase in carbon starving cultures of *A. nidulans* and the deletion of *engA* confirmed the prominent role of EngA in cell wall degradation and autolysis. However, gene deletion did not abolish the β-1,3-glucanase enzyme production (Szilágyi *et al.*, 2010a). Nuero *et al.* (1993) purified three enzymes with β-1,3-glucanase activity from autolyzing cultures of *A. nidulans* confirming the existence of several enzymes with β-1,3-glucanase activity in carbon-depleted cultures of *A. nidulans*.

High α-1,3-glucanase activities are characteristic for carbon starving cultures of *A. nidulans* (Zonneveld, 1974). Correspondingly, genes encoding enzymes with putative α-1,3-glucanase activity were up-regulated during carbon starvation in *A. niger* and *A. nidulans* (Nitsche *et al.*, 2012; Szilágyi *et al.*, 2013). One of these, *A. niger* AgnB, was characterized as α-1,3-glucanase which releases glucose from α-1,3-glucan substrates. However, no direct role for this enzyme in cell wall degradation could be identified (van Munster *et al.*, 2015b). In *A. nidulans*, which in contrast to *A. niger* has a sexual cycle, the α-1,3-glucanase MutA hydrolyses cell wall material during sexual reproduction, possibly to generate energy that is needed for the production of the cleistothecia (fruiting bodies) (Wei *et al.*, 2001; Zonneveld, 1972). The complete hydrolysis of glucans is achieved by the synergistic action of endo- and exo-glucanases. The up-regulation of high-affinity glucose transporters also supports the hypothesis that glucan monomers are taken and utilized by the surviving cells upon carbon starvation (Szilágyi *et al.*, 2013).

Chitinase ChiB is rate-limiting in biomass degradation during carbon starvation in *A. nidulans.* Deletion of *chiB* resulted in the loss of extracellular endochitinase producing capability and caused a non-autolysing phenotype, which was characterized by the lack of hyphal fragmentation and pellet disorganization in ageing carbon starving cultures (Pócsi *et al.*, 2009; Yamazaki *et al.*, 2007). Deletion of the homologous *A. niger cfcA* and *A. fumigatus chiB1* – also encoding the main chitinases induced during starvation – does not result in a change in biomass loss during starvation (Jaques *et al.*, 2003; van Munster *et al.*, 2015b); the physiological role of these enzymes thus appears to be somewhat different. However, the *A. niger* enzyme is responsible for cell wall degradation as deletion of the corresponding gene reduces hyphal fragmentation during starvation. *A. fumigatus* ChiB and *A. niger* CfcA hydrolyse chitin and chitin oligosaccharides by cleaving mainly chitobiose from the non-reducing substrate end, but can also perform a transglycosylation reaction (Escott *et al.*, 1998; Jaques *et al.*, 2003; Lü *et al.*, 2009; van Munster *et al.*, 2015b; Xia *et al.*, 2001). The *A. nidulans* ChiB appears to have similar activity (Erdei *et al.*, 2008). Chitobiose can be further hydrolysed to monomers by *N*-acetyl-β-D-glucosaminidase enzymes. The *A. niger* and *A. nidulans nagA*, both encoding an *N*-acetyl-β-D-glucosaminidase, were induced by carbon starvation (Nitsche *et al.*, 2012; Pusztahelyi *et al.*, 2006; Szilágyi *et al.*, 2013).The *A. nidulans nagA* disruptant strain was not able to grow on medium containing chitobiose as sole carbon source (Kim *et al.*, 2002), but there was no effect on autolysis. Instead, deletion reduced cell death (Shin *et al.*, 2009). Similar viability effects were observed for the *ΔengAΔchiB* strain (Szilágyi *et al.*, 2012), showing a possible link between cell wall-hydrolysing enzymes to the initiation of cell death processes in old carbon starving cultures.

High extracellular hydrolase activities were demonstrated to have a strong antifungal activity against the producer itself (Szilágyi *et al.*, 2012). It means that living fungal cells need to protect themselves against the deleterious effects of these enzymes. To shed light on the details of such self-protection remains a challenge for future research in this field. The transcriptionally regulated *mpkA* and *rlmA* genes, which encode key elements of the cell wall integrity stress signalling pathway (Fujioka *et al.*, 2007), were not up-regulated during carbon starvation. The clear overall repression of cell wall biosynthetic genes also rejects the idea that the cell wall integrity pathway would have a prominent role in self-protection (Szilágyi *et al.*, 2013). An alternative way for protection could be an increased production of the pigment melanin in the surviving cells. This could be an effective response against the deleterious effects of cell wall hydrolysing enzymes as melanin itself can inhibit autolytic chitinase and glucanase activities and, as a consequence, the enzymatic lysis of cells in *A. nidulans* (Kuo and Alexander, 1967).

Recycling responses: GH specific for conidiation

During periods of stress, such as during prolonged starvation, *Aspergillus* can survive by forming stress-resistant spores. The formation of these spores requires energy and building blocks. In the event of nutrient limitation, fungal autolysis and autophagy may make these nutrients available by recycling the cell wall or intracellular components. The regulatory machinery responsible for the initiation of sporulation is linked to the regulation of autolysis and is discussed in the following section on genetic control of the responses to carbon starvation.

Extensive morphological changes are taking place during sporulation, and the fungal cell wall needs to be adapted to enable these changes. This can be observed by changes in appearance (Adams *et al.*, 1998), as well as by the differences in carbohydrate composition of vegetative mycelium and spores (Maubon *et al.*, 2006). Spore walls of *A. niger* and *A. fumigatus* contain, compared to walls of vegetative hyphae, more mannose (25% instead of 4–6%) and galactose (12–14% instead of 4–8%), and less (*N*-acetyl)-glucosamine (6% instead of 13%) (Maubon *et al.*, 2006; van Munster *et al.*, 2013). In addition to a carbohydrate backbone, the spore cell walls contain an outer layer of melanin, which gives the spores their colour and increases its resistance to environmental stress (Latgé *et al.*, 2005). CAZymes build and remodel this cell wall during sporulation. Sporulation-specific chitin synthases (Fujiwara *et al.*, 2000; Lee *et al.*, 2004) have been identified that contribute to these cell wall changes. Recently, sporulation-specific CAZymes that are predicted to be active on the fungal cell wall, were identified by comparing the expression profiles of glycoside hydrolases in wild-type and a strain carrying a deletion of the developmental regulator *brlA* (van Munster *et al.*, 2015a). These are a glycoside transferase from GT family 1, a protein containing only a CBM14 domain, the three β-glucan acting enzymes BgtD, An02g00850 and GelG, and the chitin acting CfcI, CtcB and an AA11 member.

Transcriptome analysis shows that the sporulation-specific glycoside hydrolases are strongly induced in sporulating aerial mycelium compared to vegetative mycelium (van Munster *et al.*, 2013). Using promoter fusion strains, activities of the promoters of *cfcI* and *ctcB* were located to the conidophores. These genes encode the chitinases CfcI and CtcB. Deletion of both genes resulted in accumulation of chitin in the spore cell wall, indicating that both enzymes have a function in modification of the cell wall during sporulation (van Munster *et al.*, 2013). CfcI is capable of hydrolysing chitotriose and longer chitin oligosaccharides by cleaving off monomers, possibly in a processive mode. The enzyme acts on the reducing end of the oligosaccharide substrates. CtcB belongs to the phylogenetic subgroup of GH18 that only encodes endo-chitinases. Together these enzyme activities would thus be capable of degrading the chitin polymer to N-acetyl-glucosamine monomers (van Munster *et al.*, 2012, 2015c).

The *A. fumigatus* β-1,3-glucanosyltransferase Gel7, the orthologue of the sporulation-specific *A. niger* GelG, was found to be important during sporulation. Deletion of the *gel7* gene resulted in abnormal conidia formation. In addition, the enzyme is capable of compensating for cell wall effects generated by *gel1* and *gel2* deletion (Zhao *et al.*, 2014). Orthologues of the *A. niger* sporulation-specific An02g00850 have been characterized, the *A. fumigatus eng2* and *A. nidulans xgeA* hydrolyse soluble β-1,3-glucan with β-1,6-glucan branches as well as β-1,3:1,4-glucan (lichenan) (Bauer *et al.*, 2006; Hartl *et al.*, 2011). Deletion of *eng2* in *A. fumigatus* gave no phenotype during vegetative growth consistent with a possible (unexplored) role during sporulation (Hartl *et al.*, 2011).

Recycling responses: scouting for new C-sources

Next to recycling fungal cell components, another possibility for carbon-starved fungi is to find an external carbon source. One of the substrates that may be available is the recalcitrant cell wall from dead plants. Plant cell walls can contain the carbohydrates cellulose, hemicelluloses and pectins, as well as other components such as lignin and structural proteins. The exact plant cell wall composition is dependent on the plant species, but also varies between different plant structures and wall types. For example, the primary cell walls of grasses contain low amounts of pectin and relatively high amounts of the hemicelluloses glucuronoarabinoxylans and mixed-linkage glucans. For gymnosperms, the cell wall contains more xyloglucan, pectin and structural proteins (Sarkar *et al.*, 2009; Vogel, 2008). The structure of individual plant cell wall carbohydrates is also dependent on their origin (Scheller and Ulvskov, 2010). Plant cell walls thus constitute a carbon source that is very variable in composition, but in all cases, plant cell wall components are cross-linked and form a complex network that may be quite resistant to enzymatic degradation.

A. niger and *A. nidulans* are both metabolically versatile and are capable of growing on a large number of different carbon sources. Their genome encodes a large suite of CAZymes that are active on plant-derived polysaccharides, and a recent catalogue of the *A. niger* genome showed that its genome encodes enzymes that are predicted to be active on at least 14 different plant cell wall polysaccharides (Andersen *et al.*, 2012). The response of *A. niger* to different lignocellulosic substrates seems to require a core set of enzymes, that is complemented by a set of genes that seem differentially expressed on different substrates (Pullan *et al.*, 2014), a strategy similar to that observed in *N. crassa* (Benz *et al.*, 2014).

The expression of the genes encoding all these enzymatic activities requires a big investment by the fungus, and thus needs to be carefully regulated. This regulation is mediated by the coordinated action of transcriptional repression and inducer-dependent transcriptional activation.

In the presence of sufficient amounts of easily metabolisable carbon sources such as glucose or xylose, carbon catabolite repression (CCR) inhibits transcription of genes that encode enzymes required for complex carbon sources (Ruijter and Visser, 1997; de Vries *et al.*, 1999). Similar to other filamentous fungi and yeast, in *A. niger* and *A. nidulans* CCR is mediated by the global transcriptional repressor CreA. In nutrient rich conditions, glucose is phosphorylated after uptake by the kinases GlkA and HxkA (Flipphi *et al.*, 2003). In *A. nidulans*, similar to in *S. cerevisiae*, a kinase can phosphorylate CreA, causing it to shuttle from the nucleus to the cytoplasm under de-repressing conditions. The presence of phosphorylated sugar inhibits this kinase via RAS-signalling, allowing non-phosphorylated CreA to remain in the nucleus under repressive conditions (Brown *et al.*, 2013). There, CreA can bind to two palindromic consensus sequences (5′-SYGGRG-3) in the promoter of genes (Cubero and Scazzocchio, 1994; Takashima *et al.*, 1996) and represses gene transcription. CreA influences other transcription factors, for example the XlnR transcriptional activators of genes encoding plant polysaccharide active CAZymes (de Vries *et al.*, 1999), as well as the transcription of individual CAZyme encoding enzymes.

To be transcribed to a high level, genes encoding plant polysaccharide active enzymes require activation by inducer-dependent transcriptional regulators, such as XlnR for hemicellulase and cellulase genes, or ClbR, ClrA, ClrB for cellulase genes (Glass *et al.*, 2013).

In carbon starvation conditions, low level transcription of a set of genes that encode plant polysaccharide-degrading enzymes has been detected in *A. niger* as well as in other fungi (Coradetti *et al.*, 2012; Foreman *et al.*, 2003; Martens-Uzunova and Schaap, 2009; van Munster *et al.*, 2014; Szilágyi *et al.*, 2013). These genes include for example *A. niger* endo-α-1,5-arabinanase encoding *abnA* and *abnC*, α-arabinofuranosidase encoding *abfB* and *abfC*, exo-polygalacturonase encoding *pgaX*, *pgxB*, several genes encoding beta-glucosidases and the exo-inulinase encoding *inuE*. The encoded enzyme activities are mainly predicted to be active on terminal residues of plant polysaccharides, or their corresponding oligosaccharides. The regulatory mechanism for the up-regulation of genes during carbon starvation is not known completely. However, for the cellobiohydrolase encoding *cbhB*, its neighbouring gene encoding a GH5 cellulase and also the α-arabinofuranosidase encoding *abfB*, low level up-regulation was detected not only under carbon starvation conditions, but also in a *creA* deletion mutant grown on glucose (Delmas *et al.*, 2012). This suggests that carbon catabolite derepression via CreA is responsible for the observed up-regulation.

In the culture filtrate of carbon-starved cultures, proteins corresponding to such transcripts have also been found (Braaksma *et al.*, 2010; van Munster *et al.*, 2014, 2015a; Nitsche *et al.*, 2012). The concerted action of these enzymes on plant polysaccharides results in the release of products that include monomers as well as small linear and branched oligosaccharides. These reaction products include known inducers for lignocelluloses degrading enzymes such as arabinose (van Munster *et al.*, 2014). This supports a model where carbon starvation allows low-level up-regulation of a subset of genes, which encode enzymes that are thought to perform a scouting role. These enzymes are thought to release inducers from any substrates detected in the environment, either alone or in coordination with any constitutively expressed plant polysaccharide-degrading enzymes. These inducers may be taken up by the fungus and induce the transcription of a full set of plant carbohydrate degrading enzymes through dedicated transcription factors. Correspondingly, the early responses of *A. niger* and *N. crassa* to carbon starvation overlap with their early response to wheat straw and Avicel respectively (Coradetti *et al.*, 2012; van Munster *et al.*, 2014). This indicates that an initial phase of carbon starvation is required, during which scouting enzymes may generate inducers and dedicated transcription factors are allowed to respond to inducers. Subsequently, gene transcription is activated to generate the full set of enzymes required for the degradation of lignocellulose.

Co-regulation of conidiogenesis and cell wall hydrolase production

In the previous sections we have shown that carbon stress induces complex morphological, developmental and physiological changes both in *A. nidulans* and *A. niger*. Many of these stress responses (e.g. hyphal fragmentation and cell wall hydrolase production on one hand and conidiophore formation and survival on the other) seem to be contradictory. The diversity of the stress responses as well as their antagonistic nature suggests the existence of a complex regulatory network in which overlaps and intensive cross talks among the different signalling pathways are responsible for their proper orchestration in these multicellular organisms.

A key step of conidiation in *A. nidulans* is the activation of *brlA* (Adams *et al.*, 1998; Park and Yu, 2012). BrlA is a transcription factor which is responsible for the induction of AbaA which further induces a third transcription factor WetA (Adams *et al.*, 1998; Park and Yu, 2012). These three transcription factors (the so-called central regulatory pathway of conidiogenesis) coordinates all the conidiation-specific gene expressions during the development of conidiophores and the maturation of spores (Adams *et al.*, 1998; Park and Yu, 2012). BrlA (but not AbaA or WetA) is necessary for induction of *engA* and *chiB* during carbon stress (Pócsi *et al.*, 2009; Szilágyi *et al.*, 2010a). Deletion of or loss-of-function mutations in *brlA* inhibits the accumulation of extracellular chitinase and β-1,3-glucanase activities in the fermentation broth and hinder hyphal fragmentation in carbon starved cultures (Pócsi *et al.*, 2009; Szilágyi *et al.*, 2010a). It is assumed that in some cells induction of *brlA* is followed by the activation of the whole central regulatory pathway of conidiogenes resulting in conidiophore development and conidia formation. In other cells, BrlA does not induce *abaA* (as was observed in submerged carbon starved cultures (Pócsi *et al.*, 2009)) and these cells are responsible for cell wall hydrolase production. Identification of factors preventing the complete activation of the central regulatory pathway after *brlA* induction needs further investigations. Successful escape of hyphae into the air with the help of hydrophobins induced by BrlA is one possible regulator of conidiogenesis versus cell wall hydrolase production at least in surface cultures.

Two heterotrimeric G-protein mediated signalling pathways, the FadA and GanB pathways, negatively influence *brlA* expression (Park and Yu, 2012; Yu, 2006). Activation of the FadA(Gα)-SfaD(Gβ)-GpgA(Gγ) heterotrimeric G protein modulate cAMP-protein kinase A signalling, repress *brlA*, negatively regulate secondary metabolism and maintain vegetative growth (Park and Yu, 2012; Shimizu and Keller, 2001; Yu, 2006). Dissociation of the GanB(Gα)-SfaD(Gb)-GpgA(Gγ) complex also mediates cAMP-protein kinase A signalling, negatively regulates the central pathway of conidiogenesis and positively regulates germination and stress responses (Park and Yu, 2012; Yu, 2006). FluG is proposed to be a key positive activator of *brlA* in *A. nidulans* (Lee and Adams, 1994). Its activity somehow belongs to the extracellular formation of a diorcinol – dehydroaustinol adduct which triggers the activation of the pathway (Lee and Adams, 1994; Rodríguez-Urra *et al.*, 2012). Accumulation of FluG beyond a certain level inhibits the negative regulation imposed by SfgA on Flb (including FlbA) proteins and as a consequence activates *brlA* (Park and Yu, 2012). FlbA is a key regulator of the FadA pathway (Adams *et al.*, 1998; Park and Yu, 2012; Pusztahelyi and Pócsi, 2013), therefore activation of FluG signalling not only induces *brlA* but also decreases the negative effect of FadA signalling on *brlA* transcription (Adams *et al.*, 1998; Park and Yu, 2012; Pusztahelyi and Pócsi, 2013). Activity of FadA and GanB G-protein signalling is proposed to depend on nutritional conditions therefore, carbon starvation decreases their activity, hinders vegetative growth and allows *brlA* to be induced. Carbon stress-dependent activation of *brlA* cannot be explained exclusively by the decreased activity of heterotrimeric G-protein pathways. Mutations in the genes involved in FadA or GanB signalling had only negligible effect on chitinase production and hyphal fragmentation in carbon stressed cultures (Molnár *et al.*, 2004, 2006). Since loss of function mutation in *fluG* (but not in *flbA*) inhibited the carbon starvation induced activation of *engA* and *chiB* (Emri *et al.*, 2005; Molnár *et al.*, 2004; Pócsi *et al.*, 2009; Szilágyi *et al.*, 2010a) a carbon stress-dependent behaviour of FluG signalling is assumed.

Elements of the sexual development are also known to influence *brlA* expression, e.g. NsdD; an activator of sexual development (Lee *et al.*, 2014) and MpkB MAP kinase; an upstream regulator of NsdD (Kang *et al.*, 2013), both negatively regulate *brlA* expression. At the same time, BrlA has also been reported to represses *steA*, which encodes another activator of sexual development (Vallim *et al.*, 2000), indicating that BrlA has a central role in the developmental choices. Moreover, Kang *et al.* (2013) showed the negative effect of MpkB not only on conidiogenesis but also on cell wall hydrolase production during long cultivation times which might also be controlled via BrlA.

Atoui *et al.* (2010) demonstrated that appropriate glucose concentration can modify the light-dependent intracellular localization of VeA which is an important element of the light dependent regulation of conidiogenesis (including *brlA* and *fluG* transcription), sexual development, and secondary metabolism (Bayram and Braus, 2012; Etxebeste *et al.*, 2010; Park and Yu, 2012). This observation suggests a further possible way for carbon source-dependent modulation of differentiation, secondary metabolism and cell wall hydrolase production.

RlmA is also a negative regulator of *brlA* and represses *engA* and *chiB* transcription right after glucose depletion (Pusztahelyi and Pócsi, 2013). Since RlmA is supposed to be involved in cell wall stress response (Fujioka *et al.*, 2007), this connection between hydrolase production and cell wall stress may represent a negative feedback regulation of cell wall hydrolase production preventing overproduction of these enzymes which would be dangerous for the producer cells.

Regulation of cell wall hydrolase production depends not only on BrlA, but other regulatory mechanism including the carbon catabolite repressor CreA (Emri *et al.*, 2006). Positive regulatory effect of mono- and oligosaccharides released during cell wall hydrolysis is also emphasized (Szilágyi *et al.*, 2010a). The CreA-dependent repression of certain cell wall hydrolases (e.g. *chiB*, but not *engA*) in the presence of repressive carbon sources during carbon starvation may explain why cell wall hydrolase production depends strongly on carbon stress while in case of conidiogenesis nutritional stress is only one and probably not the most important signal for induction (Park and Yu, 2012).

Regulation of extracellular protease formation during carbon starvation

The induction of proteases has also been proposed as a key event in ageing fungal cultures (White *et al.*, 2002). The production of extracellular proteases is a part of a general stress response aimed at scavenging new energy sources for survival in carbon starving cultures. The proteases may be involved in the degradation of extracellular proteins and/or the protein content of the cell wall. For industrial relevant fungi, such as *A. niger*, this is relevant in relation to the production of (recombinant) protein in this fungus. High protease activity during starvation has a negative effect on the yield of product formation because of the unwanted proteolytic degradation of the product (Braaksma *et al.*, 2009). In *A. niger*, Nitsche *et al.* (2012) identified several proteases that are highly induced during starvation. These proteases included *pepA* and *pepB* which encode the major extracellular proteases. Regulation of protease expression in *A. niger* involves the Zn(II)2Cys6 transcription factor PrtT (Punt *et al.*, 2008), but the expression of protease encoding genes is also under control of carbon catabolite repression (van den Hombergh *et al.*, 1997). The expression of the PrtT transcription factor is also strongly induced upon starvation (Nitsche *et al.*, 2012). PrtT-homologues are only present in a few other aspergilli. Interestingly, *A. nidulans* does not have a PrtT homologue and the protease regulation seems to be fundamentally different in comparison with *A. niger.*

The genome of *A. nidulans* contains several genes encoding putative extracellular proteases which are up-regulated by carbon starvation (Katz *et al.*, 2013; Szilágyi *et al.*, 2013). These include *prtA* and *pepJ* encoding the two major extracellular proteases of this fungus (Emri *et al.*, 2009; Katz *et al.*, 1994; Szilágyi *et al.*, 2011). The p53-like transcription factor XprG was identified as a major positively acting transcription factor of extracellular protease formation in carbon starved cultures of *A. nidulans* (Katz *et al.*, 1996, 2006). Microarray data demonstrates that XprG regulates (directly or indirectly) a large number of genes related to carbon stress including secondary metabolism, development or high-affinity glucose uptake (Katz *et al.*, 2013). It also activates hydrolases other than proteases: It proved to be important in the carbon starvation dependent activation of *engA* and *chiB* most likely via the activation of BrlA (Katz *et al.*, 2013), regulates the carbon stress dependent induction of *agdB* encoding a putative α-glucosidase (Katz *et al.*, 2013) and the production of acidic phosphatase in response to phosphate limitation (Katz *et al.*, 2006). These data suggest that XprG can be a key regulator of responses to starvation (Katz *et al.*, 2013). Little is known of the regulation of XprG. It is well documented that carbon starvation induces it (Katz *et al.*, 2013; Szilágyi *et al.*, 2013). HxkC, HxkD and AtmA are negative regulators of XprG (Bernardo *et al.*, 2007). HxkC and HxkD are regulatory type hexokinases; HxkC is associated with mitochondria, while HxkD is a nuclear protein (Bernardo *et al.*, 2007). AtmA is a serine/threonine protein kinase, an orthologue of mammalian ataxia-telangiectasia mutated (ATM) kinase which acts as a redox sensor controlling mitochondrial function (Ditch and Paull, 2012). Deletion of *atmA* in *A. nidulans* has very pleiotropic effects. AtmA is involved in the DNA damage response and required for polarized hyphal growth (Malavazi *et al.*, 2006, 2007, 2008). Among carbon starving conditions AtmA is necessary for normal transcription of genes involved in autophagy (Krohn *et al.*, 2014), it may contribute in the selective clearance of damaged mitochondria which prevent accumulation of reactive oxygen species (Krohn *et al.*, 2014), genetically interacts with XprG inhibiting XprG functions (Krohn *et al.*, 2014) and it is necessary for induction of certain carbohydrate active enzymes (CAZymes) but acts independently to CreA (Brown *et al.*, 2013). CreA also negatively regulates extracellular protease formation during carbon starvation (Emri *et al.*, 2006; Katz *et al.*, 2008). It is suggested that CreA regulates *xprG* expression or interacts directly with the XprG protein (Katz *et al.*, 2008).

FluG is necessary for extracellular protease formation in carbon starved cultures; no extracellular protease activity was detected in a loss-of-function *fluG* mutant (Emri *et al.*, 2005; Szilágyi *et al.*, 2011). BrlA is also a positive regulator of protease formation, however its action depended on the experimental design: when growing cultures were transferred into carbon source free media, inactivation of *brlA* (or *fluG*) decreased markedly the extracellular protease activity. In contrast, when carbon starvation was induced simply by long time cultivation, *brlA* deletion had only a minor effect, while in the *fluG* loss-of-function mutant no proteinase activity was detected (Szilágyi *et al.*, 2011). These data suggest that FluG may act not only by regulating *brlA* transcription. Since mutations in FadA and GanB signalling have modulatory effect on protease production, it is possible that FluG, at least in parts, regulates the protease genes via FlbA and heterotrimeric G protein signalling (Molnár *et al.*, 2004, 2006). A connection between FluG and XprG is also possible but this hypothesis needs further investigation.

Although the presence of extracellular proteins is not needed for the induction of protease genes in carbon starving cultures (Katz *et al.*, 2008), the substrate of these enzymes can influence indirectly their production. High amount of organic nitrogen sources can repress the formation of proteases (as well as cell wall hydrolases) which depends on the MeaB transcription factor (Spitzmüller *et al.*, 2015; Szilágyi *et al.*, 2010b). The alkalization of pH by the release of ammonia during carbon starvation also supports the role of protein degradation in carbon-starving cultures (Emri *et al.*, 2004). It is worth mentioning that PrtA, PepJ, EngA and ChiB preserves their activity in alkaline conditions that shows that the fungus is adapted to the alkaline environment during carbon starvation (Emri *et al.*, 2009; Erdei *et al.*, 2008; Katz *et al.*, 1994; Szilágyi *et al.*, 2010a). Moreover, alkaline pH is important in the up-regulation of extracellular protease and secondary metabolite production partly by the activity of PacC (Brakhage *et al.*, 2004; Szilágyi *et al.*, 2011).

Conclusions and future trends

The lack of available carbon source can be life-threatening for fungi as maintenance of crucial cellular processes such as ion homeostasis requires ATP. A lack of ATP could consequently result in rapid cell death. Endogenous recycling (autolysis) and extracellular hydrolysis are ways to liberate energy-rich building blocks to generate ATP under carbon starvation conditions. In filamentous fungi, the energy obtained from recycling processes is not only used for maintenance, but also used to fuel asexual development (Fig. 6.1). The formation of conidiospores might assist in the propagation of the fungal species as conidia are much more resistant to unfavourable conditions (including starvation) and allow spreading of the species. Many authors have referred to carbon starvation response as autolysis (White *et al.*, 2002). However, autolysis suggests a rather uncontrolled and destructive process. In fact, the carbon starvation response seems to be a sophisticated manner to deal with the limitation

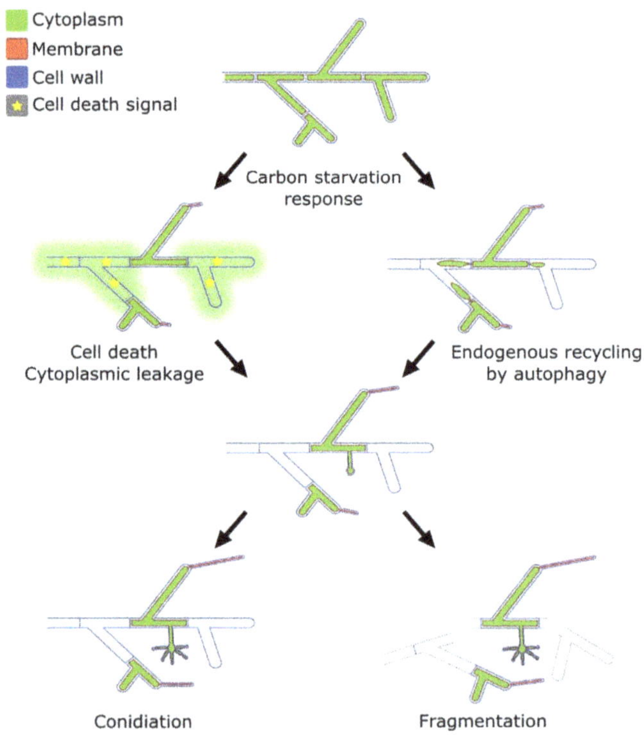

Figure 6.1 Model for carbon starvation responses in filamentous fungi. Schematic representation of morphological starvation responses during submerged carbon starvation. Upon carbon starvation, empty hyphal compartments emerge as a consequence of cell death and/or endogenous recycling of neighbouring compartments by autophagy. Cell death of compartments results in the leakage of cytoplasm into the culture broth, whereas this is prevented when cellular contents are endogenously recycled by autophagic processes. Liberated carbon sources are reused to fuel the outgrowth of thin hyphae. Towards late phases of carbon starvation, only a few surviving compartments are left in the culture broth, often bearing reproductive structures and elongated thin hyphae. Depending on the species and the cultivation conditions, the largely empty mycelial network either remains intact or is being fragmented by the activity of glycosyl hydrolases weakening the cell walls (amended and adopted from Nitsche *et al.*, 2012).

and to search for alternative sources and to maintain viability of the species via reproduction. Several studies have indicated that the process of autophagy plays an important role in the starvation response. Preliminary studies have shown that autophagy is important to delay cell death and as such allows adaptation of the fungus to the carbon starvation condition (Nitsche *et al.*, 2013). However, the carbon starvation-related phenotypes of autophagy defective mutants are subtle, indicating that additional processes or pathway are also contributing.

An important challenge for future research will be to understand the molecular mechanisms driving the multi-cellular nature of a carbon-starved mycelium. When exposed to carbon starvation, the fungal mycelium starts to differentiate and initially thin hyphae are formed (Fig. 6.1). These hyphae might be considered as scouting hyphae and might produce new enzymes with interesting biochemical properties to look for alternative carbon sources. During starvation, some compartments in the mycelium are protected from cell death, while other compartments seem to be sacrificed. It is currently unknown whether the cell content of empty compartments leaks into the medium or that neighbouring compartments can directly take up the cell content via the septal pore. Further understanding the destination and function of different cell types (thin hyphae, sacrificed compartment, surviving hyphae and hyphae that form conidiospores) may be assisted by recent tool developments for single cell/compartment transcriptomics and by identifying molecular signals and proteins related to programmed cell death or cell surviving. Identification of these molecular signals to promote fungal cell death or protect cells from dying in response to starvation conditions are also very interesting targets for antifungal drug development.

Acknowledgements

The authors would like to thank Thomas Jørgensen and Benjamin Nitsche for sharing unpublished results and helpful discussions. We thank Paul Daly for proofreading the manuscript. AB and AR were supported by grants of the SenterNovem IOP Genomics project (IGE07008) and by the Kluyver Centre for Genomics of Industrial Fermentation and the Netherlands Consortium for Systems Biology, which are part of the Netherlands Genomics Initiative/Netherlands Organization for Scientific Research (KC4.3). JvM was supported by grant IGE07008 from the Dutch SenterNovem IOP Genomics project and by grant BB/G01616X/1 from the UK Biotechnology and Biological Sciences Research Council (BBSRC) Sustainable Bioenergy Centre (BSBEC) under the Lignocellulosic Conversion to Ethanol (LACE) program. IP and TE were supported by the Hungarian Scientific Research Fund (OTKA K100464, K112181) and by the SROP-4.2.2.B-15/1/KONV-2015-0001 project. The project has been supported by the European Union, co-financed by the European Social Fund.

References

Adams, T.H., Wieser, J.K., and Yu, J.H. (1998). Asexual sporulation in *Aspergillus nidulans*. Microbiol. Mol. Biol. Rev. *62*, 35–54.

Andersen, M.R., Giese, M., de Vries, R.P., and Nielsen, J. (2012). Mapping the polysaccharide degradation potential of *Aspergillus niger*. BMC Genomics *13*, 313.

Atoui, A., Kastner, C., Larey, C.M., Thokala, R., Etxebeste, O., Espeso, E.A., Fischer, R., and Calvo, A.M. (2010). Cross-talk between light and glucose regulation controls toxin production and morphogenesis in *Aspergillus nidulans*. Fungal Genet. Biol. 47, 962–972.

Bartoszewska, M., Kiel, J.A.K.W., Bovenberg, R.A.L., Veenhuis, M., and van der Klei, I.J. (2011). Autophagy deficiency promotes β-lactam production in *Penicillium chrysogenum*. Appl. Environ. Microbiol. *77*, 1413–1422.

Bauer, S., Vasu, P., Persson, S., Mort, A.J., and Somerville, C.R. (2006). Development and application of a suite of polysaccharide-degrading enzymes for analyzing plant cell walls. Proc. Natl. Acad. Sci. U.S.A. *103*, 11417–11422.

Bayram, O.~~, and Braus, G.H.~~ (2012). Coordination of secondary metabolism and development in fungi: the velvet family of regulatory proteins. FEMS Microbiol. Rev. *36*, 1–24.

Benz, J.P., Chau, B.H., Zheng, D., Bauer, S., Glass, N.L., and Somerville, C.R. (2014). A comparative systems analysis of polysaccharide-elicited responses in *Neurospora crassa* reveals carbon source-specific cellular adaptations. Mol. Microbiol. *91*, 275–299.

Bernardo, S.M.H., Gray, K.-A., Todd, R.B., Cheetham, B.F., and Katz, M.E. (2007). Characterization of regulatory non-catalytic hexokinases in *Aspergillus nidulans*. Mol. Genet. Genomics *277*, 519–532.

Bleichrodt, R.-J., Hulsman, M., Wösten, H.A.B., and Reinders, M.J.T. (2015). Switching from a unicellular to multicellular organization in an *Aspergillus niger* hypha. MBio *6*, e00111–e00115.

Braaksma, M., Smilde, A.K., van der Werf, M.J., and Punt, P.J. (2009). The effect of environmental conditions on extracellular protease activity in controlled fermentations of *Aspergillus niger*. Microbiology *155*, 3430–3439.

Braaksma, M., Martens-Uzunova, E.S., Punt, P.J., and Schaap, P.J. (2010). An inventory of the *Aspergillus niger* secretome by combining *in silico* predictions with shotgun proteomics data. BMC Genomics *11*, 584.

Brakhage, A.A., Spröte, P., Al-Abdallah, Q., Gehrke, A., Plattner, H., and Tüncher, A. (2004). Regulation of penicillin biosynthesis in filamentous fungi. Adv. Biochem. Eng. Biotechnol. *88*, 45–90.

Brown, N.A., de Gouvea, P.F., Krohn, N.G., Savoldi, M., and Goldman, G.H. (2013). Functional characterisation of the non-essential protein kinases and phosphatases regulating *Aspergillus nidulans* hydrolytic enzyme production. Biotechnol. Biofuels *6*, 91.

Bushell, M.E., and Bull, A.T. (1999). Sporulation at minimum specific growth rate in *Aspergillus nidulans* chemostat culture predicted using protein synthesis efficiency estimations. J. Basic Microbiol. *39*, 293–298.

Choi, C.-J., Ju, H.-J., Park, B.-H., Qin, R., Jahng, K.-Y., Han, D.-M., and Chae, K.-S. (2005). Isolation and characterization of the *Aspergillus nidulans eglC* gene encoding a putative beta-1,3-endoglucanase. Fungal Genet. Biol. *42*, 590–600.

Coradetti, S.T., Craig, J.P., Xiong, Y., Shock, T., Tian, C., and Glass, N.L. (2012). Conserved and essential transcription factors for cellulase gene expression in ascomycete fungi. Proc. Natl. Acad. Sci. U.S.A. *109*, 7397–7402.

Cubero, B., and Scazzocchio, C. (1994). Two different, adjacent and divergent zinc finger binding sites are necessary for CREA-mediated carbon catabolite repression in the proline gene cluster of *Aspergillus nidulans*. EMBO J. *13*, 407–415.

Dechant, R., and Peter, M. (2008). Nutrient signals driving cell growth. Curr. Opin. Cell Biol. *20*, 678–687.

De Groot, P.W.J., Ram, A.F., and Klis, F.M. (2005). Features and functions of covalently linked proteins in fungal cell walls. Fungal Genet. Biol. *42*, 657–675.

Delmas, S., Pullan, S.T., Gaddipati, S., Kokolski, M., Malla, S., Blythe, M.J., Ibbett, R., Campbell, M., Liddell, S., Aboobaker, A., *et al.* (2012). Uncovering the genome-wide transcriptional responses of the filamentous fungus *Aspergillus niger* to lignocellulose using RNA sequencing. PLoS Genet. *8*, e1002875.

Ditch, S., and Paull, T.T. (2012). The ATM protein kinase and cellular redox signaling: beyond the DNA damage response. Trends Biochem. Sci. *37*, 15–22.

Dyer, P.S., and O'Gorman, C.M. (2012). Sexual development and cryptic sexuality in fungi: insights from *Aspergillus* species. FEMS Microbiol. Rev. *36*, 165–192.

Dyer, P.S., Ingram, D.S., and Johnstone, K. (1992). The control of sexual morphogenesis in the Ascomycotina. Biol. Rev. *67*, 421–458.

Ellström, M., Shah, F., Johansson, T., Ahrén, D., Persson, P., and Tunlid, A. (2015). The carbon starvation response of the ectomycorrhizal fungus *Paxillus involutus*. FEMS Microbiol. Ecol. *91*, fiv027.

Emri, T., Molnár, Z., Pusztahelyi, T., and Pócsi, I. (2004). Physiological and morphological changes in autolyzing *Aspergillus nidulans* cultures. Folia Microbiol. (Praha). *49*, 277–284.

Emri, T., Molnár, Z., Pusztahelyi, T., Varecza, Z., and Pócsi, I. (2005). The fluG-BrlA pathway contributes to the initialisation of autolysis in submerged *Aspergillus nidulans* cultures. Mycol. Res. *109*, 757–763.

Emri, T., Molnár, Z., Veres, T., Pusztahelyi, T., Dudás, G., and Pócsi, I. (2006). Glucose-mediated repression of autolysis and conidiogenesis in *Emericella nidulans*. Mycol. Res. *110*, 1172–1178.

Emri, T., Molnár, Z., Szilágyi, M., and Pócsi, I. (2008). Regulation of autolysis in *Aspergillus nidulans*. Appl. Biochem. Biotechnol. *151*, 211–220.
Emri, T., Szilágyi, M., László, K., M-Hamvas, M., and Pócsi, I. (2009). PepJ is a new extracellular proteinase of *Aspergillus nidulans*. Folia Microbiol. (Praha). *54*, 105–109.
Erdei, E., Pusztahelyi, T., Miskei, M., Barna, T., and Pócsi, I. (2008). Characterization and heterologous expression of an age-dependent fungal/bacterial type chitinase of *Aspergillus nidulans*. Acta Microbiol. Immunol. Hung. *55*, 351–361.
Escott, G.M., Hearn, V.M., and Adams, D.J. (1998). Inducible chitinolytic system of *Aspergillus fumigatus*. Microbiology *144*, 1575–1581.
Etxebeste, O., Garzia, A., Espeso, E.A., and Ugalde, U. (2010). *Aspergillus nidulans* asexual development: making the most of cellular modules. Trends Microbiol. *18*, 569–576.
Feng, Y., He, D., Yao, Z., and Klionsky, D.J. (2014). The machinery of macroautophagy. Cell Res. *24*, 24–41.
Flipphi, M., van de Vondervoort, P.J.I., Ruijter, G.J.G., Visser, J., Arst, H.N., and Felenbok, B. (2003). Onset of carbon catabolite repression in *Aspergillus nidulans*: Parallel involvement of hexokinase and glucokinase in sugar signaling. J. Biol. Chem. *278*, 11849–11857.
Fontaine, T., Hartland, R.P., Beauvais, A., Diaquin, M., and Latge, J.P. (1997). Purification and characterization of an endo-1,3-beta-glucanase from *Aspergillus fumigatus*. Eur. J. Biochem. *243*, 315–321.
Foreman, P.K., Brown, D., Dankmeyer, L., Dean, R., Diener, S., Dunn-Coleman, N.S., Goedegebuur, F., Houfek, T.D., England, G.J., Kelley, A.S., *et al.* (2003). Transcriptional regulation of biomass-degrading enzymes in the filamentous fungus *Trichoderma reesei*. J. Biol. Chem. *278*, 31988–31997.
Fujioka, T., Mizutani, O., Furukawa, K., Sato, N., Yoshimi, A., Yamagata, Y., Nakajima, T., and Abe, K. (2007). MpkA-dependent and -independent cell wall integrity signaling in *Aspergillus nidulans*. Eukaryot. Cell *6*, 1497–1510.
Fujiwara, M., Ichinomiya, M., Motoyama, T., Horiuchi, H., Ohta, A., and Takagi, M. (2000). Evidence that the *Aspergillus nidulans* class I and class II chitin synthase genes, *chsC* and *chsA*, share critical roles in hyphal wall integrity and conidiophore development. J. Biochem. *127*, 359–366.
Glass, N.L., Schmoll, M., Cate, J.H.D., and Coradetti, S. (2013). Plant cell wall deconstruction by ascomycete fungi. Annu. Rev. Microbiol. *67*, 477–498.
Han, K., Lee, D., Kim, J., Kim, M., Kim, W., Park, Y., Kim, H., and Han, D. (2003). Environmental factors affecting development of *Aspergillus nidulans*. J. Microbiol. *41*, 34–40.
Hartl, L., Gastebois, A., Aimanianda, V., and Latgé, J.-P. (2011). Characterization of the GPI-anchored endo β-1,3-glucanase Eng2 of *Aspergillus fumigatus*. Fungal Genet. Biol. *48*, 185–191.
van den Hombergh, J.P., van de Vondervoort, P.J., Fraissinet-Tachet, L., and Visser, J. (1997). *Aspergillus* as a host for heterologous protein production: the problem of proteases. Trends Biotechnol. *15*, 256–263.
Horiuchi, H. (2009). Functional diversity of chitin synthases of *Aspergillus nidulans* in hyphal growth, conidiophore development and septum formation. Med. Mycol. 47(Suppl. 1), S47–S52.
Jaques, A.K., Fukamizo, T., Hall, D., Barton, R.C., Escott, G.M., Parkinson, T., Hitchcock, C.A., and Adams, D.J. (2003). Disruption of the gene encoding the ChiB1 chitinase of *Aspergillus fumigatus* and characterization of a recombinant gene product. Microbiology *149*, 2931–2939.
Jaroszuk-Scisel, J., Kurek, E., Slomka, A., Janczarek, M., and Rodzik, B. (2011). Activities of cell wall degrading enzymes in autolyzing cultures of three *Fusarium culmorum* isolates: growth-promoting, deleterious and pathogenic to rye (*Secale cereale*). Mycologia *103*, 929–945.
Jørgensen, T.R., Nitsche, B.M., Lamers, G.E., Arentshorst, M., van den Hondel, C.A., and Ram, A.F. (2010). Transcriptomic insights into the physiology of *Aspergillus niger* approaching a specific growth rate of zero. Appl. Environ. Microbiol. *76*, 5344–5355.
Kamada, Y., Funakoshi, T., Shintani, T., Nagano, K., Ohsumi, M., and Ohsumi, Y. (2000). Tor-mediated induction of autophagy via an Apg1 protein kinase complex. J. Cell Biol. *150*, 1507–1513.
Kang, J.Y., Chun, J., Jun, S.-C., Han, D.-M., Chae, K.-S., and Jahng, K.Y. (2013). The MpkB MAP kinase plays a role in autolysis and conidiation of *Aspergillus nidulans*. Fungal Genet. Biol. *61*, 42–49.
Kasuga, T., and Glass, N.L. (2008). Dissecting colony development of *Neurospora crassa* using mRNA profiling and comparative genomics approaches. Eukaryot. Cell 7, 1549–1564.
Katz, M.E., Rice, R.N., and Cheetham, B.F. (1994). Isolation and characterization of an *Aspergillus nidulans* gene encoding an alkaline protease. Gene *150*, 287–292.
Katz, M.E., Flynn, P.K., vanKuyk, P.A., and Cheetham, B.F. (1996). Mutations affecting extracellular protease production in the filamentous fungus *Aspergillus nidulans*. Mol. Gen. Genet. *250*, 715–724.

Katz, M.E., Gray, K.-A., and Cheetham, B.F. (2006). The *Aspergillus nidulans xprG* (*phoG*) gene encodes a putative transcriptional activator involved in the response to nutrient limitation. Fungal Genet. Biol. *43*, 190–199.

Katz, M.E., Bernardo, S.M., and Cheetham, B.F. (2008). The interaction of induction, repression and starvation in the regulation of extracellular proteases in *Aspergillus nidulans*: evidence for a role for CreA in the response to carbon starvation. Curr. Genet. *54*, 47–55.

Katz, M.E., Braunberger, K., Yi, G., Cooper, S., Nonhebel, H.M., and Gondro, C. (2013). A p53-like transcription factor similar to Ndt80 controls the response to nutrient stress in the filamentous fungus, *Aspergillus nidulans*. F1000Research *2*, 72.

Kelly, R., Register, E., Hsu, M.J., Kurtz, M., and Nielsen, J. (1996). Isolation of a gene involved in 1,3-beta-glucan synthesis in *Aspergillus nidulans* and purification of the corresponding protein. J. Bacteriol. *178*, 4381–4391.

Kikuma, T., Ohneda, M., Arioka, M., and Kitamoto, K. (2006). Functional analysis of the ATG8 homologue Aoatg8 and role of autophagy in differentiation and germination in *Aspergillus oryzae*. Eukaryot. Cell *5*, 1328–1336.

Kim, S., Matsuo, I., Ajisaka, K., Nakajima, H., and Kitamoto, K. (2002). Cloning and characterization of the *nagA* gene that encodes beta-n-acetylglucosaminidase from *Aspergillus nidulans* and its expression in *Aspergillus oryzae*. Biosci. Biotechnol. Biochem. *66*, 2168–2175.

Krijgsheld, P., Altelaar, A.F.M., Post, H., Ringrose, J.H., Müller, W.H., Heck, A.J.R., and Wösten, H.A.B. (2012). Spatially resolving the secretome within the mycelium of the cell factory *Aspergillus niger*. J. Proteome Res. *11*, 2807–2818.

Krijgsheld, P., Bleichrodt, R., van Veluw, G.J., Wang, F., Müller, W.H., Dijksterhuis, J., and Wösten, H.A.B. (2013). Development in *Aspergillus*. Stud. Mycol. *74*, 1–29.

Krohn, N.G., Brown, N.A., Colabardini, A.C., Reis, T., Savoldi, M., Dinamarco, T.M., Goldman, M.H.S., and Goldman, G.H. (2014). The *Aspergillus nidulans* ATM kinase regulates mitochondrial function, glucose uptake and the carbon starvation response. G3 (Bethesda) *4*, 49–62.

Kuo, M.J., and Alexander, M. (1967). Inhibition of the lysis of fungi by melanins. J. Bacteriol. *94*, 624–629.

Lahoz, R., Reyes, F., and Perez Leblic, M.I. (1976). Lytic enzymes in the autolysis of filamentous fungi. Mycopathologia *60*, 45–49.

Latgé, J.-P. (2007). The cell wall: a carbohydrate armour for the fungal cell. Mol. Microbiol. *66*, 279–290.

Latgé, J.P., Mouyna, I., Tekaia, F., Beauvais, A., Debeaupuis, J.P., and Nierman, W. (2005). Specific molecular features in the organization and biosynthesis of the cell wall of *Aspergillus fumigatus*. Med. Mycol. *43*(Suppl. 1), S15–S22.

Lee, B.N., and Adams, T.H. (1994). The *Aspergillus nidulans fluG* gene is required for production of an extracellular developmental signal and is related to prokaryotic glutamine synthetase I. Genes Dev. *8*, 641–651.

Lee, J.I., Choi, J.H., Park, B.C., Park, Y.H., Lee, M.Y., Park, H.M., and Maeng, P.J. (2004). Differential expression of the chitin synthase genes of *Aspergillus nidulans, chsA, chsB,* and *chsC*, in response to developmental status and environmental factors. Fungal Genet. Biol. *41*, 635–646.

Lee, M.-K., Kwon, N.-J., Choi, J.M., Lee, I.-S., Jung, S., and Yu, J.-H. (2014). NsdD is a key repressor of asexual development in *Aspergillus nidulans*. Genetics *197*, 159–173.

Levin, A.M., de Vries, R.P., Conesa, A., de Bekker, C., Talon, M., Menke, H.H., van Peij, N.N.M.E., and Wösten, H.A.B. (2007). Spatial differentiation in the vegetative mycelium of *Aspergillus niger*. Eukaryot. Cell *6*, 2311–2322.

Lü, Y., Yang, H., Hu, H., Wang, Y., Rao, Z., and Jin, C. (2009). Mutation of Trp137 to glutamate completely removes transglycosyl activity associated with the *Aspergillus fumigatus* AfChiB1. Glycoconj. J. *26*, 525–534.

Malavazi, I., Semighini, C.P., Kress, M.R., Harris, S.D., and Goldman, G.H. (2006). Regulation of hyphal morphogenesis and the DNA damage response by the *Aspergillus nidulans* ATM homolog AtmA. Genetics *173*, 99–109.

Malavazi, I., Savoldi, M., da Silva Ferreira, M.E., Soriani, F.M., Bonato, P.S., de Souza Goldman, M.H., and Goldman, G.H. (2007). Transcriptome analysis of the *Aspergillus nidulans* AtmA (ATM, Ataxia-Telangiectasia mutated) null mutant. Mol. Microbiol. *66*, 74–99.

Malavazi, I., Lima, J.F., de Castro, P.A., Savoldi, M., de Souza Goldman, M.H., and Goldman, G.H. (2008). Genetic interactions of the *Aspergillus nidulans atmA* ATM homolog with different components of the DNA damage response pathway. Genetics *178*, 675–691.

Martens-Uzunova, E.S., and Schaap, P.J. (2009). Assessment of the pectin degrading enzyme network of *Aspergillus niger* by functional genomics. Fungal Genet. Biol. 46(Suppl. 1), S170–S179.

Maubon, D., Park, S., Tanguy, M., Huerre, M., Schmitt, C., Prévost, M.C., Perlin, D.S., Latgé, J.P., and Beauvais, A. (2006). AGS3, an α(1-3)glucan synthase gene family member of *Aspergillus fumigatus*, modulates mycelium growth in the lung of experimentally infected mice. Fungal Genet. Biol. *43*, 366–375.

McIntyre, M., Berry, D.R., and McNeil, B. (2000). Role of proteases in autolysis of *Penicillium chrysogenum* chemostat cultures in response to nutrient depletion. Appl. Microbiol. Biotechnol. *53*, 235–242.

Molnár, Z., Mészáros, E., Szilágyi, Z., Rosén, S., Emri, T., and Pócsi, I. (2004). Influence of fadAG203R and deltaflbA mutations on morphology and physiology of submerged *Aspergillus nidulans* cultures. Appl. Biochem. Biotechnol. *118*, 349–360.

Molnár, Z., Emri, T., Zavaczki, E., Pusztahelyi, T., and Pócsi, I. (2006). Effects of mutations in the GanB/RgsA G protein mediated signalling on the autolysis of *Aspergillus nidulans*. J. Basic Microbiol. *46*, 495–503.

Mouyna, I., Sarfati, J., Recco, P., Fontaine, T., Henrissatz, B., and Latge, J.P. (2002). Molecular characterization of a cell wall-associated beta(1–3)endoglucanase of *Aspergillus fumigatus*. Med. Mycol. *40*, 455–464.

van Munster, J.M., van der Kaaij, R.M., Dijkhuizen, L., and van der Maarel, M.J.E.C. (2012). Biochemical characterization of *Aspergillus niger* CfcI, a glycoside hydrolase family 18 chitinase that releases monomers during substrate hydrolysis. Microbiology *158*, 2168–2179.

van Munster, J.M., Nitsche, B.M., Krijgsheld, P., van Wijk, A., Dijkhuizen, L., Wösten, H.A., Ram, A.F., and van der Maarel, M.J.E.C. (2013). Chitinases CtcB and CfcI modify the cell wall in sporulating aerial mycelium of *Aspergillus niger*. Microbiology *159*, 1853–1867.

van Munster, J.M., Daly, P., Delmas, S., Pullan, S.T., Blythe, M.J., Malla, S., Kokolski, M., Noltorp, E.C.M., Wennberg, K., Fetherston, R., *et al.* (2014). The role of carbon starvation in the induction of enzymes that degrade plant-derived carbohydrates in *Aspergillus niger*. Fungal Genet. Biol. *72*, 34–47.

van Munster, J.M., Nitsche, B.M., Akeroyd, M., Dijkhuizen, L., van der Maarel, M.J.E.C., and Ram, A.F.J. (2015a). Systems approaches to predict the functions of glycoside hydrolases during the life cycle of *Aspergillus niger* using developmental mutants Δ*brlA* and Δ*flbA*. PLoS One *10*, e0116269.

van Munster, J.M., Dobruchowska, J.M., Veloo, R., Dijkhuizen, L., and van der Maarel, M.J.E.C. (2015b). Characterization of the starvation-induced chitinase CfcA and α-1,3-glucanase AgnB of *Aspergillus niger*. Appl. Microbiol. Biotechnol. *99*, 2209–2223.

van Munster, J.M., Sanders, P., ten Kate, G.A., Dijkhuizen, L., and van der Maarel, M.J.E.C. (2015c). Kinetic characterization of *Aspergillus niger* chitinase CfcI using a HPAEC-PAD method for native chitin oligosaccharides. Carbohydr. Res. *407*, 73–78.

Nitsche, B.M., Jørgensen, T.R., Akeroyd, M., Meyer, V., and Ram, A.F. (2012). The carbon starvation response of *Aspergillus niger* during submerged cultivation: Insights from the transcriptome and secretome. BMC Genomics *13*, 380.

Nitsche, B.M., Burggraaf-Van Welzen, A.-M., Lamers, G., Meyer, V., and Ram, A.F.J. (2013). Autophagy promotes survival in aging submerged cultures of the filamentous fungus *Aspergillus niger*. Appl. Microbiol. Biotechnol. *97*, 8205–8218.

Nuero, O.M., Alfonso, C., Del Amo, F., and Reyes, F. (1993). Study of β-1,3-glucanase activity during autolysis of *Aspergillus nidulans* by FPLC ion-exchange chromatography. Lett. Appl. Microbiol. *17*, 104–108.

Park, H.-S., and Yu, J.-H. (2012). Genetic control of asexual sporulation in filamentous fungi. Curr. Opin. Microbiol. *15*, 669–677.

Pel, H.J., de Winde, J.H., Archer, D.B., Dyer, P.S., Hofmann, G., Schaap, P.J., Turner, G., de Vries, R.P., Albang, R., Albermann, K., *et al.* (2007). Genome sequencing and analysis of the versatile cell factory *Aspergillus niger* CBS 513.88. Nat. Biotechnol. *25*, 221–231.

Perez-Leblic, M.I., Reyes, F., Martinez, M.J., and Lahoz, R. (1982). Cell wall degradation in the autolysis of filamentous fungi. Mycopathologia *80*, 147–155.

Pinar, M., Pantazopoulou, A., and Peñalva, M.A. (2013). Live-cell imaging of *Aspergillus nidulans* autophagy: RAB1 dependence, Golgi independence and ER involvement. Autophagy *9*, 1024–1043.

Pócsi, I., Leiter, E., Kwon, N.-J., Shin, K.-S., Kwon, G.-S., Pusztahelyi, T., Emri, T., Abuknesha, R.A., Price, R.G., and Yu, J.-H. (2009). Asexual sporulation signalling regulates autolysis of *Aspergillus nidulans* via modulating the chitinase ChiB production. J. Appl. Microbiol. *107*, 514–523.

Pollack, J.K., Li, Z.J., and Marten, M.R. (2008). Fungal mycelia show lag time before re-growth on endogenous carbon. Biotechnol. Bioeng. *100*, 458–465.

Pullan, S.T., Daly, P., Delmas, S., Ibbett, R., Kokolski, M., Neiteler, A., van Munster, J.M., Wilson, R., Blythe, M.J., Gaddipati, S., *et al.* (2014). RNA-sequencing reveals the complexities of the transcriptional response to lignocellulosic biofuel substrates in *Aspergillus niger*. Fungal Biol. Biotechnol. *1*, 3.

Punt, P.J., Schuren, F.H.J., Lehmbeck, J., Christensen, T., Hjort, C., and van den Hondel, C.A.M.J.J. (2008). Characterization of the *Aspergillus niger* prtT, a unique regulator of extracellular protease encoding genes. Fungal Genet. Biol. *45*, 1591–1599.

Pusztahelyi, T., and Pócsi, I. (2013). Functions, cooperation, and interplays of the vegetative growth signaling pathway in the Aspergilli. J. Mycol. *2013*, 1–11.

Pusztahelyi, T., Molnár, Z., Emri, T., Klement, E., Miskei, M., Kerékgyárto, J., Balla, J., and Pócsi, I. (2006). Comparative studies of differential expression of chitinolytic enzymes encoded by *chiA*, *chiB*, *chiC* and *nagA* genes in *Aspergillus nidulans*. Folia Microbiol. (Praha). *51*, 547–554.

Richie, D.L., Fuller, K.K., Fortwendel, J., Miley, M.D., McCarthy, J.W., Feldmesser, M., Rhodes, J.C., and Askew, D.S. (2007). Unexpected link between metal ion deficiency and autophagy in *Aspergillus fumigatus*. Eukaryot. Cell *6*, 2437–2447.

Rodríguez-Urra, A.B., Jiménez, C., Nieto, M.I., Rodríguez, J., Hayashi, H., and Ugalde, U. (2012). Signaling the induction of sporulation involves the interaction of two secondary metabolites in *Aspergillus nidulans*. ACS Chem. Biol. *7*, 599–606.

Ruijter, G.J., and Visser, J. (1997). Carbon repression in Aspergilli. FEMS Microbiol. Lett. *151*, 103–114.

Sarkar, A., Funk, A.N., Scherlach, K., Horn, F., Schroeckh, V., Chankhamjon, P., Westermann, M., Roth, M., Brakhage, A.A., Hertweck, C., *et al.* (2012). Differential expression of silent polyketide biosynthesis gene clusters in chemostat cultures of *Aspergillus nidulans*. J. Biotechnol. *160*, 64–71.

Sarkar, P., Bosneaga, E., and Auer, M. (2009). Plant cell walls throughout evolution: towards a molecular understanding of their design principles. J. Exp. Bot. *60*, 3615–3635.

Scheller, H.V., and Ulvskov, P. (2010). Hemicelluloses. Annu. Rev. Plant Biol. *61*, 263–289.

Schrickx, J.M., Krave, A.S., Verdoes, J.C., van den Hondel, C.A., Stouthamer, A.H., and van Verseveld, H.W. (1993). Growth and product formation in chemostat and recycling cultures by *Aspergillus niger* N402 and a glucoamylase overproducing transformant, provided with multiple copies of the *glaA* gene. J. Gen. Microbiol. *139*, 2801–2810.

Shimizu, K., and Keller, N.P. (2001). Genetic involvement of a cAMP-dependent protein kinase in a G protein signaling pathway regulating morphological and chemical transitions in *Aspergillus nidulans*. Genetics *157*, 591–600.

Shin, K.S., Kwon, N.J., Kim, Y.H., Park, H.S., Kwon, G.S., and Yu, J.H. (2009). Differential roles of the chib chitinase in autolysis and cell death of *Aspergillus nidulans*. Eukaryot. Cell *8*, 738–746.

Shoji, J.Y., and Craven, K.D. (2011). Autophagy in basal hyphal compartments: A green strategy of great recyclers. Fungal Biol. Rev. *25*, 79–83.

Shoji, J.Y., Arioka, M., and Kitamoto, K. (2006). Possible involvement of pleiomorphic vacuolar networks in nutrient recycling in filamentous fungi. Autophagy *2*, 226–227.

Skromne, I., Sánchez, O., and Aguirre, J. (1995). Starvation stress modulates the expression of the *Aspergillus nidulans brlA* regulatory gene. Microbiology *141*, 21–28.

de Sousa Lima, P., Casaletti, L., Bailão, A.M., de Vasconcelos, A.T.R., da Rocha Fernandes, G., and de Almeida Soares, C.M. (2014). Transcriptional and proteomic responses to carbon starvation in *Paracoccidioides*. PLoS Negl. Trop. Dis. *8*, e2855.

Spitzmüller, Z., Kwon, N.-J., Szilágyi, M., Keserű, J., Tóth, V., Yu, J.-H., Pócsi, I., and Emri, T. (2015). γ-Glutamyl transpeptidase (GgtA) of *Aspergillus nidulans* is not necessary for bulk degradation of glutathione. Arch. Microbiol. *197*, 285–297.

Stephan, J.S., Yeh, Y.-Y., Ramachandran, V., Deminoff, S.J., and Herman, P.K. (2009). The Tor and PKA signaling pathways independently target the Atg1/Atg13 protein kinase complex to control autophagy. Proc. Natl. Acad. Sci. U.S.A. *106*, 17049–17054.

Szilágyi, M., Kwon, N.-J., Dorogi, C., Pócsi, I., Yu, J.-H., and Emri, T. (2010a). The extracellular β-1,3-endoglucanase EngA is involved in autolysis of *Aspergillus nidulans*. J. Appl. Microbiol. *109*, 1498–1508.

Szilágyi, M., Pócsi, I., Forgács, K., and Emri, T. (2010b). MeaB-dependent nutrition sensing regulates autolysis in carbon starved *Aspergillus nidulans* cultures. Indian J. Microbiol. *50*, 104–108.

Szilágyi, M., Kwon, N.-J., Bakti, F., M-Hamvas, M., Jámbrik, K., Park, H., Pócsi, I., Yu, J.-H., and Emri, T. (2011). Extracellular proteinase formation in carbon starving *Aspergillus nidulans* cultures – physiological function and regulation. J. Basic Microbiol. *51*, 625–634.

Szilágyi, M., Anton, F., Forgács, K., Yu, J.-H., Pócsi, I., and Emri, T. (2012). Antifungal activity of extracellular hydrolases produced by autolysing *Aspergillus nidulans* cultures. J. Microbiol. *50*, 849–854.

Szilágyi, M., Miskei, M., Karányi, Z., Lenkey, B., Pócsi, I., and Emri, T. (2013). Transcriptome changes initiated by carbon starvation in *Aspergillus nidulans*. Microbiology *159*, 176–190.

Takashima, S., Iikura, H., Nakamura, A., Masaki, H., and Uozumi, T. (1996). Analysis of Cre1 binding sites in the *Trichoderma reesei cbh1* upstream region. FEMS Microbiol. Lett. *145*, 361–366.

Twumasi-Boateng, K., Yu, Y., Chen, D., Gravelat, F.N., Nierman, W.C., and Sheppard, D.C. (2009). Transcriptional profiling identifies a role for BrlA in the response to nitrogen depletion and for StuA in the regulation of secondary metabolite clusters in *Aspergillus fumigatus*. Eukaryot. Cell *8*, 104–115.

Vallim, M.A., Miller, K.Y., and Miller, B.L. (2000). Aspergillus SteA (sterile12-like) is a homeodomain-C2/H2-Zn+2 finger transcription factor required for sexual reproduction. Mol. Microbiol. *36*, 290–301.

Vogel, J. (2008). Unique aspects of the grass cell wall. Curr. Opin. Plant Biol. *11*, 301–307.

de Vries, R.P., Visser, J., and de Graaff, L.H. (1999). CreA modulates the XlnR-induced expression on xylose of *Aspergillus niger* genes involved in xylan degradation. Res. Microbiol. *150*, 281–285.

Wei, H., Scherer, M., Singh, A., Liese, R., and Fischer, R. (2001). *Aspergillus nidulans* alpha-1,3 glucanase (mutanase), *mutA*, is expressed during sexual development and mobilizes mutan. Fungal Genet. Biol. *34*, 217–227.

White, S., McIntyre, M., Berry, D.R., and McNeil, B. (2002). The autolysis of industrial filamentous fungi. Crit. Rev. Biotechnol. *22*, 1–14.

Xia, G., Jin, C., Zhou, J., Yang, S., Zhang, S., and Jin, C. (2001). A novel chitinase having a unique mode of action from *Aspergillus fumigatus* YJ-407. Eur. J. Biochem. *268*, 4079–4085.

Xiong, Y., Coradetti, S.T., Li, X., Gritsenko, M.A., Clauss, T., Petyuk, V., Camp, D., Smith, R., Cate, J.H.D., Yang, F., *et al.* (2014). The proteome and phosphoproteome of *Neurospora crassa* in response to cellulose, sucrose and carbon starvation. Fungal Genet. Biol. *72*, 21–33.

Yamazaki, H., Yamazaki, D., Takaya, N., Takagi, M., Ohta, A., and Horiuchi, H. (2007). A chitinase gene, *chiB*, involved in the autolytic process of *Aspergillus nidulans*. Curr. Genet. *51*, 89–98.

Yamazaki, H., Tanaka, A., Kaneko, J., Ohta, A., and Horiuchi, H. (2008). *Aspergillus nidulans* ChiA is a glycosylphosphatidylinositol (GPI)-anchored chitinase specifically localized at polarized growth sites. Fungal Genet. Biol. *45*, 963–972.

Yu, J.-H. (2006). Heterotrimeric G protein signaling and RGSs in *Aspergillus nidulans*. J. Microbiol. *44*, 145–154.

Zhao, W., Li, C., Liang, J., and Sun, S. (2014). The *Aspergillus fumigatus* β-1,3-glucanosyltransferase Gel7 plays a compensatory role in maintaining cell wall integrity under stress conditions. Glycobiology *24*, 418–427.

Zonneveld, B.J. (1972). Morphogenesis in *Aspergillus nidulans*. The significance of a alpha-1, 3-glucan of the cell wall and alpha-1,3-glucanase for cleistothecium development. Biochim. Biophys. Acta *273*, 174–187.

Zonneveld, B.J. (1974). Alpha-1,3 glucan synthesis correlated with alpha-1,3 glucanase synthesis, conidiation and fructification in morphogenetic mutants of *Aspergillus nidulans*. J. Gen. Microbiol. *81*, 445–451.

Genetics and Physiology of Sulfur Metabolism in *Aspergillus*

7

Andrzej Paszewski, Jerzy Brzywczy, Marzena Sieńko and Sebastian Piłsyk

Abstract

Aspergillus nidulans, as many other fungi, can utilize sulfate as a sulfur source through a highly energy consuming sulfate assimilation pathway producing sulfide which is then incorporated into cysteine. When cysteine is in excess this pathway is repressed by the hierarchical sulfur metabolite repression (SMR) system, in which the SCF ubiquitin ligase activated by cysteine brings about degradation of the MetR transcription factor specific for sulfate assimilation pathway genes. Mutational impairment of SMR causes a permanent derepression of the sulfate assimilation pathway and leads to a direct synthesis of homocysteine from sulfide as an alternative route of synthesis of sulfur amino acids. This derepression results in overproduction of sulfide and sulfur-containing organic compounds associated with a considerable dissipation of ATP and NADPH. Transcriptomic analysis of mutants dysregulated in SMR has revealed considerable remodelling of cellular metabolism manifested by changes in the levels of several hundred transcripts. Among the up-regulated genes the most remarkable are those encoding proteins normally involved in responses to various environmental stresses. Expression of genes encoding enzymes involved in homocysteine metabolism is independent of SMR and is activated by this amino acid, which prevents its accumulation above a toxic level.

Introduction

The rapidly growing number of sequenced fungal genomes enables comparative analysis of diverse metabolic pathways and their regulation. Sulfur metabolism is one of the four major metabolic branches, besides those of carbon, nitrogen and phosphate. Despite being an important metabolic process, sulfur metabolism is a relatively poorly understood part of the basal metabolic processes in the cell. Numerous fungi can utilize inorganic sulfate as a sulfur source for the synthesis of cysteine which serves as a precursor of methionine, both crucial components of proteins. Apart from their use as building blocks of proteins, the two amino acids are also used for the synthesis of other sulfur-containing compounds including glutathione, *S*-adenosylmethionine, some vitamins and secondary metabolites, e.g. β-lactam antibiotics. *Aspergillus nidulans, Neurospora crassa* and *Saccharomyces cerevisiae* are the most extensively studied fungal species with respect to sulfur metabolism, showing numerous similarities but also significant differences in the enzymatic pathways used and their regulatory circuits (for a review see Marzluf, 1997; Paszewski *et al.*, 2000). Among these model organisms, *A. nidulans* appears particularly useful because it uses the richest repertoire of routes for the biosynthesis of sulfur amino acids and their interconversions. Numerous physiological and genetic aspects of sulfur metabolism in *A. nidulans* have been addressed in earlier reviews (Paszewski *et al.*, 1994, 2000, 2009). The aim of this report is to recapitulate earlier data and highlight novel findings.

Synthesis of cysteine and methionine

Physiological and genetic studies have led to the elucidation of sulfur amino acid metabolic pathways in *A. nidulans* and *N. crassa,* the model filamentous fungi. An outline of sulfur metabolism in *A. nidulans* is shown in Fig. 7.1. Recent sequencing projects have allowed the identification of orthologous genes encoding enzymes and regulatory proteins involved in sulfur metabolism in aspergilli, *Penicillium rubens* and *N. crassa,* which are listed in Table 7.1.

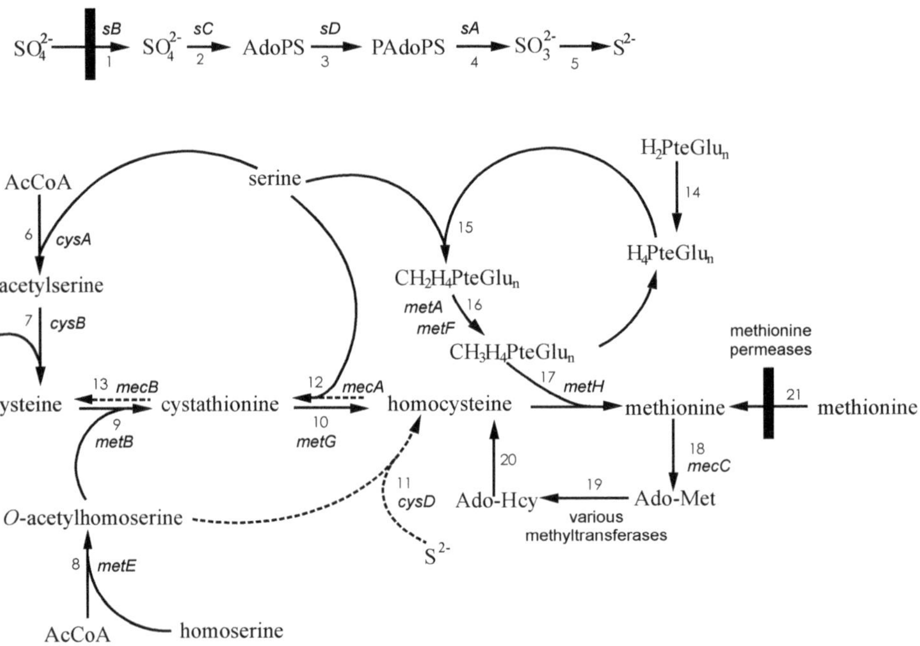

Figure 7.1 Sulfur metabolism in *Aspergillus nidulans*. Enzymatic steps: 1, sulfate permease; 2, ATP-sulfurylase; 3, AdoPS-kinase; 4, PadoPS-reductase; 5, sulfite reductase; 6, serine transacetylase; 7, cysteine synthase (*O*-acetylserine sulfhydrylase); 8, homoserine transacetylase; 9, cystathionine γ-synthase; 10, cystathionine β-lyase; 11, homocysteine synthase; 12, cystathionine β-synthase; 13, cystathionine γ-lyase; 14, dihydrofolate reductase; 15, serine hydroxymethyltransferase; 16, methylenetetrahydrofolate reductases; 17, methionine synthase; 18, *S*-adenosylmethionine synthase; 19, various methyltransferases; 20, *S*-adenosylhomocysteine hydrolase; 21, methionine permeases. Names of genes encoding particular enzymes are indicated in italics, whenever available. Metabolites: AdoMet, *S*-adenosylmethionine; AdoHcy, *S*-adenosylhomocysteine; AdoPS, adenosine 5'-phosphosulfate; PadoPS, adenosine 3′-phosphate 5′-phosphosulfate; $H_2PteGlu_n$, dihydrofolylpolyglutamate; $H_4PteGlu_n$, tetrahydrofolylpolyglutamate; $CH_2H_4PteGlu_n$, methylenetetrahydrofolylpolyglutamate; $CH_3H_4PteGlu_n$, 5-methyltetrahydrofolylpolyglutamate.

Table 7.1 Orthologous genes encoding enzymes and regulatory proteins involved in sulfur metabolism in four species of *Aspergillus*, *P. rubens* and *N. crassa*

	ORF ID/gene name					
Activity	*A. nidulans*	*A. fumigatus*	*A. niger*	*A. oryzae*	*P. rubens*	*N. crassa*
Amino acid transporters						
High affinity methionine permease	AN1631/*mupA*	Afu4g09040	An04g03940	AO090023000646	Pc18g03480	NCU04942
						NCU07754
Methionine transporter	AN12483	Afu7g00730	An08g07770	AO090038000414	Pc12g00960	NCU02195
Methionine transporter	AN2560	Afu3g15020	An09g01090	AO090701000425	Pc12g07390	-
Neutral amino acids transporter	AN10825	Afu6g04990	An15g00330	AO090701000039	Pc12g04390	NCU03815/*nap-1*
	AN8538		An12g09900	AO090003001519		
Sulfate transporters						
Sulfate permease	AN2730/*sB*	Afu1g05020	An15g04600	AO090003000798	Pc16g10020/*SutB*	NCU03235/*cys-13*
	-				Pc21g19360/*SutA*	NCU04433/*cys-14*
	-				Pc12g01540	-
High affinity sulfate transporter 1, vacuolar	AN3157	Afu3g13470	An02g09350	AO090012000782	Pc13g15010	NCU09642
Sulfate transporter family protein	AN3665	Afu4g12440	An01g07860	AO090009000456	Pc22g05460	NCU02632
Sulfate assimilation						
ATP sulfurylase	AN4769/*sC*	Afu3g06530/*sC*	An11g09790/*sC*	AO090020000349/sC	Pc20g07710/MET3	NCU01985/*cys-11*
Adenylyl-sulfate kinase	AN1194/*sD*	Afu1g10820	An08g02990	AO090038000333	Pc13g08270/KAPS	NCU09896/*cys-1*
PAPS reductase	AN4770/*sA*	Afu3g06540/*met16*	An11g09780	AO090020000347	Pc20g03220/*ParA*	NCU02005/*cys-5*
Sulfite reductase α-subunit	AN1752	Afu6g08920	An11g06670/*met10*	AO090001000571	Pc12g08580	NCU04077/*cys-2*
Sulfite reductase β-subunit	AN7600	Afu2g15590	An15g05770	AO090012000271	Pc21g02880	NCU05238/*cys-4*

Table 7.1 Continued

	ORF ID/gene name					
Activity	*A. nidulans*	*A. fumigatus*	*A. niger*	*A. oryzae*	*P. rubens*	*N. crassa*
Cysteine synthesis						
Serine *O*-acetyltransferase	AN8565/*cysA*	Afu5g08600		AO090120000210	Pc22g16570	NCU00536
Homoserine *O*-acetyltransferase	AN2229/*metE*	Afu5g07210/*met2*	An17g00630	AO090701000235	Pc21g18210	NCU07001/*met-5*
Cysteine synthase	AN8057/*cysB*	Afu5g02180	An02g10750	AO090102000276	Pc21g14890	NCU06452/*cys-17*
Cysteine synthase	AN1513/*cysF*	Afu8g05200	An16g07520	AO090005000626	Pc20g10940	NCU02564/*cys-12*
Cysteine synthase with rhodanese domain	AN7113	Afu4g03930	An14g00960	AO090011000336	Pc13g05990 Pc20g05400	NCU03788
Transulfuration						
Cystathionine β-synthase	AN5820/*mecA*	Afu2g07620	An05g00160	AO090011000931	Pc13g05320	NCU08216/*met-11*
Cystathionine γ-lyase	AN1446/*mecB*	Afu8g04340	An16g08720	AO090103000051	Pc21g05430	NCU09230/*cys-16*
Homocysteine synthase	AN8277/*cysD*	Afu5g04250/*cysD*	An09g06710	AO090102000630	Pc12g05420	NCU01652
Cystathionine β-lyase	AN7051/*metG*	Afu4g03950	An14g00930	AO090011000323	Pc13g06020	NCU07987/*met-2*
Cystathionine γ-synthase	AN3456/*metB*	Afu3g05480	An11g11160	AO090020000019	Pc20g08350	NCU02430/*met-7*

Methionine metabolism						
Methylenetetrahydrofolate reductase 2	AN8215/*metA*	Afu3g05480	An11g11160	AO090020000019	Pc21g15910	NCU09545/*met-13*
Methylenetetrahydrofolate reductase 1	AN5883/*metF*	Afu2g11300	An02g03270	AO090026000520	Pc15g00480	NCU07690/*met-12*
Methionine synthase	AN4443/*metH*	Afu4g07360/*metH*	An04g01750/*met6*	AO090023000837	Pc22g18630	NCU06512/*met-8*
S-Adenosylmethionine synthase	AN1222/*mecC*	Afu1g10630	An08g02700/*eth1*	AO090038000357	Pc16g04380	NCU02657/*eth-1*
S-Adenosyl-L-homocysteine hydrolase	AN1263	Afu1g10130	An08g01960	AO090038000417	Pc16g05080	NCU07930/*cys-18*
Regulatory factors						
E3 ubiquitin ligase complex SCF, F-box protein	AN6359/*sconB*	Afu2g14110	An02g04800	AO090026000133	Pc12g12850	NCU08563/*scon-2*
E3 ubiquitin ligase complex SCF, Skp1 subunit	AN2302/*sconC*	Afu5g06060/*skpA*	EHA22420[1]	AO090009000653	ORF ID not assigned	NCU08991/*scon-3*
Positive sulfur transcription regulator/bZIP TF	AN4361/*metR*	Afu4g06530	EHA20381[1]	AO090023000937/*metR*	Pc22g17390	NCU03536/*cys-3*
Second sulfur transcription regulator/bZIP TF	AN5218/*metZ*	Afu6g07530	An07g08880	AO090005001668	Pc18g06160	–

– GenBank access numbers

The sulfate assimilation pathway provides sulfide for cysteine synthesis. It begins with sulfate permease, responsible for the uptake of sulfate, which is then activated and reduced to sulfide. In *A. nidulans*, a single sulfate permease from the SulP family is encoded by the *sB* gene (Arst, 1968; Pilsyk *et al.*, 2007). Mutations in this gene lead to auxotrophy which can be complemented, besides the *sB* gene, by another gene found in a genomic library constructed from a Japanese isolate of *A. nidulans*. This gene was designated *astA* (alternative sulfate transporter) and was found to be a pseudogene in *A. nidulans* laboratory strains of Glasgow origin (Pilsyk *et al.*, 2007). AstA represents a type of sulfate transporter, belonging to a large but poorly characterized family of allantoate permeases from the Major Facilitator Superfamily (Pilsyk *et al.*, 2007). The closest homologues of *astA* are found in evolutionarily distant fungi which are frequently pathogenic to plants and belong to the subphylum *Pezizomycotina*. The expression of *astA*, similarly to *sB*, is tightly regulated by sulfur sources in the growth medium (Pilsyk *et al.*, 2007). A similar regulation of the *astA* homologue has recently been observed in the plant pathogen *Fusarium sambucinum* (Piłsyk *et al.*, 2015a). The occurrence of *astA* orthologues is a unique feature of particular *Aspergillus* strains, *Neosartorya fischeri* (*A. fischerianus*), *A. sydowii* and *A. versicolor*, as other sequenced *Aspergillus* sp. have only one gene encoding sulfate permease, orthologous to SB. Thus, redundant sulfate uptake proteins seem an exception rather than a rule among aspergilli.

Sulfate is activated by ATP-sulfurylase and APS-kinase (Fig. 7.1, steps 2 and 3) and then reduced by sulfate reductase and sulfite reductase (Fig. 7.1, steps 4 and 5). Sulfide – the final product of the sulfate assimilation pathway – is incorporated into cysteine in a reaction with *O*-acetylserine (OAS) catalysed by cysteine synthase (encoded by the *cysB* gene, Fig. 7.1, step 7). Methionine is then synthesized from the cysteine *via* cystathionine and homocysteine (Fig. 7.1, steps 9, 10 and 17). The latter can also be synthesized directly from *O*-acetylhomoserine and sulfide in a reaction catalysed by homocysteine synthase (CysD) (Fig. 7.1, step 11). This enzyme takes part in the synthesis of sulfur amino acids in *Yarrowia lipolytica*, *Kluyveromyces lactis*, *Schizosaccharomyces pombe* and *Trichosporon cutaneum* (Brzywczy and Paszewski, 1993; Morzycka and Paszewski, 1979). Interestingly, the pathway involving homocysteine synthase is the only route of cysteine synthesis in *S. cerevisiae* (Cherest and Surdin-Kerjan, 1992). On the other hand, *Hansenula polymorpha*, which belongs to the *Saccharomycetaceae* family, is an exceptional species lacking homocysteine synthase (Sohn *et al.*, 2014).

Homocysteine is a substrate for synthesis of methionine but can also be converted to cysteine by the cystathionine β-synthase (CBS) (*mecA*) and cystathionine γ-lyase (CGL) (*mecB*) (Fig. 7.1, steps 12 and 13), thus, homocysteine synthase provides an alternative pathway for the synthesis of sulfur amino acids. CBS and CGL are elements of the reverse transsulfuration pathway leading from methionine to cysteine *via* *S*-adenosylmethionine and *S*-adenosylhomocysteine (Fig. 7.1, steps 18, 19 and 20). The alternative pathway of cysteine synthesis is normally not operating since *metB* or *metG* mutants are methionine auxotrophs. The physiological importance of this pathway is only manifested in mutants impaired in the sulfur metabolism repression system (SMR) (see below). Under these conditions, lesions in cystathionine synthesis and cleavage steps (*metB* and *metG* – Fig. 7.1, steps 9 and 10) are bypassed. *S. pombe* lacks both cystathionine β-synthase and cystathionine γ-lyase (Brzywczy and Paszewski, 1994) so it can synthesize cysteine only from OAS.

Since *A. nidulans cysD* mutants are prototrophs growing at the wild-type rate the role of homocysteine synthase under normal conditions is not clear. As *met* mutants, except *metE*, can grow on 5'-methylthioadenosine (a by-product of polyamine synthesis) or methylthiourea, it appears that homocysteine synthase could be involved in recycling methylthio groups for methionine synthesis. Brzywczy and coworkers (1993) showed that other thiols like mercaptoethanol, thioglycerol and thioglycolic acid are substrates for this enzyme, so it could possibly play a role in detoxification of mercaptans.

Methionine synthesis and folate metabolism are strictly linked as methyltetrahydrofolate (CH_3-THF) is the only methyl group donor for methionine synthase (MS), encoded by the *A. nidulans metH* gene. Two genes – *metA* and *metF* (Fig. 7.1, step 14) – encode isoforms of methylenetetrahydrofolate reductase (MTHFR) which provides methyltetrahydrofolate (CH_3-THF) (Sieńko *et al.*, 2007). The MetA protein is responsible for only 10–15% of total MTHFR activity but detailed kinetics of reactions carried out by the both MTHFR isozymes has not been determined. If such data were available they could explain why mutations impairing the SMR system are capable of suppressing mutations in *metA*, but not those in *metF* (Sieńko *et al.*, 2007).

Two mutations, *metH2* and *metH10*, affecting methionine synthase (MS) (Fig. 7.1, step 17) have been characterized by (Kacprzak *et al.*, 2003). Both mutants accumulate cystathionine and its higher homologue – homolanthionine (Paszewski and Grabski, 1975). *metH2* has a point mutation in the coding region while *metH10* has a mutation located at the 5'-splice site of the intron. The latter results in a low level of mature *metH* transcript (Kacprzak *et al.*, 2003), which may lead to a low level of MS as compared to the wild type strain. Interestingly, the growth defect of *metH10* is rescued not only by supplementation with methionine but also with choline or betaine. Since choline synthesis consumes a bulk of methionine methyl groups (Mudd and Datko, 1986) its supplementation reduces the demand for methionine leaving more of this amino acid for other metabolic requirements. Hence, choline shows a sparing effect. The growth stimulation by betaine was unexpected since it had earlier been assumed that choline could be converted to betaine but not vice versa. However, we found that *A. nidulans* makes choline from betaine both *in vivo* and *in vitro* (Kacprzak *et al.*, 2003).

Sulfur metabolite repression

Upon addition of a favourable reduced sulfur source like methionine to the growth medium the energy-consuming sulfate assimilation pathway is shut off almost completely. When radioactive sulfate and cold methionine are simultaneously present in the medium only traces of radioactivity are found in sulfur-containing compounds isolated from the mycelium (Natorff *et al.*, 1993; Paszewski *et al.*, 1994). Methionine supplementation simultaneously renders *A. nidulans* resistant to chromate and selenate – toxic analogues of sulfate, which are taken up by sulfate permease. This regulation, known as sulfur metabolite repression (SMR) (Paszewski *et al.*, 1994), was revealed by isolation of mutations suppressing lesions in the *metB* and *metG* genes affecting the cysteine-to-methionine pathway (Fig. 7.1, steps 9 and 10) (Fig. 7.2). These suppressors affecting the main pathway of cysteine synthesis (*cysA*, *cysB*) were isolated first by Gajewski and Litwinska (1968). The *cysA* and *cysB* genes could not be easily identified in another way since the *cysA* and *cysB* mutants are prototrophs utilizing the alternative pathway of cysteine synthesis. Thus, mutations in two genes are required to cause cysteine prototrophy, e.g. *cysA*, *mecA* double mutants can grow on *O*-acetylserine or cysteine, while *cysB*, *mecB* can only be rescued by supplementation with cysteine. Moreover,

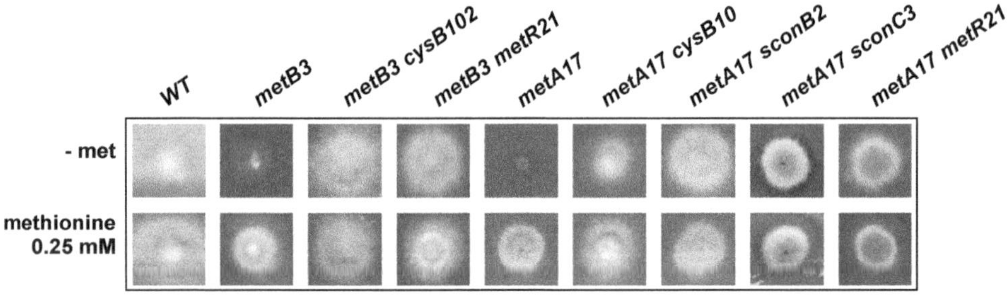

Figure 7.2 Examples of *cysB*, *sconB*, *sconC* and *metR* mutations suppressing growth defects of methionine mutants.

several enzymes are present in *A. nidulans* mycelia which exhibit cysteine synthase activity *in vitro* (Brzywczy *et al.*, 2007) (Table 7.1) so an impairment of cysteine synthase could not be easily identified by a biochemical method either.

The second class of *met* suppressors are *scon* genes (Fig. 7.2) (Natorff *et al.*, 1998; Piotrowska *et al.*, 2000) encoding components of the SCF complex (Skp1-Cullin-F-box) belonging to a large family of ubiquitin-protein ligases present in many organisms (for a review see Willems *et al.*, 2004). This complex comprises four constant components (SconC is one of them) and one variable one belonging to the WD40 protein family which determines the specificity of the complex towards the protein to be degraded or inactivated. One of the variable subunits is encoded by the *A. nidulans sconB* gene, so the corresponding complex is designated SCF^{SconB} (Fig. 7.3).

The SCF^{SconB} complex, the MetR transcription factor and cysteine constitute the sulfur metabolite repression system (SMR). Cysteine is a low molecular weight effector molecule activating the SCF^{SconB} complex, which negatively regulates the activity of MetR or targets it for degradation (Fig. 7.3). MetR is a positive regulator of transcription of some sulfur metabolism genes, in particular homocysteine synthase and those of the sulfate assimilation pathway (Natorff *et al.*, 2003). Thus, in the absence of MetR, these enzymes remain on a repressed level. They are also repressed in wild type *Aspergillus* grown in the presence of methionine due to its efficient conversion to cysteine by the reverse transsulfuration pathway (see Fig. 7.1). In effect, the level of cysteine rises about six-fold on methionine-supplemented media (Wróbel *et al.*, 2009). Notably, a much lower repression was observed in the *mecC* mutant partially impaired in *S*-adenosylmethionine synthetase so that synthesis of cysteine from methionine was limited (Paszewski and Grabski, 1974; Pieniążek *et al.*, 1973). Menant and coworkers (2006) demonstrated that cysteine but not methionine triggers degradation of Met4, which is an orthologue of MetR in yeast. If cysteine synthesis is blocked by mutation, e.g. *cysB*, such strain can make cysteine only by the less efficient alternative pathway and contains four times less of this amino acid than the wild type strain (Wróbel *et al.*, 2009). In effect, SCF^{SconB} is not active resulting in a constant induction of MetR-dependent genes and consequent activation of the alternative pathway of cysteine synthesis. The same situation takes place in the *sconB* and *sconC* mutants defective in the SCF^{SconB} complex. Since in these two mutants both pathways of cysteine synthesis are active these strains accumulate sulfur-containing compounds like glutathione, cysteine and *S*-adenosylmethionine (Natorff *et al.*, 1993; Wróbel *et al.*, 2009).

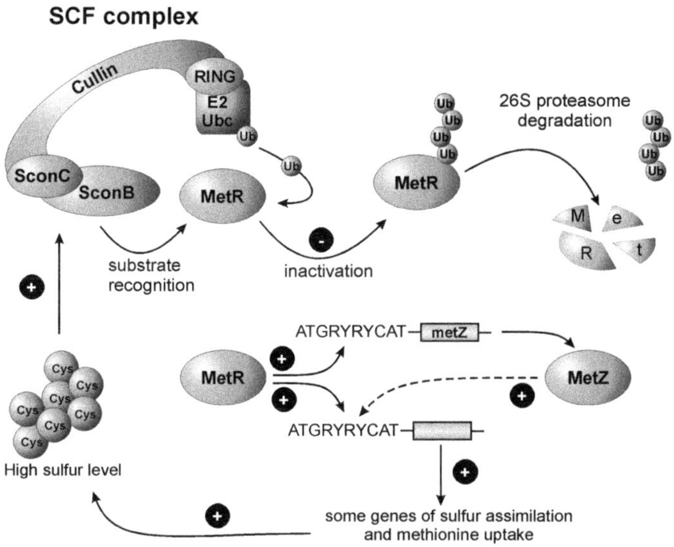

Figure 7.3 Model for the sulfur regulatory system in *A. nidulans*. SconB (F-box protein) and SconC (Skp1 orthologue) interact forming ubiquitin ligase complex (SCF^{SconB}). At high concentration of cysteine (a low-molecular-weight effector of this complex) SCF^{SconB} is responsible for ubiquitination and consequent proteasomal degradation of a positive sulfur transcription regulator – MetR (bZIP protein). Gene coding for a second sulfur transcription regulator (*metZ*) and *sconB* are also activated by MetR. Dashed line in the case of *metZ* indicates that its mode of action needs elucidation. SCF complex (Skp, Cullin, F-box containing complex) is a multi-protein E3 ubiquitin ligase complex.

In *Neurospora*, orthologues of *sconB* and *sconC* are encoded by the *scon-2* (Paietta, 1990) and *scon-3* (Sizemore and Paietta, 2002) genes, respectively. They negatively regulate the Cys-3 transcriptional activator, an orthologue of MetR. Thus, *scon-2* and *scon-3* mutations have the same regulatory effects as their *Aspergillus* counterparts. However, it has not been established yet, if these mutations can suppress lesions in genes encoding enzymes of the cysteine-to-homocysteine pathway, i.e. whether *Neurospora* can make cysteine and methionine by the alternative route.

Dysregulation of sulfur metabolism in cys and scon mutants induces stress responses

cysB, *sconB* and *sconC* mutations cause a major dysregulation of sulfur metabolism in the cell. Derepression of the sulfate assimilation pathway observed in these mutants (Fig. 7.1, steps 1, 2, 4 and 5) (Natorff *et al.*, 1993) leads to overproduction of sulfide with simultaneous dissipation of ATP and NADPH resulting in considerable changes in the pools of thiol compounds (Wróbel *et al.*, 2009). All these changes seemed likely to have strong direct and indirect global effects on gene expression and cellular metabolism. Since all three mutants are sulfur prototrophs they appeared particularly suitable for studying these effects since no supplementation of media with organic sulfur sources, which could influence the strains' metabolism, was needed.

Transcriptomic studies of the *cysB*, *sconB* and *sconC* mutants has revealed at least 2-fold up-regulation of several hundred transcripts as compared to the wild type strain (Sieńko *et al.*, 2014), many of them involved in carbohydrate and energy metabolism, along with those encoding enzymes of the glyoxylate bypass and the GABA shunt. The most striking finding among the up-regulated genes was a high occurrence of those typically related to environmental stress responses including heat and oxidative shocks and osmotic stress. Notably, the two-component signal transduction system, including stress receptor histidine kinases, is the functional category most enriched among genes up-regulated in the *cysB*, *sconB* and *sconC* mutants and this category. The activation of stress responses in these mutants must be evoked by indigenous agents of which sulfide appears to be a primary one since the transcript level of the gene encoding sulfide:quinone reductase (SQR) – an enzyme involved in sulfide detoxification – is over 2000-fold up-regulated in the mutants (Sieńko *et al.*, 2014). Interestingly, accumulation of SQR is also observed in a proteomic analysis of the *cysB*, *sconB* and *sconC* mutants (Sieńko *et al.*, unpublished). Besides SQR, other oxidoreductases, including those involved in oxidative stress response, also accumulated in the mutants studied supporting the hypothesis about induction of redox stress.

Transcription factors controlling sulfur metabolism

metR has been identified as the gene coding for a transcription factor needed for activation of several sulfur metabolism genes (Fig. 7.3) (Natorff *et al.*, 2003). *metR*-deficient mutants can grow on methionine-supplemented media, which indicates that this gene is not essential and the MetR protein may be specifically required for activation of sulfur metabolism genes. *metR* mutants show very low levels of arylsulfatase, sulfite reductase, ATP sulfurylase and homocysteine synthase (Natorff *et al.*, 2003). Interestingly, unlike other methionine auxotrophs, *metR* strains require supplementation of complete medium with additional methionine to grow. This points to an impairment of a permease specific for this amino acid, which then has to be taken up by a general amino acid permease, competing with other amino acids.

The *metR* gene is interrupted by a single 495-bp intron, unusually long for filamentous fungi (Natorff *et al.* 2003). An orthologue of *metR* in *N. crassa* is the well characterized *cys-3* gene encoding a positive activator of sulfur metabolism-related genes (Burton and Metzenberg, 1972). *Neurospora* genome sequencing has revealed a putative 5′ exon separated from the main part of the *cys-3* gene by a single 852-bp-long intron, similarly to its *Aspergillus* counterpart. The MetR and Cys-3 proteins have similar bZIP domains and a chimeric MetR/Cys-3-protein bearing the b-ZIP domain derived from Cys-3 can complement methionine auxotrophy of the *Aspergillus metR* mutant (Natorff *et al.*, 2003). These similarities suggest that both transcription factors may recognize similar binding sites in promoters of sulfur metabolism genes. Cys-3 has been demonstrated to recognize the consensus sequence ATGRYRYCAT (Li and Marzluf, 1996; Paietta, 2008) while the MetR binding site consensus has not been established experimentally though the complementation study mentioned above suggests that it should be similar to ATGRYRYCAT.

However, regulation of expression of the *metR* and *cys-3* genes differs considerably. Transcription of *metR* is not regulated by SMR and its transcript is abundant in *metR* point mutants even under repressive conditions. In contrast, the *N. crassa cys-3* gene requires the Cys-3 protein for efficient transcription (Fu and Marzluf, 1990).

Novel alleles of *metR* were isolated in a genetic screen for dominant suppressors in a diploid strain homozygous for the *metB* mutation (Brzywczy *et al.*, 2011). The *metR20, metR21* and *metR22* mutations are suppressors of *met* mutations (Fig. 7.2) showing elevated activities of sulfur metabolism enzymes along with the levels of their transcripts. These features indicate that these *metR* alleles constitute a new class of methionine suppressor mutations which appear to inactivate the SMR system possibly by making MetR resistant to Scon-dependent degradation. A similar phenotype was reported earlier for the *Neurospora cys-3* K105Q mutant having a prolonged lifetime of Cys-3 leading to its accumulation (Tao and Marzluf, 1998).

Brzywczy and coworkers (2011) also found that the *sG8* mutation (now named *metR18*), characterized previously as affecting sulfide reductase activity (Nadolska-Lutyk and Paszewski, 1988), is a frameshift affecting the C-terminal part of *metR*. Thus, *metR18* represents a third class of mutations affecting the *metR* gene, besides null mutants and the dominant suppressors described above. Its phenotype suggests that the C-terminal domain may be essential for activation of some sulfur metabolism-related genes like those encoding sulfite reductase subunits.

By applying a BLAST similarity search to the *A. nidulans* genome with the MetR amino acid sequence as a query we found a b-ZIP protein encoded by the AN5218 open reading frame showing 26% identity with the whole sequence of MetR and 77% identity within its b-ZIP domain (Piłsyk *et al.*, 2015b). The novel gene, designated *metZ*, has a single intron of 570 bp located in a similar position as that in the *metR* gene. Interestingly, orthologues of *metZ* can be found among fungi of the *Eurotiales* order only. The *A. nidulans metZ* gene, like the *Neurospora cys-3* (Tao and Marzluf, 1998), is strongly regulated by SMR and its expression depends on the MetR protein (Piłsyk *et al.*, 2015b). Since attempts to disrupt *metZ* were unsuccessful its physiological role was studied by fusing it to a MetR-independent promoter. Such construct was introduced to a *metR* deletion strain and the transformants obtained did not require supplementation with methionine on complete medium but they still required methionine or cysteine for growth on minimal medium. In addition, these transformants were sensitive to chromate (taken up by sulfate permease, whose expression is tightly dependent on MetR) even in the presence of methionine at a repressing concentration. This indicates that MetZ activates a gene encoding a sulfate permease as well as a permease specific for methionine (and possibly cysteine) which is not expressed in *metR* strains. Similarly, a methionine-specific permease is induced by Cys-3 in *Neurospora* (Burton and Metzenberg, 1972). Thus, overexpression of MetZ partially complements the *metR* mutant defects.

Transcriptomic analysis of this transformant has revealed that overexpression of *metZ* leads to elevated transcription of genes coding for permeases and enzymes incorporating sulfide into sulfur amino acids while transcription of other sulfate assimilation genes is decreased (Piłsyk *et al.*, 2015b). Thus, the MetZ protein may partially substitute for MetR in the activation of sulfur metabolism genes. These two transcription factors bearing almost identical b-ZIP domains are potentially able to form a heterodimer, which could contribute to a more precise control of sulfur metabolism in *Eurotiales*.

Other regulatory systems in sulfur metabolism

Growth of *A. nidulans mecA* mutants, which are impaired in cystathionine β-synthase and cannot metabolize homocysteine to cystathionine (Fig. 7.1), is strongly inhibited by methionine (Pieniążek *et al.*, 1973; Sieńko *et al.*, 2007). Such effects have also been observed in *cys4* mutants affecting cystathionine β-synthase in *S. cerevisiae* (Jakubowski, 1991). An excess of methionine is harmful to these mutants, because it is readily metabolized to homocysteine which can be converted further to homocysteine thiolactone in an error-editing reaction catalysed by methionyl-tRNA synthetase (Jakubowski, 2006). The highly reactive homocysteine thiolactone causes protein *N*-homocysteinylation through the formation of amide bonds with ε-amino groups of lysine, a modification that alters or impairs protein function. Homocysteine, like other thiols, becomes oxidized in the presence of molecular oxygen and transition metal catalysts, eventually leading to the formation of reactive oxygen species (ROS) including hydrogen peroxide. Additionally, in the presence of nitric oxide, the oxide anion can form a powerful oxidant peroxynitrite (Loscalzo, 1996). An elevated production of ROS induces oxidative stress and, consequently, the responses observed in *cys* and *scon* mutants.

One would expect *A. nidulans cysB* mutants to accumulate homocysteine because they have a derepressed sulfate assimilation pathway and higher availability of *O*-acetylhomoserine for homocysteine synthase than the wild type strain. However, even if homocysteine is overproduced it is readily converted to cystathionine and, to a lesser extent, further to cysteine (Paszewski and Grabski, 1974; Wróbel *et al.*, 2009). A low cysteine level ensures constant derepression of the sulfate assimilation pathway. Cystathionine is also accumulated in the *sconB* strain but neither this mutant nor *cysB* accumulate more homocysteine than wild type does – 0.28, 0.22 and 0.25 nmoles of homocysteine per mg protein in the wild type, *sconB* and *cysB* strains, respectively (Paszewski and Wróbel, unpublished results). Since *sconB* also accumulates *S*-adenosylmethionine (Natorff *et al.*, 1993) it appears that *A. nidulans* can effectively metabolize an excess of homocysteine. This detoxification can be achieved, at least in part, by homocysteine-mediated activation of genes encoding enzymes directly or indirectly involved in its metabolism: cystathionine b-synthase (*mecA*), cystathionine γ-lyase (*mecB*), *S*-adenosylmethionine synthase (*mecC*), two methylenetetrahydrofolate reductases (*metA, metF*), *S*-adenosylhomocysteine hydrolase (AN1263), and folylpolyglutamate synthase (AN4384) (Paszewski *et al.*, 2009; Sieńko *et al.*, 2009) (Fig. 7.1). This mechanism seems to lower the intracellular homocysteine level. We designated this coordinately regulated group of genes a 'homocysteine regulon' (Sieńko *et al.*, 2007) which represents a sulfur metabolism regulatory system different from SMR.

The *metE* gene encoding homoserine transacetylase appears to be regulated in still another way than the two regulatory systems described earlier since its transcript's level rises strongly upon methionine limitation (Grynberg *et al.*, 2001). Transcription of *metG* encoding cystathionine β-lyase is not appreciably regulated by sulfur sources in growth medium (Sieńko and Paszewski, 1999). *metB*, the second gene in the cysteine-to-homocysteine transsulfuration pathway (Fig. 7.1, step 9), seems not to be regulated by sulfur sources either, as indicated by transcriptomic analysis of the *cysB* and *scon* mutants (Sieńko *et al.*, 2014). Thus, some areas of sulfur metabolism, e.g. the main homocysteine biosynthesis pathway, are not subject to regulation.

Future trends

Despite the considerable knowledge of the sulfur metabolic pathways and their regulatory circuits, many more molecular aspects of both have to be elucidated. Several genes still lack a functional assignment. For instance, there is a puzzling cysteine synthase-like family of genes, which are conserved from *Aspergillus* through *S. cerevisiae* but their physiological role remains to be found.

The sulfur metabolic repression (SMR) system involving the SCF^{SconB} complex and MetR transcription factor acting in a hierarchical fashion is, at the moment, the best known sulfur regulatory system in *A. nidulans*. The exact mechanism of its functioning is not fully elucidated yet, but by analogy with the *S. cerevisiae* system, ubiquitination of MetR by SCF^{SconB} activated by cysteine, leading to MetR degradation, appears to be the most likely mode of sulfate assimilation repression. The interaction between MetR and the newly discovered MetZ transcription factor seems possible but has not been experimentally demonstrated yet. In general, the role of MetZ in the regulation of sulfur metabolism in *Eurotiales* is only beginning to be recognized. Its occurrence may be another significant difference between *Eurotiales* and *Sordariales* besides their different regulation of *metR* vs. *cys-3* expression. As stated in the preceding section, other sulfur regulatory circuits function in *A. nidulans* besides SMR, like the homocysteine regulon and the apparent regulation of *metE* gene expression by methionine, with unknown molecular mechanisms.

The availability of full genomic sequences for a growing number of fungi should facilitate resolving these and other issues in the genetics of fungal sulfur metabolism. Further proteomic and metabolomic analyses should provide a better insight into the integration of this metabolism with the rest of cellular processes.

Web resources

Aspergillus Genome Database (AspGD): http://www.aspergillusgenome.org/

Aspergillus Comparative Database from Broad Institute: http://www.broadinstitute.org/annotation/genome/aspergillus_group/MultiHome.html

Neurospora crassa Database from Broad Institute: http://www.broadinstitute.org/annotation/genome/neurospora/MultiHome.html

Penicillium chrysogenum (*P. rubens*) *v1.0* database from Department of Energy Joint Genome Institute (DOE JGI): http://genome.jgi.doe.gov/Pench1/Pench1.home.html

Saccharomyces Genome Database (SGD): http://www.yeastgenome.org/

References

Arst, H.N. Jr. (1968). Genetic analysis of the first steps of sulphate metabolism in *Aspergillus nidulans*. Nature *219*, 268–270.

Brzywczy, J., Kacprzak, M.M., and Paszewski, A. (2011). Novel mutations reveal two important regions in *Aspergillus nidulans* transcriptional activator MetR. Fungal Genet. Biol. *48*, 104–112.

Brzywczy, J., Natorff, R., Sieńko, M., and Paszewski, A. (2007). Multiple fungal enzymes possess cysteine synthase activity *in vitro*. Res. Microbiol. *158*, 428–436.

Brzywczy, J., and Paszewski, A. (1993). Role of O-acetylhomoserine sulfhydrylase in sulfur amino acid synthesis in various yeasts. Yeast *9*, 1335–1342.

Brzywczy, J., and Paszewski, A. (1994). Sulfur amino acid metabolism in *Schizosaccharomyces pombe*: occurrence of two O-acetylhomoserine sulfhydrylases and the lack of the reverse transsulfuration pathway. FEMS Microbiol. Lett. *121*, 171–174.

Brywczy, J., Yamagata, S., and Paszewski, A. (1993). Comparative studies in *O*-actylhomoserine sulfhydrylase: physiological role and characterization of the *Aspergillus nidulans* enzyme. Acta Biochim. Pol. *40*, 421–428.

Burton, E.G., and Metzenberg, R.L. (1972). Novel mutation causing derepression of several enzymes of sulfur metabolism in *Neurospora crassa*. J. Bacteriol. *109*, 140–151.

Cherest, H., and Surdin-Kerjan, Y. (1992). Genetic analysis of a new mutation conferring cysteine auxotrophy in *Saccharomyces cerevisiae*: updating of the sulfur metabolism pathway. Genetics *130*, 51–58.

Fu, Y.H., and Marzluf, G.A. (1990). *cys-3*, the positive-acting sulfur regulatory gene of *Neurospora crassa*, encodes a sequence-specific DNA-binding protein. J. Biol. Chem. *265*, 11942–11947.

Gajewski, W., and Litwińska, J. (1968). Methionine loci and their suppressors in *Aspergillus nidulans*. Mol Gen Genet. *102*, 210–220.

Grynberg, M., Piotrowska, M., Pizzinini, E., Turner, G., and Paszewski, A. (2001). The *Aspergillus nidulans metE* gene is regulated by a second system independent from sulphur metabolite repression. Biochim. Biophys. Acta *1519*, 78–84.

Jakubowski, H. (1991). Proofreading *in vivo*: editing of homocysteine by methionyl-tRNA synthetase in the yeast *Saccharomyces cerevisiae*. EMBO J. *10*, 593–598.

Jakubowski, H. (2006). Pathophysiological consequences of homocysteine excess. J. Nutr. *136*, 1741S-1749S.

Kacprzak, M.M., Lewandowska, I., Matthews, R.G., and Paszewski, A. (2003). Transcriptional regulation of methionine synthase by homocysteine and choline in *Aspergillus nidulans*. Biochem. J. *376*, 517–524.

Li, Q., and Marzluf, G.A. (1996). Determination of the *Neurospora crassa* CYS 3 sulfur regulatory protein consensus DNA-binding site: amino-acid substitutions in the CYS3 bZIP domain that alter DNA-binding specificity. Curr. Genet. *30*, 298–304.

Loscalzo, J. (1996). The oxidant stress of hyperhomocyst(e)inemia. J. Clin. Invest. *98*, 5–7.

Marzluf, G.A. (1997). Molecular genetics of sulfur assimilation in filamentous fungi and yeast. Annu. Rev. Microbiol. *51*, 73–96.

Menant, A., Baudouin-Cornu, P., Peyraud, C., Tyers, M., and Thomas, D. (2006). Determinants of the ubiquitin-mediated degradation of the Met4 transcription factor. J. Biol. Chem. *281*, 11744–11754.

Morzycka, E., and Paszewski, A. (1979). Two pathways of cysteine biosynthesis in *Saccharomycopsis lipolytica*. FEBS Lett. *101*, 97–100.

Mudd, S.H., and Datko, A.H. (1986). Methionine methyl group metabolism in lemna. Plant Physiol. *81*, 103–114.

Nadolska-Lutyk, J., and Paszewski, A. (1988). A new gene controlling sulphite reductase in *Aspergillus nidulans*. Genet. Res. *51*, 1–3.

Natorff, R., Balińska, M., and Paszewski, A. (1993). At least four regulatory genes control sulphur metabolite repression in *Aspergillus nidulans*. Mol. Gen. Genet. *238*, 185–192.

Natorff, R., Piotrowska, M., and Paszewski, A. (1998). The *Aspergillus nidulans* sulphur regulatory gene *sconB* encodes a protein with WD40 repeats and an F-box. Mol. Gen. Genet. *257*, 255–263.

Natorff, R., Sieńko, M., Brzywczy, J., and Paszewski, A. (2003). The *Aspergillus nidulans metR* gene encodes a bZIP protein which activates transcription of sulphur metabolism genes. Mol. Microbiol. *49*, 1081–1094.

Paietta, J.V. (1990). Molecular cloning and analysis of the *scon-2* negative regulatory gene of *Neurospora crassa*. Mol. Cell. Biol. *10*, 5207–5214.

Paietta, J.V. (2008). DNA-binding specificity of the CYS3 transcription factor of *Neurospora crassa* defined by binding-site selection. Fungal Genet. Biol. *45*, 1166–1171.

Paszewski, A., Brzywczy, J., and Natorff, R. (1994). Sulphur metabolism. Prog. Ind. Microbiol. *29*, 299–319.

Paszewski, A., and Grabski, J. (1974). Regulation of *S*-amino acids biosynthesis in *Aspergillus nidulans*. Role of cysteine and-or homocysteine as regulatory effectors. Mol. Gen. Genet. *132*, 307–320.

Paszewski, A., and Grabski, J. (1975). Homolanthionine in fungi: accumulation in the methionine-requiring mutants of *Aspergillus nidulans*. Acta Biochim. Pol. *22*, 263–268.

Paszewski, A., Natorff, R., Piotrowska, M., Brzywczy, J., Sieńko, M., Grynberg, M., Pizzinini, E., and Turner, G. (2000). Regulation of sulfur aminoacid biosynthesis in *Aspergillus nidulans*: physiological and genetical aspects. In Sulfur Nutrition and Sulfur Assimilation in Higher Plants, C. Brunold, H. Rennenberg, L.J. De Kok, I. Stulen, and J.-C. Davidian, eds. (Paul Haupt, Bern, Switzerland), pp. 93–105.

Paszewski, A., Natorff, R., Sieńko, M., Brzywczy, J., and Lewandowska, I. (2009). Sulfur metabolism in *Aspergillus nidulans*: pathways and regulation. In Sulfur Metabolism in Plants, A. Sirko, L.J. De Kok, S. Haneklaus, M.J. Hawkesford, H. Rennenberg, K. Saito, E. Schnug, and I. Stulen, eds. (Backhuys Publishers, Leiden, the Netherlands), pp. 279–283.

Pieniążek, N., Stepień, P.P., and Paszewski, A. (1973). An *Aspergillus nidulans* mutant lacking cystathionine -synthase: identity of L-serine sulfhydrylase with cystathionine -synthase and its distinctness from *O*-acetyl-L-serine sulfhydrylase. Biochim. Biophys. Acta *297*, 37–47.

Pilsyk, S., Natorff, R., Sienko, M., and Paszewski, A. (2007). Sulfate transport in *Aspergillus nidulans*: A novel gene encoding alternative sulfate transporter. Fungal Genet. Biol. *44*, 715–725.

Piłsyk, S., Natorff, R., Gawińska-Urbanowicz, H., and Kruszewska, J.S. (2015a). *Fusarium sambucinum astA* gene expressed during potato infection is a functional orthologue of *Aspergillus nidulans astA*. Fungal Biol. *119*, 509–517.

Piłsyk, S., Natorff, R., Sieńko, M., Skoneczny, M., Paszewski, A., and Brzywczy, J. (2015b). The *Aspergillus nidulans metZ* gene encodes a transcription factor involved in regulation of sulfur metabolism in this fungus and other *Eurotiales*. Curr. Genet. *61*, 115–125.

Piotrowska, M., Natorff, R., and Paszewski, A. (2000). *sconC*, a gene involved in the regulation of sulphur metabolism in *Aspergillus nidulans*, belongs to the SKP1 gene family. Mol. Gen. Genet. *264*, 276–282.

Sieńko, M., Natorff, R., Owczarek, S., Olewiecki, I., and Paszewski, A. (2009). *Aspergillus nidulans* genes encoding reverse transsulfuration enzymes belong to homocysteine regulon. Curr. Genet. *55*, 561–570.

Sieńko, M., Natorff, R., Skoneczny, M., Kruszewska, J., Paszewski, A., and Brzywczy, J. (2014). Regulatory mutations affecting sulfur metabolism induce environmental stress response in *Aspergillus nidulans*. Fungal Genet. Biol. *65*, 37–47.

Sieńko, M., Natorff, R., Zielinski, Z., Hejduk, A., and Paszewski, A. (2007). Two *Aspergillus nidulans* genes encoding methylenetetrahydrofolate reductases are up-regulated by homocysteine. Fungal Genet. Biol. *44*, 691–700.

Sieńko, M., and Paszewski, A. (1999). The *metG* gene of *Aspergillus nidulans* encoding cystathionine beta-lyase: cloning and analysis. Curr. Genet. *35*, 638–646.

Sizemore, S.T., and Paietta, J.V. (2002). Cloning and characterization of *scon-3+*, a new member of the *Neurospora crassa* sulfur regulatory system. Eukaryot. Cell *1*, 875–883.

Sohn, M.J., Yoo, S.J., Oh, D.B., Kwon, O., Lee, S.Y., Sibirny, A.A., and Kang, H.A. (2014). Novel cysteine-centered sulfur metabolic pathway in the thermotolerant methylotrophic yeast *Hansenula polymorpha*. PLoS One *9*, e100725.

Tao, Y., and Marzluf, G.A. (1998). Synthesis and differential turnover of the CYS3 regulatory protein of *Neurospora crassa* are subject to sulfur control. J. Bacteriol. *180*, 478–482.

Willems, A.R., Schwab, M., and Tyers, M. (2004). A hitchhiker's guide to the cullin ubiquitin ligases: SCF and its kin. Biochim. Biophys. Acta *1695*, 133–170.

Wróbel, M., Lewandowska, I., Bronowicka-Adamska, P., and Paszewski, A. (2009). The level of sulfane sulfur in the fungus *Aspergillus nidulans* wild type and mutant strains. Amino Acids *37*, 565–571.

Production of Feruloyl Esterases by *Aspergillus* Species

8

Miia R. Mäkelä, Luis Alexis Jiménez Barboza,
Ronald P. de Vries and Kristiina S. Hildén

Abstract

Ferulic acid esterases are enzymes capable of releasing ester-linked hydroxycinnamic acids from certain polysaccharide moieties. These carbohydrate polymers, xylan and pectin, are ubiquitous in nature, and are mostly found as major constituents of plant cell walls. Plant cell walls are the most common renewable carbon source on earth. They encompass many different types of biopolymer complexes harnessed together such as cellulose, hemicelluloses, pectin, lignin and proteins. Together these compounds form lignocellulose, which is one of the most resilient materials found in nature.

Ferulic acid is one of the possible linkages between plant cell wall polymers and therefore is a major factor in the recalcitrance of cell wall against microbial attack. Saprobic plant cell wall-degrading fungi therefore produce feruloyl esterases that are capable of removing ferulic acid from the polysaccharides in order to weaken the integrity of the cell wall.

In this study we evaluated the variation in feruloyl esterase production of a set of *Aspergillus* species, a fungal genus with many industrial applications, during growth on plant biomass in the presence and absence of ferulic acid.

Introduction

Hydroxycinnamic acid residues in plant cell walls

Plant cell walls contain different amounts of hydroxycinnamic acid derivatives such as ferulic acid, *p*-coumaric acid and sinapic acid, usually esterified or etherified to polymeric components within the lignocellulose matrix (Ishii, 1997). They are found in both the primary and secondary cell walls (Carnachan and Harris, 2000; Ishii, 1997), mainly in xylem vessels, epidermis and sclerenchyma (Buanafina, 2009), and are involved in linking cell wall polymers. The interaction of polymeric components in a typical secondary cell wall of a gramineous species is shown in Fig. 8.1.

Ferulates and diferulates are important structural components of graminaceous cell walls, where the amount of ferulic acid can reach up to 3% of the cell wall dry weight (Kroon *et al.*, 1999). In arabinoxylans and glucuronoarabinoxylans, hydroxycinnamic acid residues (mostly feruloyl and *p*-coumaroyl) are O-5-linked to a usually terminal α-L-arabinose side chain (de Vries and Visser, 2001). The covalent ester linkage is formed between the carboxylic acid of the phenolic compound and the hydroxyl group of the sugar. Furthermore,

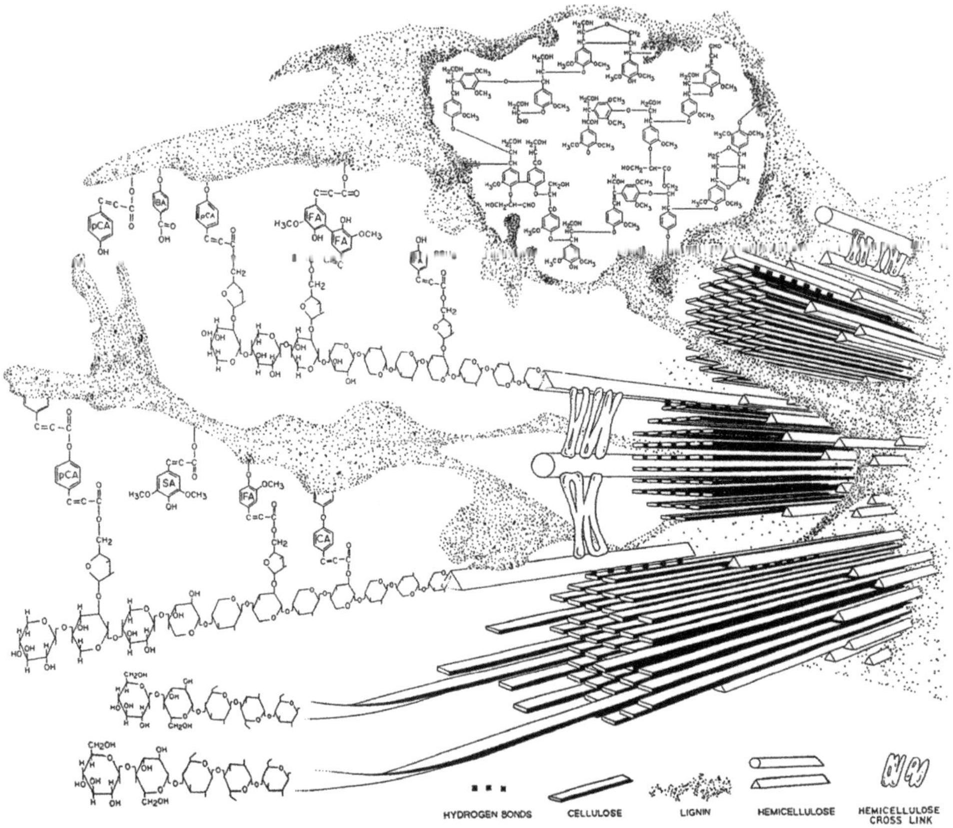

Figure 8.1 Interaction of different components of lignocellulose in a typical secondary cell wall of a gramineous species. FA, ferulic acid; CA, caffeic acid; pCA, *p*-coumaric acid; SA, sinapic acid. Reprinted and modified from Bidlack *et al.* (1992) with permission from the Oklahoma Academy of Science.

oxidative coupling carried out by hydrogen peroxide and peroxidases can create an array of dehydrodiferulate dimers (Fig. 8.2). Diferulates in arabinoxylans have mostly been detected in areas with high arabinose substitutions whereas monomeric ferulates prefer regions where arabinose polymers are scarce. According to Waldron and Faulds (2007), this can be explained by the fact that diferulate bridges require the molecule to be flexible and coiled in order to allow both ferulates to be physically in close proximity. In addition, Ishii (1997) suggested that feruloylated xyloglucan, like arabinoxylan, could perform this type of cross-linkages with other ferulate-linked polysaccharides.

Even though ferulate monomers are more common than diferulates (Buanafina, 2009), Fry (2004) proposed that this type of phenolic coupling has been underestimated and not restricted to dimers, but also include trimers and probably higher oligoferulates that can sometimes be more prevalent. Moreover, oxidative coupling of different hydroxycinnamic acids is a feasible possibility, as well as the existence of joint ferulate–*p*-coumarate dimers within the plant cell wall (Fry, 2004). Additionally, sinapate–ferulate dimers have been reported among other new ferulate dehydrodimers and trimers (Ralph *et al.*, 2004). This

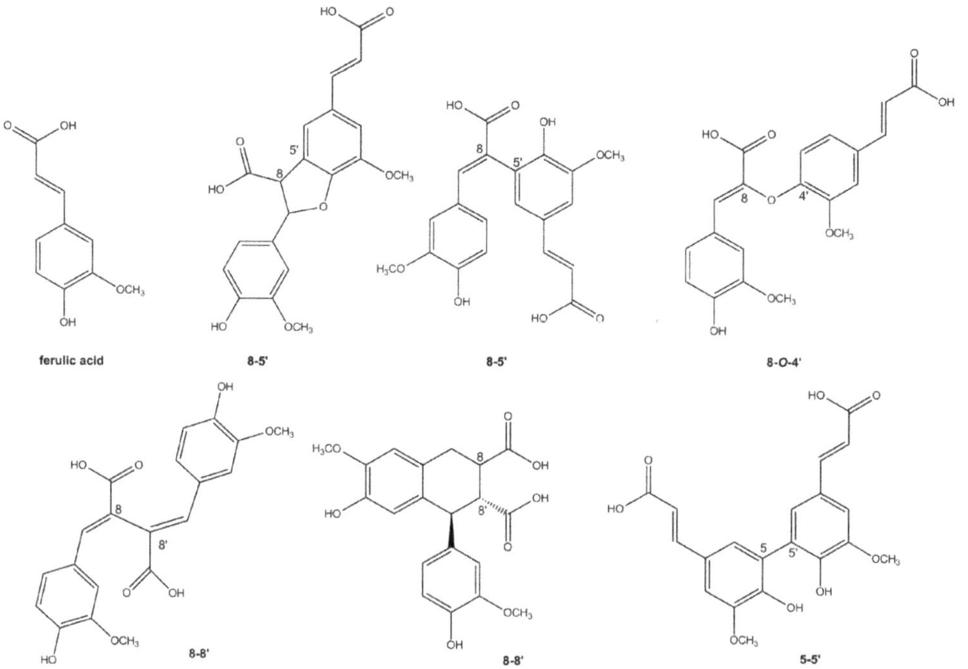

Figure 8.2 Ferulic acid and diferulic acid structures identified in plant cell walls. Based on (Mäkelä *et al.*, 2015).

exemplifies the extent of ferulate-coupling reactions as major cross-linking agents in plant cell walls.

Lignin can be connected to ferulate residues through ether linkage between the hydroxyl group of a ferulate and a phenolic lignin monomer (Buanafina, 2009). As bifunctional molecules, with carboxylic acid and hydroxyl groups, ferulates can form both ether and ester linkages, enabling the creation of ferulate–polysaccharide–lignin complexes (Buanafina, 2009; de Vries and Visser, 2001; Kondo *et al.*, 1990; Sun *et al.*, 2002). Moreover, Sun *et al.* (2002) reported for the first time that ferulic acid is able to connect different lignin fragments via ester and/or ether bridges.

In dicots, ferulic acid is esterified mostly to pectic polysaccharide chains, specifically, through O-6-linkage to a galactan residue and through O-2-linkage to an arabinose residue (de Vries and Visser, 2001).

Even though the existence of cinnamic acids crosslinking in cell walls of gymnosperms has been unclear, Carnachan and Harris (2000) showed the existence of hydroxycinnamic acids such as ferulic and *p*-coumaric acid in cell walls of 41 gymnosperm species from Pinaceae, Taxodiaceae, Cupressaceae, Araucariaceae and Taxaceae families. These species represent all four existing classes of gymnosperms (Carnachan and Harris, 2000).

Feruloylated plant cell wall polysaccharides have a tighter structure than non-feruloylated polysaccharides, presumably due to the cross-linkages that they create with other polymeric constituents (Waldron and Faulds, 2007). Moreover, cell wall digestibility by microorganisms appears to be dependent on the degree of feruloylation. Plant cell walls are more resistant to microbial degradation when the amount of ferulate cross-linkages increases (de

Vries and Visser, 2001; Grabber *et al.*, 1998). Finally, it has been shown that the number of feruloyl polysaccharide ester linkages increases as the cell matures and therefore constrains cell extensibility and growth (Ishii, 1997).

Classification of feruloyl esterases (FAEs)

Feruloyl esterase (FAE) is the commonly used name for esterases releasing hydroxycinnamic acids from plant cell walls. Other names that are or have been used for such enzymes are ferulic acid esterases, cinnamoyl esterases or (hydroxy)cinnamic acid esterases (Benoit *et al.*, 2008). FAEs catalyse the hydrolysis of the carboxylic ester linkages found between various hydroxycinnamic acids and plant cell wall polymers, including some of the diferulic acid linkages that are present in plant cell walls (de Vries and Visser, 2001). Therefore FAEs are considered to be important accessory enzymes produced by fungi to aid in the biodegradation of plant cell wall polymers.

A common feature of all known FAEs is the conserved amino acid catalytic triad serine–histidine–aspartic acid located in their active site. However, differences in adjacent amino acid sequences forming surface loops and additional domains grant FAEs with the ability to accommodate and process distinct aromatic substrates with different affinities (Udatha *et al.*, 2012).

FAEs are a subclass of carboxylic acid esterases with the enzyme commission (EC) number 3.1.1.73. However, beyond this the classification of this group of enzymes has not been unambiguous. FAEs feature both catalytic and substrate promiscuity and therefore conventional classification approaches fail to indicate their broad mode of action (Udatha *et al.*, 2012). Most enzymes involved in plant biomass degradation have been assigned to specific families of the Carbohydrate Active Enzyme database (CAZy, www.cazy.org) (Lombard *et al.*, 2014). While some FAEs can be found in Carbohydrate Esterase (CE) family 1, this is only a small subset of the FAEs characterized so far, and e.g. two of the best-characterized FAEs, FaeA and FaeB from *Aspergillus niger* (Benoit *et al.*, 2006a; de Vries *et al.*, 1997, 1999a, 2000, 2002; de Vries and Visser, 1999; Faulds and Williamson, 1994, 1995; Kroon *et al.*, 1996, 1997), are not members of this family. Several classifications of FAEs have been published over the years, and will be summarized here.

One FAE classification is based on their capacity to hydrolyse synthetic methyl ester model substrates, i.e. methyl caffeate (MC), methyl ferulate (MF), methyl *p*-coumarate (M*p*C) and methyl sinapate (MS). Crepin *et al.* (2004) proposed a classification where FAEs were organized into four functional sub-families (A, B, C and D) based on their primary amino acid sequence, their specificity for model hydroxycinnamic acid methyl esters and their ability to release oligoferulates from complex plant cell wall substrates.

However, this classification did not take into account all the available data on their activity against natural substrates. For instance, *A. niger* FaeA and FaeB were considered to be xylan- and pectin-specific, respectively (Crepin *et al.*, 2004), while other studies demonstrated that these esterases can remove ferulic acid from both xylan and pectin (de Vries *et al.*, 2000, 2002). A more refined classification of FAEs was reported by Benoit *et al.* (2008), acknowledging phylogenetic analysis from available fungal genomes. Within this scheme, seven sub-families of FAEs were proposed (SF1–7) containing only three biochemically characterized types (A, B and C), but excluding the type D FAEs. This classification also clarifies the large differences between the different FAE subfamilies as they appear to have no common ancestor, but have derived from various other types of esterases. *A. niger* FaeA

Table 8.1 Candidate feruloyl esterases in six *Aspergillus* species (based on Benoit *et al.*, 2015)

Species	SF1	SF2	SF3	SF4	SF5	SF6	SF7
A. nidulans	1	0	0	0	1	0	0
A. niger	1	0	0	2	1	0	1
A. terreus	2	0	1	0	3	0	1
A. oryzae	2	0	2	1	3	0	1
A. clavatus	1	0	0	0	4	0	0
A. fischeri	2	0	0	0	2	0	0

SF = subfamily.

and FaeB are related to lipases and tannases, respectively, while CE1 FAEs are related to acetyl xylan esterases (Benoit *et al.*, 2008).

The availability of fungal genome sequences also enables a more detailed comparison of their diversity with respect to the presence of genes encoding specific enzymes, such as FAEs. The genus *Aspergillus* is a particularly interesting group to evaluate, due to the history of use of FAEs (see above), an already considerable set of species with a genome sequence (Benoit *et al.*, 2013), and the ongoing effort to genome sequence all species of this genus.

An earlier study identified candidate FAEs in the genomes of *A. niger*. *A. nidulans* and *A. oryzae* (Coutinho *et al.*, 2009). Based on these data, more recently the diversity of eight *Aspergillus* species with respect to plant biomass degradation related genes was evaluated (Benoit *et al.*, 2015). This revealed a significant diversity in candidate FAE encoding genes, based on the subfamily classification of Benoit *et al.* (2008) (Table 8.1).

Biotechnological applications of FAEs

Hydroxycinnamates released by FAEs are regarded as potentially important bioactive components of the human diet. They possess important antioxidative (Yuan *et al.*, 2005) and anti-inflammatory properties (Murakami *et al.*, 2002). Ferulic acid can remove free radicals and reactive oxygen species from cells exposed to air and UV radiation, providing additional benefits against cancer and age-related affections (Graf, 1992). Similar beneficial features were demonstrated for caffeic acid, *p*-coumaric acid and sinapic acid (Kikuzaki *et al.*, 2002).

By applying FAE-based technologies, high-value phenolic compounds can potentially be obtained from by-products of agricultural and food industry. Significant amounts of ferulic acid and *p*-coumaric acid were released when steam exploded wheat straw was treated with *A. niger* FaeA and FaeB (Benoit *et al.*, 2008). Coffee industry residues are difficult to dispose of (Pandey *et al.*, 2000), yet they contain high amounts of phenolic compounds wherein chlorogenic acid was found to be the main constituent (42.2%) of coffee pulp (Ramirez, 1988). Hydrolysis experiments utilizing *A. niger* FaeB successfully released 100% and 73% of the chlorogenic acid and *p*-coumaric acid from coffee pulp, respectively (Benoit *et al.*, 2006b). Similar results were reported from other cinnamate-rich by-products such as apple marc (Benoit *et al.*, 2006b). Moreover, ferulic acid is an important flavour precursor as it can be converted to vanillin (4-hydroxy-3-methoxybenzaldehyde), one of the most used flavours in the food industry and the main component of vanilla. Ferulic acid for vanillin production can be obtained by enzymatic hydrolysis using FAEs from agricultural residues such as wheat bran (Di Gioia *et al.*, 2007), sugar beet pulp, rice bran oil (Zheng *et al.*, 2007) and other by-products.

FAEs have a positive synergistic effect on saccharification of plant biomass (e.g. for biofuel production) when combined with other hydrolases. For example, Yu *et al.* (2003) demonstrated how the combination of *A. niger* FaeA and *Trichoderma reesei* cellulases and xylanases practically doubled the release of reducing sugars from grinded oat hulls compared to the cellulase treatment alone. Similarly, Tabka *et al.* (2006) reported a significant enhancement in the enzymatic hydrolysis of steam exploded wheat straw when *A. niger* FaeA, and *T. reesei* xylanases and cellulases were used in conjunction, yielding more than 80% of the maximum amount of glucose recovery in the hydrolysate. A more detailed study on synergy in plant biomass degradation demonstrated the positive effects of FAE on release of all monomeric plant biomass components (de Vries *et al.*, 2000).

The kappa number is used to measure the amount of chemicals required during the pulp bleaching. Efficient delignification and a low kappa number for wheat straw pulp was reported by Record *et al.* (2003) using the homologously overexpressed FaeA from *A. niger* combined with xylanase and laccase treatment. Moreover, comparing alternative enzymatic-bleaching treatments with a subsequent peroxide stage, Sigoillot *et al.* (2005) reported a more significant kappa number reduction and a higher percentage of ISO brightness of flax pulp fibres using *A. niger* FAE compared to the xylanase treatment. Likewise, using *A. niger* FaeA in a totally chlorine-free bleaching process of oilseed flax straw, Tapin *et al.* (2006) achieved not only a low kappa number and high pulp brightness, but also recovery of phenolic compounds including ferulic acid, *p*-coumaric acid, vanillic acid and vanillin.

Induction of the production of FAEs

Production of FAEs was observed in cultures with ferulic acid containing plant biomass and free ferulic acid was demonstrated to have a positive influence on the level of FAE activity (Faulds *et al.*, 1997). More detailed studies into the expression of the genes encoding FaeA and FaeB from *A. niger* provided information about the molecular mechanism underlying FAE production. Expression of *faeA* is controlled by (at least) three independent regulatory systems (de Vries and Visser, 1999). It is one of the genes regulated by the (hemi-)cellulolytic transcriptional activator XlnR (van Peij *et al.*, 1998), while it is also induced by the presence of ferulic acid independently of XlnR (de Vries and Visser, 1999), indicating the presence a ferulic acid-responsive transcriptional regulator. Finally, the carbon catabolite repressor protein CreA suppresses the expression of *faeA* in the presence of monomeric sugars (de Vries and Visser, 1999), in a concentration dependent manner (de Vries *et al.*, 1999b).

Expression of *faeB* is also repressed by CreA, but is not activated by XlnR (de Vries *et al.*, 2002). Both *faeA* and *faeB* are expressed in the presence of ferulic acid, but it is not clear if this is mediated by the same regulatory system. The other aromatic compounds that induce expression of *faeA* or *faeB* differ, suggesting either specific modulation of their expression, or the presence of two independent regulatory systems (de Vries *et al.*, 2002). Interestingly, there is a direct correlation between the aromatic compounds that induce the expression of the two genes (de Vries *et al.*, 2002) and the methyl esters of these compounds the corresponding enzymes are active on (Kroon *et al.*, 1997). This suggests a highly fine-tuned regulatory system that responds specifically to the presence of real substrates of these two enzymes.

In this study we analysed the production of FAE activity of 14 aspergilli during growth on wheat bran and sugar beet pulp, two plant biomass substrates that both contain ferulic acid, but linked to xylan and pectin, respectively. In addition, we explored the influence of adding free ferulic acid to the culture on the level of FAE activity. With the ongoing genome sequencing efforts of the *Aspergillus* genus, these data will provide a biological data set that can be used to query genomic differences in FAE encoding genes.

Materials and methods

Fungal strains and cultivation

Eight *Aspergillus* strains (*A. niger* CBS513.88, *A. niger* ATCC1015, *A. niger* N402, *A. clavatus* CBS513.65, *A. fischeri* CBS544.65, *A. nidulans* FGSCA4, *A. oryzae* Rib40, *A. terreus* NIH2624), representing six different species, were obtained from CBS-KNAW Fungal Biodiversity Centre, the Netherlands). The strains were maintained on malt extract (2% w/v) agar (2% w/v) plates and stored at 4°C. To induce spore formation, strains were grown on malt agar plates at 30°C in the presence of light.

Fungal strains were grown in 250 ml Erlenmeyer flasks in 50 ml liquid minimal medium (de Vries *et al.*, 2004), which contained 0.1% (v/v) of Tween 80 and 2% (w/v) of either wheat bran (WB) or sugar beet pulp (SBP) with or without 0.13 mM ferulic acid (FA) supplementation.

Liquid cultures were inoculated with 10^6 spores/ml and incubated at 30°C with agitation (250 rpm). Each *Aspergillus* strain was cultivated as three biological replicates.

Culture filtrate samples were taken on day 3 and used for enzyme assays. Feruloyl esterase activities were determined spectrophotometrically at 37°C in 100 mM 3-(*N*-morpholino)propanesulfonic acid (MOPS, Sigma-Aldrich) buffer (pH 6.0) using methyl caffeate (MC), methyl ferulate (MF), methyl *p*-coumarate (M*p*C) and methyl sinapate (MS) as substrates (APIN Chemicals Ltd). Absorbance was monitored for 5 min at 322 nm for MC (ε_{322} = 14,720/M/cm), 320 nm for MF (ε_{320} = 29,680/M/cm), 308 nm for M*p*C (ε_{308} = 20,390/M/cm) and MS (ε_{320} = 15,890/M/cm).

Results

FAE activity during growth on wheat bran and sugar beet pulp

FAE activity profiles of eight *Aspergillus* strains varied strongly in the studied liquid cultures (Fig. 8.3). Differences were observed regarding the carbon source used as well as the substrate that was hydrolysed in the activity assays. In general, the studied *Aspergillus* strains produced higher FAE activities in the WB than in the SBP cultures.

The highest FAE activities were detected in the WB cultures of *A. nidulans* and they showed activity against MF, M*p*C and MC (Fig. 8.3A). When compared to SBP cultures, *A. nidulans* produced three-fold and two-fold higher activity against MF and MC, respectively, in the WB cultures. Also a more diverse set of FAE activities was detected during the growth of *A. nidulans* in WB as no activity against M*p*C was detected in the SBP cultures (Fig. 8.3B). FA supplementation increased activity against M*p*C, but had adverse effect with MF and MC. However, no activity against MS was observed in the *A. nidulans* cultures.

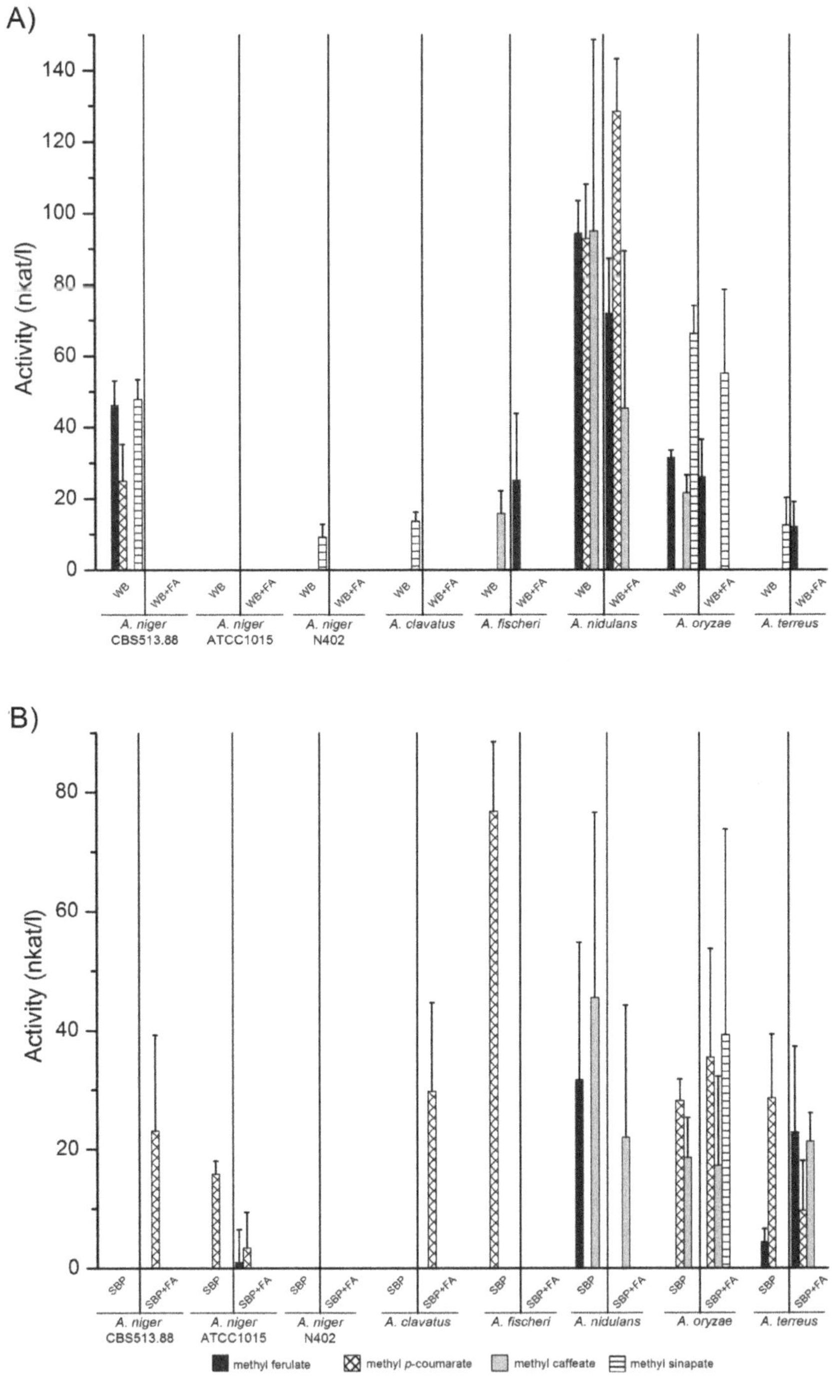

Figure 8.3 Activities produced against four different methyl ester model substrates (methyl ferulate, methyl *p*-coumarate, methyl caffeate and methyl sinapate) by eight *Aspergillus* strains. The activities were detected after three days of growth in liquid cultures supplemented with (A) 2% wheat bran (WB) or 2% wheat bran and 0.13 mM ferulic acid (WB+FA) or (B) 2% sugar beet pulp (SBP) or 2% sugar beet pulp supplemented with 0.13 mM ferulic acid (SBP+FA).

Activity profiles of *A. nidulans* and *A. oryzae* were similar in both WB and SBP cultures when MC was used as FAE substrate. For other substrates, the activity profiles differed between these two species. *A. oryzae* produced FAE activity in WB cultures with or without FA supplementation when MF and MS were used as substrates (Fig. 8.3A). However, in FA-supplemented cultures of *A. oryzae* activity against MC was not observed. In SBP cultures, *A. oryzae* produced FAE activity against M*p*C and MC, and in SBP cultures with FA supplementation activity against MS was also detected (Fig. 8.3B).

Differences between FAE activity profiles were observed between the three *A. niger* strains. *A. niger* CBS513.88 showed FAE activity against a wide range of substrates, MF, M*p*C and MS, when cultivated in WB (Fig 8.3A). *A. niger* N402 showed activity only when MS was used as a substrate in WB cultures. While no activity was observed in the WB cultures of *A. niger* ATCC1015, it produced activity against M*p*C when cultivated in SBP and also low activities (< 4 nkat/l) against MF and M*p*C were detected in the SBP cultures amended with FA (Fig 8.3B). In contrast, *A. niger* N402 did not produce any FAE activity in SBP cultures. These data suggest that the studied *Aspergillus* strains produce different FAEs since high variation observed in the substrate specificity of orthologous enzymes towards the methyl ester substrates is unlikely.

Both *A. niger* CBS513.88 and *A. clavatus* produced FAE activity only in the FA-supplemented SBP cultures with M*p*C as substrate (Fig. 8.3B). In contrast, *A. clavatus* FAE activity was detected only in the WB cultures with MS as a substrate similarly to *A. niger* N402 (Fig. 8.3A). Also *A. fischeri* showed narrow FAE substrate profile. Low FAE activity (< 30 nkat/l) against MC and MF was detected in WB- and FA-supplemented WB cultures, respectively. In SBP cultures, high activity (> 70 nkat/l) was detected when M*p*C was used as substrate.

In SBP cultures of *A. terreus* FA induced MF activity and also activity against MC was detected (Fig. 8.3B). Similarly to *A. fischeri* activity against MF was detected from FA-supplemented WB cultures of *A. terreus*.

Discussion

Interestingly, despite the similarity in genome content, *A. niger* CBS 513.88, ATCC1015 and N402 strains showed different specificities against the methyl ester substrates assayed in this study. The different specificity of fungal FAEs is well documented and is mostly dependent on the phenolic ring substitutions of the substrate (Faulds and Williamson, 1994). In this work, after three days of growth *A. niger* CBS513.88 produced activity towards MF, M*p*C and MS, and *A. niger* N402 towards MS in WB medium, and *A. niger* ATCC1015 towards MF and M*p*C in SBP cultures. In line with our results, the highest FAE activity for FaeA (FAE-III) from another strain of *A. niger* (CBS 120.49) has been detected against MS (156 U/mg protein) and no activity has been observed against MC (Faulds and Williamson, 1994). This could imply that FaeA may be the main FAE isoenzyme produced by *A. niger* CBS 513.88 and N402 strains or that the yet uncharacterized FAEs prefer MS as a substrate. It has been proposed that MC can be used as a marker substrate to assess the presence of other FAEs, such as FaeB of *A. niger* (Brézillon *et al.*, 1996). However, FaeA (FAE-III) possess very low specific activity against M*p*C (0.07 U/mg protein), which suggests that M*p*C activity detected for *A. niger* CBS513.88 and ATCC1015 strains in this study may be due to the production of other types of FAEs in the studied culture conditions.

In this study, FAE activity against MF was detected when *A. niger* CBS513.88 was cultured with WB as the carbon source, and very low activity when *A. niger* ATCC1015 was grown in SBP cultures supplemented with FA. Most of this activity is probably due to the production of FaeA of *A. niger* (Brézillon *et al.*, 1996). Additional evidence of the action of FaeA is the FAE activity against MS detected in the WB liquid cultures of *A. niger* CBS513.88. Also the production of FaeA (FAE-III) by *A. niger* CBS120.49 has been reported to be favoured when the fungus grew in cereal-derived carbon sources (Faulds and Williamson, 1994). MF has been detected to be a relatively poor substrate for FaeA and FaeB which are the two most studied *A. niger* FAEs (Ralet *et al.*, 1994). Moreover, in accordance with our results obtained from FA-supplemented SBP cultures of *A. niger* ATCC1015, very low extracellular FAE activity against MF was reported when *A. niger* CS 180 was grown in liquid cultures with SBP as the carbon source (Brézillon *et al.*, 1996).

The available carbon sources affected the overall FAE activity of *A. niger* strains studied here. In line with these results, FAE activity of *A. niger* CBS120.49 was also dependent on the carbon source used in the culture medium (Faulds and Williamson, 1994). When compared to WB, a mixture of WB and maize bran as a carbon source was reported to increase the production of FAE by *A. niger* ATCC16404 (Ou *et al.*, 2011). Moreover, when *A. niger* has been grown in pectin-rich materials, an increase in the FAE activity was observed (Faulds and Williamson, 1993, 1994). Brézillon *et al.* (1996) has detected from four to six times higher levels of FAE activity in SBP culture medium of *A. niger* CS 180 than in xylan-based cultures. It has been shown that liquid cultures of *A. niger* CBS120.49, which contained 1% oat spelt xylan and 1% glucose as carbon source, exhibited the highest FAE and xylanase activity after day four of cultivation (Faulds and Williamson, 1994), likely due to initial repression of the production caused by the presence of glucose. The accumulation of FAE activity at later stages of cultivation of *A. niger* strains in our study (data not shown) is not likely due to initial carbon repression, but may indicate increments in the production of other hydrolytic enzymes that allow the fungi to better assimilate the carbon source.

FAE activity of other *Aspergillus* species

Different substrate specificity profiles were obtained from the liquid cultures of *A. clavatus*, *A. fischeri*, *A. nidulans*, *A. oryzae* and *A. terreus*. *A. nidulans* was a strong producer of FAE activity in all culture conditions. The potential of *A. nidulans* as plant biomass degrader is evident due to the wide range of carbohydrate-active enzyme (CAZyme) encoding genes present in its genome (Coutinho *et al.*, 2009). The FAE activity data suggests that *A. nidulans* activity profile corresponds to type A and B FAEs. Whole genome sequencing has revealed at least two putative FAE encoding genes for *A. nidulans* (Olivares-Hernandez *et al.*, 2010), but it lacks a orthologue of *A. niger* FaeA (Benoit *et al.*, 2008). The only characterized FAE from *A. nidulans* so far is type B FAE with an optimum activity at 45°C and pH 7.0 (Shin and Chen, 2007).

A. oryzae also showed a wide FAE substrate specificity suggesting that it is able to produce type A, B and C FAEs. High activity against MS suggests the production of type A FAE in WB and in FA-supplemented WB and SBP cultures after three days of growth. Two FAEs have been characterized from *A. oryzae*: AoFaeC with activity towards MC, MF, M*p*C and MS and AoFaeB with activity towards the three methyl esters but not MS (Koseki *et al.*, 2009). It is most likely that also other types of FAEs will be found in *A. oryzae* as nine putative FAE encoding genes have been detected in its genome (Table 8.1; Benoit *et al.*, submitted).

A. terreus produced the highest FAE activity in SBP cultures both with and without FA supplementation. The ability of *A. terreus* to produce FAEs in different types of plant biomass is supported by the previous study where production of FAE activity was tested in various agricultural crop residues (Kumar *et al.*, 2011). The most effective FAE production was detected with maize bran as substrate, followed by WB, coconut husk and rice husk. Recently, three FAEs have been characterized from *A. terreus*: AtFAE-1, AtFAE-2 and AtFAE-3 (Kumar *et al.*, 2013). AtFAE-1 was determined to be a type A FAE whereas the other two were revealed to belong to the type C FAEs (Kumar *et al.*, 2013). Our results showed that *A. terreus* produced activity towards all the four FAE substrates studied, which is in line with the biochemical characteristics of AtFAEs. In addition, recombinant FAE ATEG 06644 from *A. terreus* has shown to be active on a wide range of methyl ester substrates, including MC, MF, M*p*C and MS (Mäkelä *et al.*, unpublished).

A. fischeri produced FAE activity against MF, M*p*C and MC suggesting that this species secretes at least type A and B FAEs. Low FAE activity detected only against MS and M*p*C in the cultures of *A. clavatus* could reflect the weak ability of the species to thrive in some of these substrates. For example, *A. clavatus* is known to grow poorly in pectin-rich materials (Benoit *et al.*, 2015). It is also possible that some of the studied species harbour new types of FAE encoding genes in their genome, with different specificities and regulatory mechanisms not explored before. Recombinant AcFAE (XP_001274884) from *A. clavatus* has been shown to exhibit hydrolytic activity on MC, M*p*C and MF, which suggests that this particular FAE was not produced in the culture conditions studied here (Damasio *et al.*, 2013).

Influence of ferulic acid on the FAE activity of the cultures

Several studies indicate that the availability of aromatic compounds such as FA in the culture medium increases the extracellular FAE activity production of *Aspergillus* species (de Vries *et al.*, 2002; de Vries and Visser, 1999; Faulds *et al.*, 1997; Koseki *et al.*, 2007). However, in this study addition of 0.13 mM FA to the liquid media caused no increase in FAE activities. This may be due to the fact that aromatic compounds such as FA are also to some extent toxic to *Aspergillus* species (Guiraud *et al.*, 1995; Mäkelä *et al.*, 2015) or, alternatively, FA may adsorb to the plant biomass substrate and therefore not be sensed by the fungi. Even though WB and SBP contain up to 1% of dry weight of FA (Faulds *et al.*, 1997; Hartley and Jones, 1977; Rombouts and Thibault, 1986), extracellular FAE activity is not only dependent on the presence of FA. FAE activity in the liquid cultures of *A. niger* CS 180 (CMICC298302) was observed both with and without FA addition (Brézillon *et al.*, 1996; Kroon *et al.*, 1996). This strain is a different lineage but was confirmed to be a true *A. niger* (R.P. de Vries, unpublished results). Moreover, increase in concentration of feruloylated oligosaccharides seemed not to induce the production of extracellular FAE activity by *A. niger* CS 180, thus suggesting that the presence of certain non-aromatic, hemicellulose or pectin components is far more important for the induction of FAEs than FA. This was confirmed at the gene expression level for *A. niger faeA* that was much higher expressed on xylose than on FA (de Vries and Visser, 1999).

Similar studies with different *Aspergillus* species concluded that FA is not strictly required to induce FAE activity (Faulds *et al.*, 1997; Koseki *et al.*, 2007). However, FAE production was stimulated in the cultures amended with 0.01% and 0.05% FA (Faulds *et al.*, 1997). Subsequently, complex medium supplemented with 0.03% FA has been shown to induce the expression of *faeA* gene in *A. niger* (de Vries and Visser, 1999). A similar experiment reported that the highest levels of *faeB* transcripts were observed in the presence of 0.03% of caffeic acid and *p*-coumaric acid (de Vries *et al.*, 2002). Likewise, Koseki *et al.* (2007) detected an increase in the production of extracellular xylan-active FAE of *Aspergillus kawachii* when the fungus was cultivated in medium supplemented with 0.5 mM FA. In addition, when the stability of FA in the culture medium of *A. niger* CBS 120.49 was explored, the levels of FA decreased quickly and were not detectable after five days of cultivation (Faulds *et al.*, 1997). Consequently, the low levels of FA utilized in this study may explain the lack of expected induction of FAE activity in the WB and SBP liquid cultures.

Furthermore, the aromatic compounds inducing the production of FaeA by *A. niger* are different from those inducing FaeB thus highlighting the complexity of gene regulation in response to aromatic compounds (de Vries *et al.*, 2002). This suggests that the aromatic regulatory system that triggers FAE production in *Aspergillus* species is very complex. The mechanism by which FA induces FAE activity is still not well understood and has yet to be elucidated (Faulds *et al.*, 1997).

Conclusion

FAEs are one of the most diverse types of plant biomass acting enzymes. They appear to have originated by modifications of different types of enzymes (e.g. tannase, lipase, acetyl xylan esterase) (Benoit *et al.*, 2008) and the FAE gene content in fungal genomes differs strongly. The data in this study demonstrate that, in addition to the genomic variation, the production of orthologous enzymes in various *Aspergillus* species differs significantly. This suggests a variable dependence on this enzyme for efficient plant biomass degradation for different species. The currently ongoing genome project for the whole *Aspergillus* genus will provide a unique opportunity to track the origin of individual FAE encoding genes. However, as indicated by our data, without experimental data about the production of the enzymes it will be hard to predict the importance of FAEs for the individual species during growth on plant biomass.

References

Benoit, I., Asther, M., Sulzenbacher, G., Record, E., Marmuse, L., Parsiegla, G., Gimbert, I., Asther, M., and Bignon, C. (2006a). Respective importance of protein folding and glycosylation in the thermal stability of recombinant feruloyl esterase A. FEBS Lett. *580*, 5815–5821.

Benoit, I., Culleton, H., Zhou, M., DiFalco, M., Aguilar-Osorio, G., Battaglia, E., Bouzid, O., Brouwer, C.P.J.M., El-Bushari, H.B.O., Coutinho, P.M., *et al.* (2015). Closely related fungi employ diverse enzymatic strategies to degrade plant biomass. Biotech Biofuels *8*, 107.

Benoit, I., Danchin, E.G., Bleichrodt, R.J., and de Vries, R.P. (2008). Biotechnological applications and potential of fungal feruloyl esterases based on prevalence, classification and biochemical diversity. Biotechnol. Lett. *30*, 387–396.

Benoit, I., Malavazi, I., Goldman, G., Baker, S.E., and de Vries, R.P. (2013). *Aspergillus* – Genomics of a cosmopolitan fungus. In: Genomics of soil- and plant-associated fungi, B.A. Horwitz, P.K. Mukherjee, M. Mukherjee, and C.P. Kubicek, eds. (Springer, Heidelberg, Germany), pp. 89–126.

Benoit, I., Navarro, D., Marnet, N., Rakotomanomana, N., Lesage-Meessen, L., Sigoillot, J.C., Asther, M., and Asther, M. (2006b). Feruloyl esterases as a tool for the release of phenolic compounds from agro-industrial by-products. Carbohydr. Res. *341*, 1820–1827.

Bidlack, J., Malone, M., and Benson, R. (1992). Molecular structure and component integration of secondary cell walls in plants. Proc. Okla. Acad. Sci. *72*, 51–56.

Brézillon, C., Kroon, P.A., Faulds, C.B., Brett, G.M., and Williamson, G. (1996). Novel ferulic acid esterases are induced by growth of *Aspergillus niger* on sugar beet pulp. Appl. Microbiol. Biotechnol. *45*, 371–376.

Buanafina, M. (2009). Feruloylation in grasses: current and future perspectives. Mol. Plant *2*, 861–872.

Carnachan, S.M., and Harris, P.J. (2000). Ferulic acid is bound to the primary cell walls of all gymnosperm families. Biochem. Syst. Ecol. *28*, 865–879.

Coutinho, P.M., Andersen, M.R., Kolenova, K., vanKuyk, P.A., Benoit, I., Gruben, B.S., Trejo-Aguilar, B., Visser, H., van Solingen, P., Pakula, T., Seiboth, B., Battaglia, E., Aguilar-Osorio, G., de Jong, J.F., Ohm, R.A., Aguilar, M., Henrissat, B., Nielsen, J., Stalbrand, H., and de Vries, R.P. (2009). Post-genomic insights into the plant polysaccharide degradation potential of *Aspergillus nidulans* and comparison to *Aspergillus niger* and *Aspergillus oryzae*. Fungal Genet. Biol. 46(Suppl. 1), S161–S169.

Crepin, V.F., Faulds, C.B., and Connerton, I.F. (2004). Functional classification of the microbial feruloyl esterases. Appl. Microbiol. Biotechnol. *63*, 647–652.

Damasio, A.R., Braga, C.M., Brenelli, L.B., Citadini, A.P., Mandelli, F., Cota, J., de Almeida, R.F., Salvador, V.H., Paixao, D.A., Segato, F., Mercadante, A.Z., de Oliveira Neto, M., do Santos, W.D., and Squina, F.M. (2013). Biomass-to-bio-products application of feruloyl esterase from *Aspergillus clavatus*. Appl. Microbiol. Biotechnol. *97*, 6759–6767.

Di Gioia, D., Sciubba, L., Setti, L., Luziatelli, F., Ruzzi, M., Zanichelli, D., and Fava, F. (2007). Production of biovanillin from wheat bran. Enzyme Microb. Technol. *41*, 498–505.

Faulds, C.B., de Vries, R.P., Kroon, P.A., Visser, J., and Williamson, G. (1997). Influence of ferulic acid on the production of feruloyl esterases by *Aspergillus niger*. FEMS Microbiol. Lett. *157*, 239–244.

Faulds, C.B., and Williamson, G. (1993). Ferulic acid esterase from *Aspergillus niger*: purification and partial characterization of two forms from a commercial source of pectinase. Biotechnol Appl Biochem *17*, 349–359.

Faulds, C.B., and Williamson, G. (1994). Purification and characterization of a ferulic acid esterase (FAE-III) from *Aspergillus niger*: specificity for the phenolic moiety and binding to microcrystalline cellulose. Microbiology *140*, 779–787.

Faulds, C.B., and Williamson, G. (1995). Release of ferulic acid from wheat bran by a ferulic acid esterase (FAE-III) from *Aspergillus niger*. Appl. Microbiol. Biotechnol. *43*, 1082–1087.

Fry, S. (2004). Oxidative coupling of tyrosine and ferulic acid residues: Intra- and extra-protoplasmic occurrence, predominance of trimers and larger products, and possible role in inter-polymeric cross-linking. Phytochem. Rev. *3*, 97–111.

Grabber, J.H., Ralph, J., and Hatfield, R.D. (1998). Ferulate cross-links limit the enzymatic degradation of synthetically lignified primary walls of maize. J. Agric. Food Chem. *46*, 2609–2614.

Graf, E. (1992). Antioxidant potential of ferulic acid. Free Radic. Biol. Med. *13*, 435–448.

Guiraud, P., Steiman, R., Seigle-Murandi, F., and Benoit-Guyod, J.L. (1995). Comparison of the toxicity of various lignin-related aromatic compounds towards selected fungi perfecti and fungi imperfecti. Ecotoxicol. Environ. Safety *32*, 29–33.

Hartley, R.D., and Jones, E.C. (1977). Phenolic components and degradability of cell walls of grass and legume species. Phytochem. *16*, 1531–1534.

Ishii, T. (1997). Structure and functions of feruloylated polysaccharides. Plant Sci. *127*, 111–127.

Kikuzaki, H., Hisamoto, M., Hirose, K., Akiyama, K., and Taniguchi, H. (2002). Antioxidant properties of ferulic acid and its related compounds. J. Agric. Food Chem. *50*, 2161–2168.

Kondo, T., Mizuno, K., and Kato, T. (1990). Cell wall-bound *p*-coumaric and ferulic acids in Italian ryegrass. J. Plant Sci. *70*, 495–499.

Koseki, T., Hori, A., Seki, S., Murayama, T., and Shiono, Y. (2009). Characterization of two distinct feruloyl esterases, AoFaeB and AoFaeC, from *Aspergillus oryzae*. Appl. Microbiol. Biotechnol. *83*, 689–696.

Koseki, T., Mimasaka, N., Hashizume, K., Shiono, Y., and Murayama, T. (2007). Stimulatory effect of ferulic acid on the production of extracellular xylanolytic enzymes by *Aspergillus kawachii*. Biosci. Biotechnol. Biochem. *71*, 1785–1787.

Kroon, P.A., Faulds, C.B., Brézillon, C., and Williamson, G. (1997). Methyl phenylalkanoates as substrates to probe the active sites of esterases. Eur. J. Biochem. *248*, 245–251.

Kroon, P.A., Faulds, C.B., and Williamson, G. (1996). Purification and characterisation of a novel esterase induced by growth of *Aspergillus niger* on sugar-beet pulp. Biotechnol. Appl. Biochem. *23*, 255–262.

Kroon, P.A., Garcia-Conesa, M.T., Fillingham, I.J., Hazlewood, G.P., and Williamson, G. (1999). Release of ferulic acid dehydrodimers from plant cell walls by feruloyl esterases. J. Sci. Food Agric. *79*, 428–434.

Kumar, C.G., Kamle, A., and Kamal, A. (2013). Purification and biochemical characterization of feruloyl esterases from *Aspergillus terreus* MTCC 11096. Biotechnol. Prog. *29*, 924–932.
Kumar, C.G., Kamle, A., Mongolla, P., and Joseph, J. (2011). Parametric optimization of feruloyl esterase production from *Aspergillus terreus* strain GA2 isolated from tropical agro-ecosystems cultivating sweet sorghum. J. Microbiol. Biotechnol. *21*, 947–953.
Lombard, V., Golaconda Ramulu, H., Drula, E., Coutinho, P.M., and Henrissat, B. (2014). The carbohydrate-active enzymes database (CAZy) in 2013. Nucleic Acids Res. *42*, D490–495.
Mäkelä, M.R., Marinović, M., Nousiainen, P., Liwanag, A.J.M., Benoit, I., Sipilä, J., Hatakka, A., de Vries, R.P., and Hildén, K.S. (2015). Aromatic metabolism of filamentous fungi in relation to the presence of aromatic compounds in plant biomass. Adv. Appl. Microbiol. *91*, 63–137.
Murakami, A., Nakamura, Y., Koshimizu, K., Takahashi, D., Matsumoto, K., Hagihara, K., Taniguchi, H., Nomura, E., Hosoda, A., Tsuno, T., Maruta, Y., Kim, H.W., Kawabata, K., and Ohigashi, H. (2002). FA15, a hydrophobic derivative of ferulic acid, suppresses inflammatory responses and skin tumor promotion: comparison with ferulic acid. Cancer Lett. *180*, 121–129.
Olivares-Hernandez, R., Sunner, H., Frisvad, J.C., Olsson, L., Nielsen, J., and Panagiotou, G. (2010). Combining substrate specificity analysis with support vector classifiers reveals feruloyl esterase as a phylogenetically informative protein group. PLoS One *5*, e12781.
Ou, S., Zhang, J., Wang, Y., and Zhang, N. (2011). Production of feruloyl esterase from *Aspergillus niger* by solid-state fermentation on different carbon sources. Enzyme Res. *2011*, 848939.
Pandey, A., Soccol, C.R., Nigam, P., Brand, D., Mohan, R., and Roussos, S. (2000). Biotechnological potential of coffee pulp and coffee husk for bioprocesses. Biochem. Eng. J. *6*, 153–162.
van Peij, N., Gielkens, M.M.C., de Vries, R.P., Visser, J., and de Graaff, L.H. (1998). The transcriptional activator XlnR regulates both xylanolytic and endoglucanase gene expression in *Aspergillus niger*. Appl. Environ. Microbiol. *64*, 3615–3619.
Ralet, M.-C., Faulds, C.B., Williamson, G., and Thibault, J.-F. (1994). Degradation of feruloylated oligosaccharides from sugar beet pulp and wheat bran by ferulic acid esterases from *Aspergillus niger*. Carbohydr. Res. *263*, 257–269.
Ralph, J., Lundquist, K., Brunow, G., Lu, F., Kim, H., Schatz, P.F., Marita, J.M., Hatfield, R.D., Ralph, S.A., Holst Christensen, J., and Boerjan, W. (2004). Lignins: natural polymers from oxidative coupling of 4-hydroxyphenylpropanoids. Phytochem. Rev. *3*, 29–60.
Ramirez, J. (1988). Phenolic compounds in coffee pulp: Quantitative determination by HPLC. J. Sci. Food Agric. *43*, 135–144.
Record, E., Asther, M., Sigoillot, C., Pages, S., Punt, P.J., Delattre, M., Haon, M., van den Hondel, C.A., Sigoillot, J.C., Lesage-Meessen, L., and Asther, M. (2003). Overproduction of the *Aspergillus niger* feruloyl esterase for pulp bleaching application. Appl. Microbiol. Biotechnol. *62*, 349–355.
Rombouts, F.M., and Thibault, J.-F. (1986). Feruloylated pectic substances from sugar-beet pulp. Carbohydr. Res. *154*, 177–187.
Shin, H.D., and Chen, R.R. (2007). A type B feruloyl esterase from *Aspergillus nidulans* with broad pH applicability. Appl. Microbiol. Biotechnol. *73*, 1323–1330.
Sigoillot, C., Camarero, S., Vidal, T., Record, E., Asther, M., Perez-Boada, M., Martinez, M.J., Sigoillot, J.C., Asther, M., Colom, J.F., and Martinez, A.T. (2005). Comparison of different fungal enzymes for bleaching high-quality paper pulps. J. Biotechnol. *115*, 333–343.
Sun, R., Sun, X., Wang, S., Zhu, W., and Wang, X. (2002). Ester and ether linkages between hydroxycinnamic acids and lignins from wheat, rice, rye, and barley straws, maize stems, and fast-growing poplar wood. Ind. Crops Prod. *15*, 179–188.
Tabka, M.G., Herpoël-Gimberta, I., Monod, F., Asther, M., and Sigoillot, J.C. (2006). Enzymatic saccharification of wheat straw for bioethanol production by a combined cellulase xylanase and feruloyl esterase treatment. Enzyme Microbiol. Technol. *39*, 897–902.
Tapin, S., Sigoillot, J.-C., Asther, M., and Petit-Conil, M. (2006). Feruloyl esterase utilization for simultaneous processing of nonwood plants into phenolic compounds and pulp fibers J. Agric. Food Chem. *54*, 3697–3703.
Udatha, D.B., Mapelli, V., Panagiotou, G., and Olsson, L. (2012). Common and distant structural characteristics of feruloyl esterase families from *Aspergillus oryzae*. PLoS One *7*, e39473.
de Vries, R.P., Burgers, K., van de Vondervoort, P.J.I., Frisvad, J.C., Samson, R.A., and Visser, J. (2004). A new black *Aspergillus* species, *A. vadensis*, is a promising host for homologous and heterologous protein production. Appl. Environ. Microbiol. *70*, 3954–3959.

de Vries, R.P., Faulds, C.B., and Visser, J. (1999a). The *faeA* gene from *Aspergillus niger* encoding a feruloyl esterase with activity on xylan and pectin is subject to a complex system of regulation. J. Sci. Food Agric. *79*, 443–446.

de Vries, R.P., Kester, H.C.M., Poulsen, C.H., Benen, J.A.E., and Visser, J. (2000). Synergy between accessory enzymes from *Aspergillus* in the degradation of plant cell wall polysaccharides. Carbohydr. Res. *327*, 401–410.

de Vries, R.P., Kester, H.C.M., vanKuyk, P.A., and Visser, J. (2002). The *Aspergillus niger faeB* gene encodes a second feruloyl esterase involved in pectin and xylan degradation, and is specifically induced on aromatic compounds. Biochem. J. *363*, 377–386.

de Vries, R.P., Michelsen, B., Poulsen, C.H., Kroon, P.A., van den Heuvel, R.H., Faulds, C.B., Williamson, G., van den Hombergh, J.P., and Visser, J. (1997). The *faeA* genes from *Aspergillus niger* and *Aspergillus tubingensis* encode ferulic acid esterases involved in degradation of complex cell wall polysaccharides. Appl. Environ. Microbiol. *63*, 4638–4644.

de Vries, R.P., and Visser, J. (1999). Regulation of the feruloyl esterase (*faeA*) gene from *Aspergillus niger*. Appl. Environ. Microbiol. *65*, 5500–5503.

de Vries, R.P., and Visser, J. (2001). *Aspergillus* enzymes involved in degradation of plant cell wall polysaccharides. Microb. Mol. Biol. Rev. *65*, 497–522.

de Vries, R.P., Visser, J., and de Graaff, L.H. (1999b). CreA modulates the XlnR-induced expression on xylose of *Aspergillus niger* genes involved in xylan degradation. Res. Microbiol. *150*, 281–285.

Waldron, K.W., and Faulds, C. (2007). Cell wall polysaccharides composition and structure. In: Comprehensive Glycoscience, J. Kamerling, ed. (Elsevier Ltd, Norwich), pp. 181–201.

Yu, P., McKinnon, J.J., Maenz, D.D., Olkowski, A.A., Racz, V.J., and Christensen, D.A. (2003). Enzymic release of reducing sugars from oat hulls by cellulase, as influenced by *Aspergillus* ferulic acid esterase and *Trichoderma* xylanase. J. Agric. Food Chem. *51*, 218–223.

Yuan, X., Wang, J., and Yao, H. (2005). Antioxidant activity of feruloylated oligosaccharides from wheat bran. Food Chem. *90*, 759–764.

Zheng, L., Zheng, P., Sun, Z., Bai, Y., Wang, J., and Guo, X. (2007). Production of vanillin from waste residue of rice bran oil by *Aspergillus niger* and *Pycnoporus cinnabarinus*. Biores. Technol. *98*, 1115–1119.

Secondary Metabolite Formation by the Filamentous Fungus *Penicillium chrysogenum* in the Post-genomic Era

9

Marta M. Samol, Oleksandr Salo, Peter Lankhorst, Roel A.L. Bovenberg and Arnold J.M. Driessen

Abstract

The filamentous fungus *Penicillium chrysogenum* is a major industrial producer of the β-lactam antibiotic penicillin. Penicillin was discovered by Alexander Fleming in 1928, and was among the first medications to be effective against bacterial infections. Penicillins are still widely used but are now produced by high β-lactam-yielding strains that emerged from extensive classical strain improvement programs lasting for several decades. In 2008, the genome of *P. chrysogenum* was sequenced revealing an unexploited reservoir of nonribosomal peptide synthetases and polyketide synthases genes that specify potential bioactive compounds. In recent years, several pathways have been resolved that are responsible for the production of a wide variety of secondary metabolites.

Introduction

The filamentous fungus *Penicillium chrysogenum* is a major industrial producer of the β-lactam antibiotic penicillin. Penicillin was discovered by Alexander Fleming in 1928, and during the Second World War it was rapidly considered as a magic bullet in combating infectious disease. Since then it has been used as one of the mostly used antibiotics in human health care. *P. chrysogenum* belongs to the genus *Penicillium* and features a typical brush-shaped morphology of its conidiophores. Because of the importance of this fungus in the industrial production of β-lactams, its taxonomy has been studied in detail. The fungus identified by Alexander Fleming was initially classified as *P. rubrum* (Fleming, 1955) but later renamed as *P. notatum* (Raper *et al.*, 1944). Further refinement of the taxonomical status provided Fleming's isolate with the names *P. chrysogenum* (Samson *et al.*, 1977), *P. griseoroseum* or *P. rubens* (Houbraken *et al.*, 2011). Because of the phenotypic similarity between *P. chrysogenum* and *P. notatum*, both species were included into the newly established section Chrysogena (subsection Radiata) (Raper *et al.*, 1944). In 1977, *P. notatum*, *P. meleagrinum*, *P. cyaneofulvum* and six additional species were classified as synonymous to *P. chrysogenum* (Samson *et al.*, 1977). Both Fleming's isolate and the industrial β-lactam producer Wisconsin were named *Penicillium chrysogenum* as *nomen conservandum* (Houbraken *et al.*, 2012; McNeill, 2006). According to the latest phylogenetic position based on multi-gene analysis, morphological and extrolite similarities, the section Chrysogena was expanded with up to 18 species. Eight of those produce penicillins (Houbraken *et al.*, 2012).

P. chrysogenum is found in diverse environments like soil, spoiled fruits, in the air and even in the extreme environments like in Antarctic ice cores (Foreman *et al.*, 2011; Henk *et al.*, 2011). Recently, the interest towards marine derived microorganisms as a potential source of novel bioactive compounds, lead to the identification of *P. chrysogenum* species living in symbiosis with the halophilic spongy *Ircinia fasciculata* (Bringmann *et al.*, 2003) and the red algal genera *Laurencia* (Gao *et al.*, 2011). Commonly known as non-pathogenic, *P. chrysogenum* has also been associated to invasive human infections in patients with an immune system disorder (Geltner *et al.*, 2013). *P. chrysogenum* induces a dose-dependent allergic asthma-like response in mice (Chung *et al.*, 2005). This observation is of particular importance considering the presence of this fungus in common domestic environments like old buildings, at walls and in industrial dust.

P. chrysogenum has mostly been exploited for β-lactam production. The current industrial high β-lactam yielding strains emerged during an intensive classical strain improvement programme that lasted several decades. In 2008, the 32.19 Mb genome of *P. chrysogenum* Wisconsin 54-1255 was sequenced, revealing an unexplored reservoir of nonribosomal peptide synthetases and polyketide synthases genes that specify potential bioactive compounds (van den Berg *et al.*, 2008). Here we will summarize our current understanding of secondary metabolite formation by *P. chrysogenum* and provide an outlook for future applications as an industrial host for the production of secondary metabolites in general.

Fungal secondary metabolite enzymes

A large group of secondary metabolites emerge from non-ribosomally synthesized peptides and from polyketides. Enzymes involved in the production of such compounds are responsible for the biosynthesis of a diverse group of natural products with complex chemical structures and an enormous pharmaceutical potential. These secondary metabolites are synthesized by modular nonribosomal peptide synthetase (NRPS) and polyketide synthase (PKS) enzyme complexes using a conserved thiotemplate mechanism. In addition, hybrid PKS–NRPS enzymes have been identified. They constitute one of the largest proteins found within fungal cells. For instance, cyclosporine is synthesized by a NRPS with a molecular mass of 1.6 MDa (Turgay *et al.*, 1992). In the following section each of these classes is briefly described.

Non-ribosomal peptide synthetases

Non-ribosomal peptide synthetases (NRPSs) are multifunctional enzymes organized into modules that carry out the biosynthesis of a diverse range of peptides. Each module possesses several catalytic activities and incorporates diverse monomeric substrates into the final product. In addition to amino acids, NRPSs can utilize a variety of non-protein amino acids to generate peptides (von Dohren, 2004) and to date hundreds of different peptides have been identified. In contrast to bacteria in which NRPS-related biosynthetic genes are organized in an operon, fungi make use of a single NRPS to synthesize a product or product class (Weber and Leitner, 1994). The product of NRPS enzymes are further tailored by enzymes whose genes often co-localize with the major NRPS gene in a so-called gene cluster. The size of fungal NRPS enzymes is often much larger than that of the bacterial counterparts (Finking and Marahiel, 2004).

Classes of NRPS enzymes

Based on the mechanism of peptide biosynthesis, NRPS enzymes can be subdivided into three main classes: (i) linear – the number and sequence of the modules in the NRPS matches the number and order of amino acids in the peptide; (ii) iterative – the modules or domains of the synthetase are used more than once to synthetize the peptide; and (iii) nonlinear – the arrangement of the modules does not match the sequence of amino acids (Cane and Walsh, 1999). Fungal NRPSs tend to be linear although iterative enzymes have been suggested for the synthesis of some siderophores. The three key enzymatic functions that are present in each NRPS module are adenylation, thiolation and condensation. The main determinant of the substrate specificity is the adenylation (A) domain that recognizes and activates cognate amino acids in a double mode process: while ATP is consumed, the cognate amino acid is selectively bound and converted into an amino acyl adenylate intermediate. The activated substrate is then transferred to the thiol group of the 4′-phosphopantetheine on the peptidyl carrier protein (T) domain and covalently bound through a thioester linkage (Stachelhaus and Marahiel, 1995). The T-domain mediates the transfer of amino acids or peptidyl substrates through the catalytic step of NRP synthesis (Finking and Marahiel, 2004; Linne *et al.*, 2001). Prior to NRP biosynthesis, T-domains are post-translational modified by phosphopantetheinyl transferases (PPTases), which is essential to convert the inactive *apo*-T to the active *holo*-T form. The PPTase encoding gene Pc13g04050 in *P. chrysogenum* was found to be required for penicillin biosynthesis (Garcia-Estrada *et al.*, 2008). The 4′PP cofactor – which is bonded as a thioester – is transferred to a conserved serine residue of the T-domain by the PPTase in an Mg^{2+}-dependent reaction. The 4′PP cofactor acts as a flexible arm to allow the bound amino acyl or peptidyl substrate to be shuttle between the T-domains of each module during peptide elongation (Finking and Marahiel, 2004; Hur *et al.*, 2012). The structures of T-domains in the *apo* and the *holo* state both exhibit conformational diversity, which is an essential prerequisite for the directed movement of the 4′-PP cofactor and its interaction with externally acting proteins such as thioesterases and 4′-PPTases (Koglin *et al.*, 2006). Once the amino acyl substrates bound to the T-domain have undergone all necessary modifications, they are joined together by the condensation (C) domain, which is situated at the N-terminus of each elongation module. The C-domain catalyses peptide bond formation with the downstream amino acyl unit that is bound to the T-domain of the adjacent module (Stachelhaus *et al.*, 1999). Peptide bond formation presumably involves deprotonation of the amino (imino, hydroxyl) group and stereochemistry of the incoming intermediates (von Dohren, 2004). The thioester group of the upstream donor substrate undergoes a nucleophilic attack by the α-amino group of the downstream acceptor substrate, forming an amide bond and transferring the peptide intermediate from one module to the next (Hur *et al.*, 2012).

While the basic module contains the aforementioned three domains, the first module of the NRPS system is the initiation or starting module. Moreover, modules may contain additional tailoring domains, which are responsible for various modifications that create a further diversity in NRP products. For example, *Trichoderma virens* produces an 18-residue linear peptide that includes the rare amino acid aminoisobutyric acid. The N-terminus of the peptide product is acylated while the C-terminal amino acid is hydroxylated (Wiest *et al.*, 2002). An NRPS in *Tolypocladium niveum* synthesizes the cyclic undecapeptide cyclosporine (Weber *et al.*, 1994). The final product is released from the T-domain of the termination module through either hydrolysis or macrocyclization involving a thioesterase (TE) domain (Kohli *et al.*, 2001). In fungi, however, only a few NRPSs are known to release products via a TE-domain. In most cases, the peptide chain releases through the action of a terminal condensation domain (Evans *et al.*, 2011) also called a condensation-like (C(T)) domain (Gao *et al.*, 2012). With most common multimodular NRPSs, a terminal C-domain catalyses the release of product by means of inter- or intra-molecular amide bond formation (Keating *et al.*, 2001), while with monomodular NRPSs, a reductase (R) domain catalyses the reduction with NADPH to form an aldehyde (Keating *et al.*, 2001). The peptide chain release mechanism then involves a non-enzymatic cyclization by the formation of a diketopiperazine ring (Walzel *et al.*, 1997).

One striking characteristic of some NRPs is the presence of D-amino acids. D-Amino acids can be critical for functionality as they influence the peptide orientation for further modification by tailoring domains (Hur *et al.*, 2012). D-amino acids may be formed by an additional epimerization (E) domain, which converts an amino acid from the L- to the D-configuration at the thioester stage. Moreover, the E-domain only catalyses epimerization of the amino acid while it is tethered to the T-domain (Linne *et al.*, 2001). Some A-domains specifically accept D-amino acid residues that are supplied by amino acid racemases as for example D-alanine in cyclosporine and the HC-toxin (von Dohren, 2004). A further domain is the *N*-methyltransferase domain which in fungi is usually present at the C-terminus of an associated A-domain, as for instance in the cyclosporine synthetase (Weber *et al.*, 1994). Both *N*- and *C*-methyltransferases can use *S*-adenosyl methionine as the methyl donor and transfer it to the amino acid that is tethered to the 4′-phosphopantetheine arm at the T-domain. In case of the *Fusarium scirpi* enniatin synthetase, *N*-methylated substrates are recognized and directly bound to the T-domain (Hornbogen *et al.*, 2007). The *N*-methyltransferase domain of the enniatin synthetase also exhibits enantioselectivity for the L-isomers of the amino acids (Hacker *et al.*, 2000).

The non-ribosomal code

NRPS enzymes serve as a kind of assembly line for natural peptides. The sequence of each peptide is encoded within the modular structure of the respective NRPSs (von Dohren *et al.*, 1999). Substrate recognition and activation occurs at the A-domain of each module. Analysis of the structure of the phenylalanine-activating domain (PheA) of the bacterial gramicidin S initiation module (Conti *et al.*, 1997) indicated that the substrate is bound at the interface of a large N-terminal domain and a small C-terminal domain. This binding involves 8–10 residues of the active site (Stachelhaus motif) that was confirmed to play a critical role in substrate recognition (Stachelhaus *et al.*, 1999). The A-domain undergoes at least two conformational changes to properly recognize and activate the substrate for NRP biosynthesis. Substrate recognition is determined by hydrophobic interaction and shape complementarity. The size of the substrate side chain affects the specificity conferring code, which is defined by the composition of a set of critical amino acids that are part of the active site (Villiers and Hollfelder, 2009). Additionally, other factors such as van der Waals interactions influence substrate binding (Hur *et al.*, 2012).

Many techniques have been developed to unravel the code of the NRPS specificity template. This includes adenylation domain sequence alignment, pyrophosphate/ATP exchange assay to defined specificity (Otten *et al.*, 2007), and site-directed mutagenesis (Challis *et al.*, 2000). Bioinformatics tools have been developed to predict substrate specificity based on known A-domain employing the aforementioned Stachelhaus motif and taking into account the physico-chemical properties of their substrates (Rausch *et al.*, 2005). Unfortunately, none of the prediction methods that work for prokaryote NRPSs yields reliable results for the fungal enzymes, indicating a marked evolution of different types of domains with related specificities (von Dohren, 2004). To elucidate the specificity of A-domains in relation to the structural types of NRPS is still an important challenge to improve the prediction methods for fungal NRPSs (Bushley *et al.*, 2008).

Novel compounds can also be produced through a re-arrangement of NRPS modules (Stachelhaus *et al.*, 1995). This was first done on a NRPS of the *srfA* operon that is responsible for the synthesis of the lipopeptide antibiotic surfactin in *Bacillus subtilis*. Coding regions of the Phe, Orn-, and Leu-activating domains of the NRPS of the *grs* operon from *Bacillus brevis*, and the Cys and Val domain of the ACV synthetase from *P. chrysogenum* were replaced by the amino acid-activating domains from SrfA. This resulted in the production of new compounds with slightly reduced activity compared to original product (Stachelhaus *et al.*, 1995). Replacement of domains is a promising tool allowing the production of novel metabolites. However, this approach is limited by uncertainties of the exact boundaries between the domains within a module and the possible requirement for specific domain–domain interactions.

Polyketide synthases

Polyketide synthases (PKSs) are a family of multi-domain enzymes or enzyme complexes that produce polyketides, a large class of secondary metabolites, in bacteria, fungi, plants, and a few animal lineages. The mechanistic basis of the biosynthesis of polyketides share striking similarities with fatty acid biosynthesis. PKSs are subdivided into four classes based on their synthesis mechanism: (i) type I – a single protein with multiple modules that function as a linear assembly line, in which each active site is used only once, (ii) type I – iterative – a single protein with one module that is used in an iterative manner, (iii) type II – multiple proteins each with a single functional active site, that may function in an iterative manner or only once, and (iv) type III – a single protein with multiple modules that may function in an iterative manner (Watanabe and Ebizuka, 2004). The modular type I are the most complex of these four classes of PKS enzymes while the type I – iterative is the type most commonly found in fungi.

Most fungal PKSs consist of a single large polypeptide with multiple domains. There are three key domains that carry out the basic functions, i.e. the acyltransferase (AT), thiolation (T) and ketosynthase (KS) domain. The AT-domain catalyses the thioester exchange between a malonyl-CoA and a thiol group present on a pantetheinyl group attached to an acyl carrier protein (ACP) causing the displacement of CoA. Subsequently, the polyketide chain (or acetate) is transferred from the upstream ACP to an active site serine on the KS-domain while the extender unit undergoes decarboxylation. In the final step, the thioester linkage on the KS-domain undergoes a nucleophilic attack by the carbanion to yield a chain that has been extended by two carbons (Fitzgerald *et al.*, 2011; Ridley and Khosla, 2007). In contrast to the substrate-specific fatty acid synthases that accept acyl-CoA and acetyl-CoA substrates, PKSs can use other substrates such as propionyl-CoA, methylmalonyl-CoA, ethylmalonyl-CoA, and methoxymalonyl-CoA, which allow the production of a range of molecules with structural diversity. Often fungal PKS gene clusters contain additional genes that encode modifying enzymes. These allow further chemical modifications and an even greater structural diversity which is reflected in large variety of biologically active compounds.

Classes of PKS enzymes

The single-module fungal iterative type I PKSs are subdivided in three categories; non-reducing (NR), partially reducing (PR), and highly reducing (HR). The increase in reducing nature is the direct result of the incorporation of additional ketoreductase (KR), dehydratase (DH), and methylase (MT) activities into these multienzymes (Cox, 2007) In reducing iterative PKSs, the ketone group of a ketide unit can be reduced to various degrees. This process is catalysed by the ketoreductase (KR), dehydratase (DH) and enoyl reductase (ER) domains (Kroken *et al.*, 2003). The KR-domain is responsible for reducing the ketone group to a hydroxyl group, which is further reduced by the DH-domain to an enoyl group, which in turn can be reduced to an alkyl group by the ER-domain. Reducing iterative PKSs may also contain C-methyl transferases (C-MT) domains which yield further structural diversity. The utilization of alternate acyl-CoA molecules such as methyl-malonyl-CoA in this reaction may result in the introduction of branched structures into the main carbon chain (Song *et al.*, 2004).

In the iterative subclass of type I PKS, a single module is used repeatedly. The exact molecular programming of these enzymes remains enigmatic. Example of highly reducing fungal PKSs comprises the lovastatin synthases: LovB and LovF both yield products with a high degree of saturation (Sutherland *et al.*, 2001). It has been proposed that LovB catalyses the intramolecular Diels-Alder reaction needed to form lovastatin. This cyclisation likely occurs on an enzyme-bound hexaketide, which is subsequently extended to the nonaketide (Cox and Simpson, 2009). An example of a partially reducing PKS is the 6-methylsalicylate synthase from *Penicillium patulum*, one of the smallest fungal type I PKS (Dimroth *et al.*, 1970). This unimodular enzyme contains an N-terminal KS-domain followed by AT, DH, KR and ACP-domains. 6-Methylsalicylic acid (6-MSA) is synthesized from one acetate and three malonate molecules in a process that includes several repeated events (Cox and Simpson, 2009) After the third cycle, the chain undergoes cyclisation, dehydration and enolisation to yield the aromatic 6-MSA. The absence of a terminating thioesterase (TE) domain suggests that the release of the chain follows an alternative mechanism rather than the typical hydrolytic event (Staunton and Weissman, 2001).

Non-reducing (NR) PKSs produce aromatic compounds such as orsellinic acid, pentaketide-derived 1,3,6,8-tetrahydroxynaphthalene (THN), and norsolorinic acid, an intermediate in aflatoxin B biosynthesis (Brown *et al.*, 1999) Their biosynthetic modules contain KS-, AT-, and ACP-domains, but no KR- or ER-domains. These enzymes harbour two novel domains (Udwary *et al.*, 2002), i.e. the central product template domain that may be involved in the control of polyketide intermediate chain-length, folding, and stabilization, and a starter unit-ACP transacylase (sat) domain present at the extended N-terminus of these PKSs. *In silico* analysis of the sat domain revealed a similarity to known malonyl-CoA:acyl carrier protein transacylases (malonyl transferases). Just like several other fungal non-reductive PKSs, the wA PKS synthesizing naphthopyrone of *Aspergillus nidulans* or PKS1 synthesizing THN in *Colletotrichum lagenarium*, contain a terminal TE/CLC domain (Fujii *et al.*, 2000). The Claisen cyclase (CLC) domain affects the release of the polyketide product from the PKS by a resorcinol ring-forming Claisen condensation reaction (Fujii *et al.*, 2001). The CLC-domain has also been implied in controlling chain length by intercepting the polyketide intermediate from the ACP-domain during chain elongation in order to produce the shorter chain-lengths (Cox and Simpson, 2009).

Polyketide synthetase–non-ribosomal peptide synthetase hybrids

Fungal polyketide synthetase–non-ribosomal peptide synthetase (PKS–NRPS) hybrids produce a large group of structurally complex molecules. A few examples are: cytochalasins (Scherlach *et al.*, 2010), the neuritogenic and immunosuppressive agent pseurotin A (Komagata *et al.*, 1996), cyclopiazonic acid (Tokuoka *et al.*, 2008) and flavipucine (Gressler *et al.*, 2011). PKS–NRPS hybrids usually contain a highly reducing PKS, in which individual domains are programmed to function in an iterative manner. The process of biosynthesis starts with a KS-domain, which elongates an acetyl-CoA starter unit with malonyl-CoA extenders that are loaded onto the acyl carrier protein (ACP) with the help of the AT domain (Boettger and Hertweck, 2013). After each elongation step, the ketoreductase (KR), dehydratase (DH), and enoyl reductase (ER) domains mediate furtherβ-keto-processing steps (Staunton and Weissman, 2001). A characteristic trait of fungal PKS–NRPS hybrids is the lack of an intact ER domain within the PKS module (Halo *et al.*, 2008). Its catalytic function is either not essential for product formation, or is fulfilled by a trans-acting ER-domain of another protein encoded by a gene present in the biosynthetic gene cluster or elsewhere in the genome. The NRPS module is responsible for the selection and activation of specific amino acids via the adenylation domain (A), which loads the amino acyl residue onto the T-domain. After completion of the polyketide chain, the condensation (C) domain catalyses the fusion of the C-terminus PKS structure to the activated amino acid to yield an amide (Cox, 2007). There are two possible strategies to release the amide. First, the terminal reductive domain (R) may facilitate a reductive release, thereby generating a free aldehyde that undergoes a Knoevenagel condensation to form a pyrrolinone (Sims and Schmidt, 2008). Second, a tetramic acid derivative can be released directly through a Dieckmann cyclization reaction catalysed by the terminal domain (D). The characteristic primary products of the hybrid synthetase are tetramic acids (2-pyrrolidones). A typical PKS–NRPS hybrid contains approximately 10 catalytic domains, is larger than 400 kDa, and synthesizes acyltetramic acid (3-acyl-pyrrolidine-2,4-dione) (Boettger and Hertweck, 2013). Branches in the chain are usually introduced through S-adenosylmethionine-derived α-methylation of the growing polyketide chain by designated C-methyltransferase (C-MT) domains.

Bioinformatic analysis of secondary metabolite genes in fungal genomes

During recent years, large numbers of sequenced genomes of filamentous fungi have become available. This has promoted bioinformatics screens for the genomic distribution of NRPS and PKS enzymes, and a prediction of their function. Various specific web-based prediction tools (Jenke-Kodama and Dittmann, 2009) can be used. By BLAST analysis, an initial assignment of PKS, NRPS, and hybrid PKS–NRPS genes can be made (Sanchez *et al.*, 2012). More advanced genome scanning methods (Li and Durbin, 2009; Li *et al.*, 2009) allow the identification of PKS domains and a prediction of the molecule formed based on the presumed use of certain precursors (Yadav *et al.*, 2003a,b) a prediction of the amino acid specificity of A-domains of NRPSs (Rausch *et al.*, 2005) and the construction of a nonribosomal peptide (NRP) database including a prediction of their biological activity (Abdo *et al.*, 2012). It should be emphasized, however, that the prediction of the A-domain specificity works reasonable well for bacterial NRPSs, but for fungi, this method is still highly unreliable. Rapid progress in open source secondary metabolite prediction methods is further assisted by a growing number of bioinformatics tools to process the large volume of high-resolution mass spectrometers data (Scheltema *et al.*, 2011). Several of these web-based programs offer an analysis of associated genes and a prediction of their function (Khaldi *et al.*, 2010; Priebe *et al.*, 2011). A functional association of biosynthetic gene clusters with natural product formation is predicted by the software tool the Secondary Metabolite Analysis Shell (antiSMASH) (Weber *et al.*, 2015). However, still experimental methods are necessary to reveal the actual function of NRPS, PKS and the hybrid PKS–NRPS enzymes.

Secondary metabolite gene clusters and their products in *P. chrysogenum*

Using bioinformatics tools, 32 gene clusters have been identified in the *P. chrysogenum* Wisconsin genome that could be potentially involved in secondary metabolite production. Of these clusters, 10 are NRPS-associated, 20 are PKS-associated, and 2 specify hybrid PKS–NRPS enzymes (van den Berg *et al.*, 2008) (Table 9.1). In addition, there are 15 genes that appear to specify incomplete NRPS enzymes with catalytic domains similar to those found in NRPS enzymes, but lacking the condensation domain necessary for peptide formation. The identification of secondary metabolite genes is based on unique domain features of fungal enzymes that can be detected with the genome annotation tools SMURF. All the PKS and NRPS related gene clusters, the domain organization of the associated PKSs and NRPSs as well as their putative function is listed in Table 9.1. During the last few decades, many bioactive compounds have been detected in the culture broth of *P. chrysogenum* and for several of these compounds the genes involved in their biosynthesis have been identified and the products have been characterized. Interestingly, the intensive classical strain improvement program of *P. chrysogenum* eliminated pigment and other secondary metabolite formation (Corse *et al.*, 1948). Therefore, many of the metabolites described below are only produced in significant quantities by wild type-like *P. chrysogenum* strains.

β-lactams

Penicillins were the first discovered β-lactam antibiotics and have remained therapeutically and commercially important for the last six decades. This class of compounds all have in common that they harbour a conserved β-lactam ring nucleus. Benzylpenicillin (penicillin G) and phenoxymethylpenicillin (penicillin V) were the first natural penicillins introduced for clinical applications. These antibiotics exhibit a broad antimicrobial activity against bacteria by inhibiting cell wall biosynthesis. Penicillins block the formation of the peptidoglycan cross-links as the β-lactam ring covalently associates with the DD-transpeptidases as termed penicillin binding proteins (PBPs). There are three main genes involved in penicillin biosynthesis and these are clustered in the genome (van den Berg *et al.*, 2007; Weber *et al.*, 2012). These are the *pcbAB*, *pcbC* and *penDE* genes that encode the non-ribosomal

Table 9.1 Secondary metabolite gene clusters in *Penicillium chrysogenum*

Gene ID	Gene name	Function	Domain organization	Product group
Pc13g05250	*pssC*	Siderophore synthetase	$A_1TCA_2TCTCA_3TCTCT$	Ferrichrome
Pc13g14330		Tetrapeptide synthetase	$CA_1TECA_2TCA_3TCA3TCA_4TC$	–
Pc16g03850	*pssA*	Siderophore synthetase	ATCTA	Coprogen
Pc16g04690	*hcpA*	Cyclic tetrapeptide synthetase	$A_1TECA_2A_3TCA_4TECTCT$	Fungisporin
Pc21g01710		Dipeptide synthetase	A_1TCA_2T	–
Pc21g10790		Hexapeptide synthetase	$A_1TCA_2TCA_3TECA_4TCA_5TCA_6TC$	–
Pc21g12630	*chyA*	2-aminobenzamide synthetase	A_1TCA_2TC	Chrysogine
Pc21g15480	*roqA*	Histidyl-tryptophanyldiketo-piperazine synthetase	A_1TCA_2TC	Roquefortin/ Melagrin
Pc21g21390	*pcbAB*	α-Aminoadipyl-cysteinyl-valine synthetase	$A_1TCA_2TCA_3TEte$	β-lactams
Pc22g20400	*pssB*	Siderophore synthetase	ATCTC	Fusarinines
Pc06g01540		NRPS-like	A	–
Pc12g09980		NRPS-like	A*	–
Pc12g13170		NRPS-like	A*	–
Pc13g12570		NRPS-like	A	–
Pc14g01790		NRPS-like	A*	–
Pc16g09930		NRPS-like	A*te	–
Pc18g00380		NRPS-like	A	–
Pc20g02260		NRPS-like	A*	–
Pc20g02590		NRPS-like	A*	–
Pc20g09690		NRPS-like	A*	–
Pc20g12670		NRPS-like	A	–
Pc21g22530		NRPS-like	A	–
Pc21g22650		NRPS-like	A*	–
Pc22g06310		NRPS-like	A	–
Pc22g09430		NRPS-like	A*	–
Pc12g05590		Polyketide synthase	ks-at-dh-mt-kr-acp	–
Pc13g04470		Polyketide synthase	ks-at-dh-mt-er-kr-acp	–
Pc13g08690		Polyketide synthase	ks-at-dh-mt-er-kr-acp	–
Pc16g00370		6-methylsalicylic acid synthase	ks-at-kr-acp	6-MSA
Pc16g03800		Polyketide synthase	ks-at-dh-er-kr-acp	–
Pc16g04890		Polyketide synthase	ks-at-dh-mt-er-kr-acp	–
Pc16g11480		Polyketide synthase	ks-at-dh-mt-er-kr-acp	–
Pc21g00960		Polyketide synthase	ks-at-dh-mt-er-kr-acp	–
Pc21g03930		Polyketide synthase	ks-at-dh-mt-er-kr-acp	–
Pc21g03990		Polyketide synthase	ks-at-dh-er-kr-acp	–
Pc21g04840		Polyketide synthase	ks-at-dh-er-kr-acp	–

Table 9.1 Continued

Gene ID	Gene name	Function	Domain organization	Product group
Pc21g05070		Polyketide synthase	ks-at-acp-mt-te/red	–
Pc21g05080		Polyketide synthase	ks-at-dh-mt-er-kr-acp	–
Pc21g12440		Polyketide synthase	ks-at-dh-er-kr-acp	–
Pc21g12450		Polyketide synthase	ks-at-acp-te	–
Pc21g15160		Polyketide synthase	ks-at-dh-mt-er-kr-acp	–
Pc21g16000		Naphtha-γ-pyrone synthase	ks-at-acp-acp-te	naphtha-γ-pyrone
Pc22g08170		6-Methylsalicylic acid synthase	ks-at-kr-acp	6-MSA
Pc22g22850		Polyketide synthase	ks-at-acp-met-te/red	–
Pc22g23750		Polyketide synthase	ks-at-dh-mt-er-kr-acp	–
Pc12g02940		Polyketide synthase	ks-at-dh-er-kr-acp	–
Pc16g03760		Polyketide synthase	ks-at-dh-er-kr-acp	–
Pc22g09030		3-Oxoacyl-[acyl-carrier-protein]-synthase	acp	Fatty acids
Pc14g00080		Hybrid NRPS-PKS	ks-at-dh-kr-acp-acp-CA	–
Pc16g13930		Hybrid NRPS-PKS	ks-at-mt-ATd	–

Abbreviations used for domains: ks, keto-synthase; kr, ketoreductase; at, acyltransferase; mt, methyltransferase; er, enoyl reductase; acp, acyl carrier protein; te/red, thioester reductase; A, adenylation; T, thiolation; E, epimerization; te, thioesterase; C, condensation; A*, incomplete adenylation.

peptide synthetase L-δ-(α-aminoadipyl)-L-cysteinyl-D-valine synthetase (ACVS), isopenicillin N synthase (IPNS) and isopenicillin N acyltransferase (AT), respectively (Fig. 9.1). One of the main adaptations in high yielding β-lactam producers that emerged from the classical strain improvement programmeis the amplification of the penicillin biosynthetic gene cluster with in some strains up to 50 copies (Fierro *et al.*, 2006; Newbert *et al.*, 1997)

Biosynthesis starts with the NRPS mediated condensation of the amino acids α-aminoadipic acid, cysteine and valine and formation of tripeptide L-δ-(α-aminoadipyl)-L-cysteinyl-D-valine. Importantly, a peptide bond is formed between the δ-amino group of α-aminoadipic acid and the carboxyl-group of L-cysteine. In addition, the NRPS mediates the epimerization of L-valine to incorporate D-valine. β-lactam ring formation is catalysed by isopenicillin N synthase to yield isopenicillin N. Isopenicillin N is a very weak intermediate, because it does not show strong antibiotic activity. In the next step, the α-aminoadipate moiety is replaced by an alternative side chain through a transamidation reaction involving the isopenicillin N acetyltransferase (AT) which localizes to microbodies within the cell (Meijer *et al.*, 2010; Muller *et al.*, 1991). AT accepts a CoA-activated side chain like phenylacetic acid and phenoxyacetic acid to convert isopenicillin N into penicillin G and V, respectively. Such organic acids are added to the culture supernatant to direct the biosynthesis towards the aforementioned products. However, in the absence of such a feed, *P. chrysogenum* withdraws endogenous CoA-activated acids to yield compounds such as

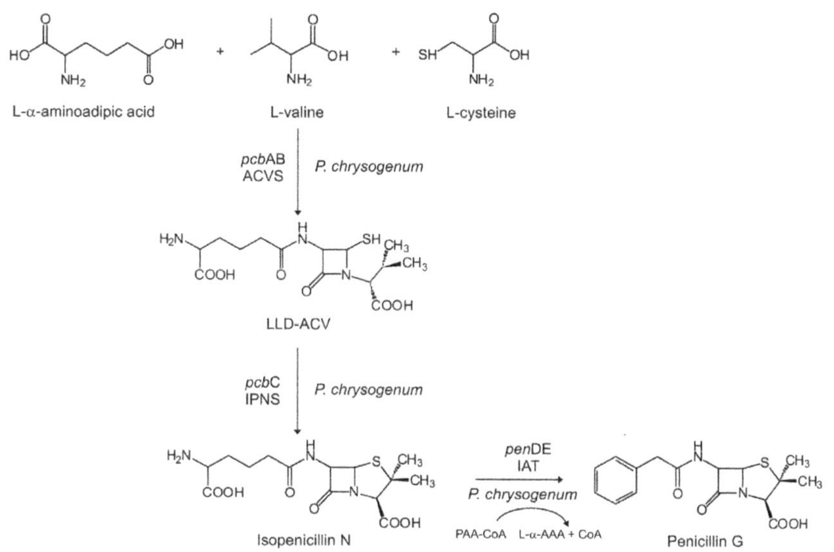

Figure 9.1 Scheme of the penicillin biosynthetic pathway. See text for details.

penicillin K and L that are less effective as the commercial used β-lactam antibiotics. The AT enzyme also harbours hydrolytic activity, and converts isopenicillin N into 6-aminopenicillanic acid (6-APA). The latter is an important precursor for the production of semi-synthetic β-lactams such as amoxicillin and ampicillin. This semisynthetic production processes has expanded the activity spectrum of this class of antibiotics.

Roquefortines

Penicillium strains produce a variety of mycotoxins termed roquefortines. This bacteriostatic compound is a well known contaminant during the production of Roquefort cheese which involves the fungus *P. roqueforti* (Scott *et al.*, 1977). The genetic origin and the enzymatic steps of the roquefortines/meleagrin biosynthesis pathway has recently been resolved in *P. chrysogenum* using a combination of metabolic profiling and structural elucidation of the products formed by a series of deletion strains of the cluster of seven genes (Ali *et al.*, 2013; Ries *et al.*, 2013). The pathway is responsible for more than 13 different metabolites that are released into the culture broth. Major abundant metabolites of this pathway are histidyl-tryptophanyldiketopiperazine (HTD), dehydrohistidyltryptophanyldiketopiperazine (DHTD), roquefortine D, roquefortine C, glandicoline A, glandicoline B and meleagrin. Specific genes could be assigned to each enzymatic reaction step (Fig. 9.2). The NRPS RoqA accepts L-histidine and L-tryptophan as substrates leading to the production of the diketopiperazine HTD. DHTD, previously suggested to be a degradation product of roquefortine C, is derived from HTD involving the cytochrome P450 oxidoreductase RoqR. The dimethylallyltryptophan synthetase RoqD prenylates both HTD and DHTD yielding directly the products roquefortine D and roquefortine C without the synthesis of the previously indicated DHTD intermediate and the involvement of RoqM. This leads to a branch in the otherwise linear pathway. Roquefortine C is subsequently converted into glandicoline B with glandicoline A as intermediates, involving

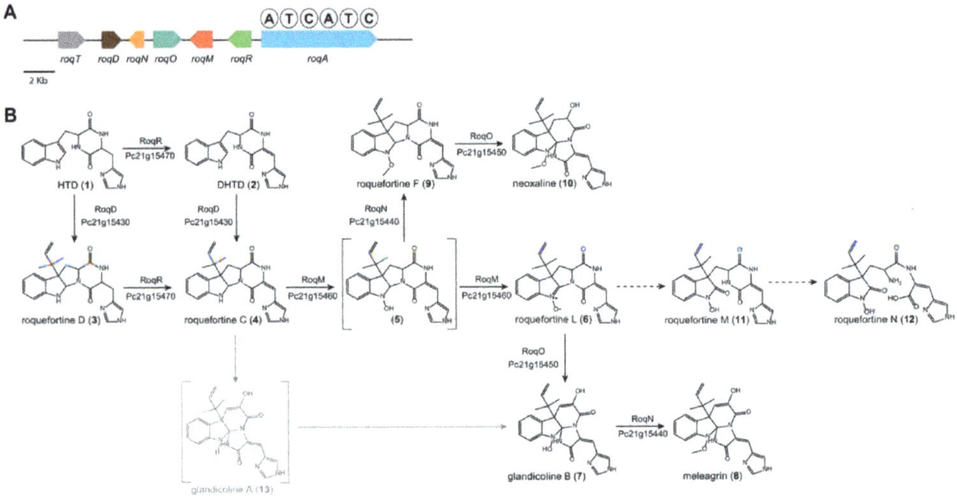

Figure 9.2 Roquefortine/meleagrin biosynthetic gene cluster and pathway. (A) Organization of the roquefortine/meleagrin biosynthetic gene cluster. (B) Proposed roquefortine/meleagrin pathway. Enzymatic catalysed reactions are indicated by solid arrows whereas chemical reactions are indicated by dashed arrows. See text for details. From (Ries *et al.*, 2013).

two monooxygenases (RoqM and RoqO) which were mixed up in an earlier attempt to elucidate the biosynthetic pathway (Garcia-Estrada *et al.*, 2011) Eventually, meleagrin is produced from glandicoline B involving a methyltransferase (RoqN). This pathway is also responsible for the production of other intermediates such as roquefortine F and L that arise from the intrinsic unspecificity of several enzymes of this pathway thereby catalysing more than one reaction, leading to excessive branching in the pathway with meleagrin and neoxaline as end products of two branches. In other *Penicillium* strains shorter pathways for roquefortine exist (Ali *et al.*, 2013; Kosalkova *et al.*, 2015; Ries *et al.*, 2013). Interestingly, *P. chrysogenum* strains that were subjected to classical strain improvement produce relatively low amounts of roquefortines and meleagrin. However, deletion of all β-lactam biosynthetic gene clusters and thus the elimination of penicillin production, results in enhanced levels of roquefortins whereas the expression of the pathway genes was not altered (Ali *et al.*, 2013; Ries *et al.*, 2013). This phenomenon is likely due to a re-diversion of nitrogen metabolism towards other NRPs such as roquefortins in the absence the amino acid drain imposed by penicillin production.

The most abundant products like roquefortine C and D exhibit neurotoxic and bacteriostatic activities. Roquefortine L is potential therapeutic agent for treatment of the ischaemic stroke and certain tumours. A *P. chrysogenum* strain lacking the *roqA* gene, processes synthetic, exogenously added HTD to yield a set of roquefortine-based secondary metabolites also produced by the wild-type strain (Ouchaou *et al.*, 2015). This was used for a mutasynthesis approach in which various synthetic HTD analogues were converted into a number of new roquefortine D derivatives, depending on the nature of the synthetic HTD added. This knowledge may be tapped for the future development of more complex semisynthetic roquefortine-based secondary metabolites.

Siderophores

The production of metal ion-binding siderophores is widely spread among filamentous fungi. The bioavailability of iron is restricted due to the low solubility of Fe^{3+} in aerobic environments. To mobilize the iron, fungi excrete high-affinity siderophores into the medium (Boukhalfa and Crumbliss, 2002). The genome of *P. chrysogenum* specifies at least three NRPS genes involved in the production of coprogen (PssA, Pc16g03850), fusarinine C (PssB, Pc22g20400) and ferrichrome (PssC, Pc13g05250) (Fig. 9.3) (Samol *et al.*, unpublished). Unlike earlier predictions, *P. chrysogenum* does not have the genes to synthesize and degrade triacetylfusarine. In addition, *P. chrysogenum* produces two novel fusarinine C-related compounds under iron deficiency conditions, which depend on both PssA and PssB for their biosynthesis. Coprogen production is dependent on PssA. There are further NRPS encoding genes that might be involved in siderophore formation as they are expressed only under iron-depleted growth conditions. Overexpression of Pc13g14330 by replacing its native promoter with the IPNS promoter resulted in almost 3-fold increase in ferrichrome biosynthesis (M. Samol, unpublished data) but its actual role is unknown. Pc21g01710 encodes a NRPS that is homologous to SidE of *A. clavatus* and *N. fischeri* that is involved biosynthesis of ferrichrome. It thus appears that *P. chrysogenum* has multiple and possibly redundant mechanisms for siderophore production.

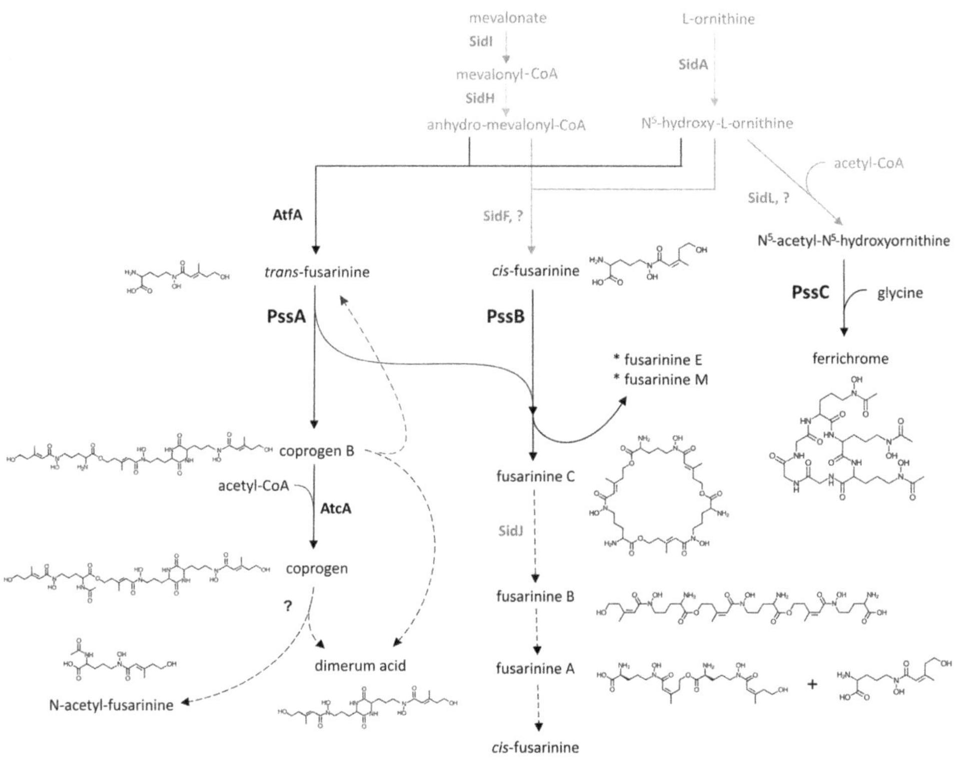

Figure 9.3 Proposed biosynthetic pathways of siderophores in *P. chrysogenum* (Black) based on the *A. fumigatus* pathways (Grey). * indicates novel siderophores with iron chelating properties and ⟶ indicates degradation routes.

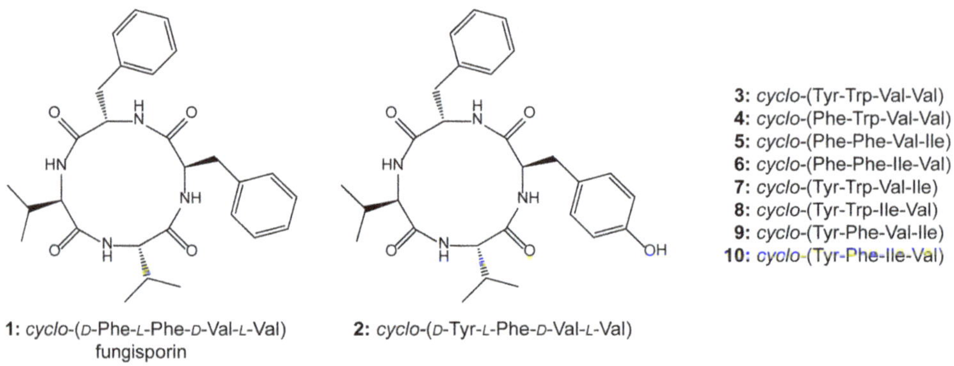

Figure 9.4 Structures of cyclic hydrophobic tetrapeptides produced by the HcpA protein. From (Ali *et al.*, 2014).

Fungisporin

A group of at least 10 cyclic hydrophobic peptides including fungisporin has been linked to a single NRPS termed HcpA (Pc16g04690) (Fig. 9.4) (Ali *et al.*, 2014). These peptides play a role in fungal colonial morphology. In addition, the fungus produces a series of linear variants of the cyclic peptides that are likely products of a proteolytic event. HcpA is a non-canonical tetra-modular NRPS with microheterogenicity of all involved adenylation domains towards their respective substrates (Ali *et al.*, 2014). In combination with the mapping of the substrate predictions for each module, a mechanism has been proposed for a 'trans-acting' adenylation domain. A fungisporin reported from *Onychocola sclerotic* was shown to be have cardiac channel blocking activities. Interestingly, the colonies of the Δ*hcpA* strain have lost the ability to produce a wrinkled surface leading to a rather smooth appearance suggesting that they fulfil a kind of hydrophobin-like role in determining the wettability of the surface. Because of their highly hydrophobic characteristics, the production of fungisporins is undesirable. For instance, in industrial application, these molecules tend to aggregate and clog tubing's of the fermentation equipment.

Sorbicillinoids

Some coloured secondary metabolites co-purify together with penicillin from the fermentation broth of *P. chrysogenum* (*P. notatum*) (Cram and Tishler, 1948). In the early days, yellow pigments like sorbicillin and penitrinic acid (see below) were found to be the main contaminants during β-lactam production. The sorbicillins are a vast group of polyketides produced by *P. chrysogenum* and have been purified and analysed in detail. Together with structurally related metabolites they belongs to the rapidly growing family of sorbicillinoids that were first identified in *P. notatum* (Maskey *et al.*, 2005). Other fungal species like *Aspergillus parasiticus* (Basaran and Demirbas, 2010), *Trichoderma* sp. USF-2690 (Abe *et al.*, 2000, 2002), and *Phaeoacremonium* NRRL 32148 (Reategui *et al.*, 2006) are also known to produce sorbicillins and derivatives. Sorbicillin production is eliminated in the early penicillin producing strains Wisconsin 54-1255 whereas the progenitor NRRL1951 produces a wide variety of sorbicillins. The unique hexaketide origin of the sorbicillins indicates the involvement of polyketide synthases in their biosynthesis. The structure of sorbicillin corresponds to a core scaffold present in about 30 monomeric and dimeric vertinoids found in a variety of fungal species and in different environments (Abe *et al.*, 2000, 2002; Basaran and Demirbas, 2010; Bringmann *et al.*, 2003; Reategui *et al.*, 2006). Novel molecules such as sorrentanone and bisvertinolone (previously also reported for *Verticillium interextum*) are produced by *P. chrysogenum*. Sorrentanone shows a moderate antimicrobial activity against *Staphylococcus* and *Enterococcus* species (Miller and Huang, 1995). The oxidized form of sorbicillin dimerizes into complex structures via Diels–Alder or Michael-type reactions which are known as bisorbicillinoids (Nicolaou *et al.*, 1999; Stocking and Williams, 2003). Both mono- and dimeric sorbicillins were extracted from the culture broth of *P. notatum*, including monomeric metabolites such as 2′,3′-dihydrosorbicillin, sohirnone A, B and C, oxosorbicillinol with the related tautomer and 7-deacetoxyyanuthone. Dimeric molecules found are bisorbicillinol, bisvertinoquinol, isorbibutenolide, rezishanones A, B, C, D and bisvertinolone (Maskey *et al.*, 2005). Many sorbicillins exhibit antimicrobial activity against *Staphylococcus aureus* and *Bacillus subtilis* (Maskey *et al.*, 2005). Recently, also new sorbicillinoids structures were identified in the BIOTECmarine program that aimed to investigate the metabolome of sponge-associated microorganisms. Sorbicillactone A and B were found to be produced by a *P. chrysogenum* isolate derived from spongy *Ircinia fasciculate* (Bringmann *et al.*, 2003). The origin of the unusual structures was elucidated through feeding experiments using ^{13}C-labelled precursors, and a biosynthetic mechanism of the sorbicillactones was proposed (Bringmann *et al.*, 2005). Sorbicillactone A is a low toxic compound with anti-HIV properties but exhibits a strong cytotoxic effect against L5178y leukaemic cells (Bringmann *et al.*, 2005). These remarkable pharmaceutical features of sorbicillactone A have led to the development of improved cultivation techniques of *P. chrysogenum* for the large-scale production of this compound (Bringmann *et al.*, 2007).

ω-Hydroxyemodin

The bright-yellow pigment ω-hydroxyemodin was purified from extracts of *P. cyclopium* (Anslow *et al.*, 1940). An unnamed compound with similar characteristics was also extracted from *P. citreoroseum* that was later characterized as *P. chrysogenum* according to the morphological, physiological and extrolites features (Houbraken *et al.*, 2012). Production of ω-hydroxyemodin and the related emodic acid was demonstrated through a systematic analysis of the extrolites produced by *Penicillium* species (Samson and Frisvad, 2004). Emodin and its derivatives affect cell transformation and DNA repair in *Salmonella typhimurium* (Westendorf *et al.*, 1990). The ability of emodic acid to generate reactive oxygen species (ROS) has been exploited for targeted photodynamic chemotherapy of cancer cells (Rahimipour *et al.*, 2001). Recently, an insecticidal activity of emodic acid has been described (Georges *et al.*, 2008). ω-hydroxy emodin is likely produced by a PKS, but the genes involved have not yet been identified.

Chrysogenin

P. chrysogenum is well known for the production of a series of compounds with a yellow colour, which are secreted in drops on top of the colonies. The production of the yellow pigment chrysogenin was proposed to be characteristic feature of *P. chrysogenum* when growing on milk (Clutterbuck *et al.*, 1932). In the early days, the typical yellow colour of the culture broth of *P. chrysogenum* was used to make a rough estimate of the antibiotic penicillin produced (Corse *et al.*, 1948). Even today, despite advanced analytical methods available, chrysogenin remains a qualitative marker for the evaluation of the fungal secondary metabolism in low penicillin producing strains (Garcia-Rico *et al.*, 2008; Macdonald *et al.*, 1963). The pigment(s), previously observed in *P. rubens* by Fleming (Fleming, 2001), was later also identified in the *P. chrysogenum* type strain (№26) and named chrysogenin (Clutterbuck *et al.*, 1932). The absorption spectrum of chrysogenin suggests the presence of a methyl polyhydroxyanthraquinone structure similar to emodin (see above) with a provisionally adopted empirical formula $C_{18}H_{22}O_6$. This suggests that it is synthesized by a PKS. No antimicrobial activity seems to be associated with the purified compound (Wolf *et al.*, 1960), and no further structural analysis has been reported.

Chrysogine

Chrysogine is a metabolite that likely is derived from the condensation of anthranilic acid and alanine through the function of di-modular NRPS (Hikino *et al.*, 1973). Recently, the biosynthetic gene cluster with six co-regulated genes including di-modular NRPS ChyA was identified (Ali *et al.*, unpublished data). The five other genes of the cluster are probably involved in tailoring the 2-aminobenzamide core, and this includes a putative amidotransferase (ChyE), oxygenase (ChyH), reductase (ChyC), desaturase (ChyM), and an asparagine synthase (ChyD). In addition, the cluster contains a local regulator ChyR. All the clustered genes are highly expressed under submerged shaking flask growth condition and co-suppressed in presence of phenylacetic acid in the medium. By gene inactivating and metabolic profiling, the production of chrysogine and many related metabolites was shown to be eliminated upon the deletion of the *chyA* gene. These metabolites have all the 2-aminobenzamide core scaffold in common.

6-Methylsalicylic acid

6-Methylsalicylic acid (6-MSA) is produced by various bacterial and fungal species (Bacha *et al.*, 2009; Fujii, 2010). This triacetic polyketide is a common product reported in *Aspergillus* and *Penicillium* species. *P. chrysogenum* contains two PKS enzymes that show homology to 6-methylsalicylic acid synthases. The cluster of genes around the PKS Pc16g00370 of *P. chrysogenum* shows similarity to a cluster of nine genes of *A. niger*, while the PR PKS Pc22g08170 is homologous to the 6-methylsalicylic acid synthase from *P. griseofulvum*. The latter is involved in patulin biosynthesis which requires additional enzymatic conversions like decarboxylation, hydroxylation and oxidation reactions. In *P. chrysogenum* these enzymatic steps are presumably provided by an aminohydrolase and two cytochrome P450 enzymes present in the cluster. However, the gene encoding the isoepoxidon dehydrogenase cannot be found in the genome of *P. chrysogenum* consistent with the notion that *P. chrysogenum* does not produce patulin. Pc22g08170 is homologous to the 6-methylsalicylic acid synthase of *Paracoccidioides brasiliensis*. The PKS enzymes have been suggested to be involved in the synthesis of prenylated and hydroxylated derivatives of 6-MSA (van den Berg *et al.*, 2008).

6-MSA synthases are usually clustered in the genome together with other modifying gene products, and the initial polyketide product may serve as a precursor for the biosynthesis of more complex products. In *A. patulum* and *A. clavatus*, the 6-MSA synthase associated gene cluster is involved in the biosynthesis of the potent mycotoxin patulin but this compound is not produced by *P. chrysogenum*. 6-MSA synthases are also linked to the formation of asperlacton and isoasperlacton in *A. westerdikiaen* (Bacha *et al.*, 2009) and is involved in the biosynthesis of terreic acid in *A. terreus* (Guo *et al.*, 2014) In *A. niger*, a 6-MSA associated cluster of ten genes is involved in the biosynthesis of the meroterpenoids yanuthones (A–E) (Holm *et al.*, 2014; Petersen *et al.*, 2014). A nearly identical cluster is present in the genome of *P. chrysogenum* including the 6-MSA encoding genes. Pc16g00370 is homologous to the aforementioned 6-MSA synthase of *A. niger*, while Pc21g16430 is homologous to the 6-MSA synthase *of A. patulum*. However, neither of the two 6-MSA PKS genes in *P. chrysogenum* are expressed under laboratory conditions (O. Salo, unpublished results).

Sesquiterpene PR-toxin

PR-toxin is a potent mycotoxin produced by *P. roqueforti*. It is also produced by *P. chrysogenum* but at significantly lower levels. This metabolite is toxic for rats and mice. It interferes with the RNA polymerase in liver cells (Moule *et al.*, 1976) but also affects as brain, heart and kidney in rats and this has been attributed to an inhibition of the HCO_3-ATPase activity in mitochondria (Hsieh *et al.*, 1986). The gene cluster for PR-toxin biosynthesis has been characterized in *P. roqueforti* (Hidalgo *et al.*, 2014). The PR-toxin is derived from the 15 carbon atoms sesquiterpene aristolochene which formed by the aristolochene synthase (encoded by Ari1). Synthesis of the PR-toxin involves a cluster of four genes. Homologous genes are found in *P. chrysogenum*, namely Pc21g06300 (Prx1, a NADPH:quinone reductase), Pc21g06320 (Prx3), Pc21g0630 (Prx4) and the aristolochene synthase Prx2 (Pc12g06310). These genes are part of a large gene cluster of eleven genes including a member of the major facilitator superfamily. This gene cluster is poorly expressed in strongly aerated cultures of *P. chrysogenum* (Harris *et al.*, 2009). When *P. chrysogenum* is grown under static culture conditions on hydrated rice medium, production of PR-toxin could be detected (Hidalgo *et al.*, 2014).

Other secondary metabolites

The yellow cultured penitrinic acid is an unusual compound with empirical formula $C_{15}H_{17}O_5N$ contains a nitrogen atom, which is not a typical feature of fungal pigments. No structural data or other properties of the penitrinic acid have been reported since the initial discovery (Stodola *et al.*, 1945). Three new citrinin analogues, penicitols A–C, and one new xanthone derivative, penixanacid A, were discovered in an extract of a mangrove-derived *P. chrysogenum* HND11-24 (Guo *et al.*, 2015)

Other metabolite genes with unknown products

For several NRPS and PKS specifying genes, the associated products have not been identified (Table 9.1). One of the main reasons is that many of these genes are not expressed under laboratory conditions. They are collectively termed as ‘silent’ genes. Pc21g10790 encodes a very large NRPS with six adenylation domains and that is homologous to putative NRPSs present in *A. oryzae* and *A. flavus* NRRL3357. However, the function of this enzyme is unknown. *P. chrysogenum* contains various genes encoding for so-called monomodular NRPSs that appear to specify incomplete enzymes. The enzymes contain an adenylation-domain but lack the condensation domain necessary for peptide formation. *P. chrysogenum* contains 15 of such NRPS-like genes of which three are highly expressed under laboratory conditions. The function of these enzymes is unknown.

P. chrysogenum also contains two hybrid PKS-NRPS genes, but these are not expressed. Pc14g00080 is homologous to the nonaketide synthase LovB of *A. terreus*, and is presumable involved in the production of lovastatin, a member of the drug class of statins. These compounds are used in combination with diet, weight loss and exercise for lowering cholesterol to reduce risk of cardiovascular disease. Pc16g13930 is homologous to PKS–NRPS hybrid enzymes found in *Fusarium heterosporum*, *Magnaporthe grisea* and *Coccidioides immitis*, but the identity of the products formed is unknown.

Also for the majority of the PKS enzymes, the actual products are unknown. The PKS enzyme Pc12g05590 is part of a large gene cluster with twenty ORFs. It shows the highest identity to the EqiS-like PKS-NRPS enzyme of *A. fumigatus*, but it does not contain the adenylation domain found in EqiS (Sims *et al.*, 2005). Pc13g04470 is also part of a large gene cluster and is homologous tor the lovastatin diketide synthase LovF of *A. terreus*. Pc16g03800 is part of a cluster of 35 genes, and appears to be associated with PssA, which is involved in the biosynthesis of the siderophore coprogen. The PKS might be involved in biosynthesis of cis-5-hydroxy-3-methylpent-2-enoic acid, the precursor of fusarinine (van den Berg *et al.*, 2008), but the deletion of this gene had no effect on coprogen or fusarinine production (M. Samol, unpublished results). Pc16g04890 and Pc16g11480 show homology to PKS enzymes of *A. niger* and *A. fumigatus*, respectively, but these proteins have no assigned function. The highly reducing PKSs Pc21g00960, Pc21g03930 and Pc21g03990 show homology to some uncharacterized PKSs of other fungi with unknown function.

There are two neighbouring, oppositely oriented genes Pc21g05070 and Pc21g05080 that encode a NR and HR PKS enzyme, respectively. They belong to a large gene cluster. The NR PKS Pc21g05070 exhibits identity to citrinin polyketide synthase of *Monascus purpureus* while the HR PKS Pc21g05080 is homologous to lovastatin diketide synthase LovF of *A. terreus* (van den Berg *et al.*, 2008). A second putative gene cluster encodes a HR PKS Pc21g12440 and NR PKS Pc21g12450. These genes are part of a large cluster that also specifies ChyA and associated genes, and that appear to be co-regulated. The functional relation between the two PKS and ChyA is unknown. The HR PKS Pc21g15160 is homologous to an unknown polyketide synthase from *Trichophyton rubrum*, while the NR PKS encoded by Pc21g16000 shows similarity to the *A. nidulans* wA and *A. fumigatus* PksP naphthopyrone synthases. These enzymes are involved in the biosynthesis of the precursor for a conidial pigment. Finally, the HR PKS encoded by Pc22g23750 gene shows homology to a type I polyketide synthase of *Colletotrichum fioriniae*.

Approaches for the activation of 'silent' secondary metabolite gene clusters

The successful development of *P. chrysogenum* as a β-lactam production strain was initially based on an optimization of process parameters, but this was rapidly followed by a more elaborate classical strain improvement programme (van den Berg *et al.*, 2007). This pre-genomic strategy, however, is not applicable to the majority of the secondary metabolite pathways that are cryptic and these require sophisticated techniques for activation (Brakhage and Schroeckh, 2011). The post genomic era offers two main approaches for the activation of the silent gene clusters potentially involved in secondary metabolite production in filamentous fungi: (i) molecular based techniques targeting local transcriptional regulators, epigenetics interference, and promoter replacement, and (ii) simulation by using different growth conditions or conditions that correspond to the natural ecosystem involving predation by for instance insects (Brakhage and Schroeckh, 2011).

Gene cluster-specific transcriptional regulators and promoter replacement

Ectopic overexpression of pathway-specific regulators under the control of an inducible promoter has led to the discovery of novel secondary metabolites produced by a PKS-NRPS in *A. nidulans* (Bergmann *et al.*, 2010) but similar methods have also been used in *P. chrysogenum*. The genome of *P. chrysogenum* contains two polyketide synthase encoding genes that are homologous to 6-MSA synthases, Pc16g00370 and Pc21g16430 (Table 9.1) that are not expressed under laboratory conditions. Promoter replacement using the *pcbC* (isopenicillin synthase) promoter resulted in the expression of Pc16g00370 and Pc21g16430 and the formation of 6-MSA and several other compounds by *P. chrysogenum* (O. Salo, unpublished data).

Epigenetic interference

Epigenetic regulation of the secondary metabolism deserves special attention. It can be defined as reversible heritable changes in gene expression that are not related to changes in the DNA sequence. Recently, this definition gained broader applications and also concerns gene regulation via a change in chromatin structure that are not necessarily heritable (Smith *et al.*, 2012). This regulation is controlled by DNA methylation or other post-translational modifications of the histones and is linked to structural changes of the chromosome. The re-establishment between the closed (heterochromatin) and open (euchromatin) regions on the chromosome was shown to cause pleotropic effects on gene expression. This phenomenon was successfully applied for the discovery of the secondary metabolite pathways encoded by the clustered genes that are mostly silenced within compact heterochromatin locations on the chromosome. For example, interfering with the function of the transcriptional regulator LaeA in *A. nidulans* has resulted in the identification and functional characterization of a gene cluster responsible for the biosynthesis of terrequinone A (Bouhired *et al.*, 2007). LaeA is a nuclear protein homologous to the histone and arginine methyltransferases and the first pleotropic regulator of the secondary metabolism described in 2004 in *A. nidulans* and *P. chrysogenum* (Kosalkova *et al.*, 2009). It functions in association with proteins like LreA, LreB, FphA, KapA, VeA, VelB and VosA that together constitute the Velvet complex (Purschwitz *et al.*, 2008; Sarikaya Bayram *et al.*, 2010; Bayram and Braus, 2012). The regulatory effect of the Velvet complex on cell development and secondary metabolism of filamentous fungi has been used as a powerful tool for the discovery of novel metabolites (Bayram *et al.*, 2008). LaeA functions through heterochromatin repression and reorganization of the heterochromatic marks on the chromosome. LaeA is named after the *Loss of aflR expression* phenotype that was observed in corresponding gene inactivation mutant of *A. nidulans* (Bok and Keller, 2004). Homologues of LaeA are also found in other aspergilli as well as in *P. chrysogenum* where its impact on the secondary metabolism was examined (Kale *et al.*, 2008; Veiga *et al.*, 2012). Studies on *A. fumigatus* show that LeaA transcriptionally deregulates 13 secondary metabolite gene clusters out of the 22 present in the genome (Perrin *et al.*, 2007). In *A. nidulans*, LaeA positively regulates sterigmatocystin and penicillin biosynthesis. A similar phenotype was shown for gliotoxin production in *A. fumigatus* (Sugui *et al.*, 2007) and lovastatin production in *A. terreus* (Bok and Keller, 2004). Studies on the role of LaeA in *P. chrysogenum*, however, lead to some controversial observations. LaeA was initially reported as an activator of penicillin biosynthetic gene cluster and was linked to morphological features like sporulation and pigmentation using a prolonged batch cultivation technique (Hoff *et al.*, 2010; Kosalkova *et al.*, 2009). Such a phenotype, however, was not observed when cells were grown under aerobic, glucose-limited conditions. It appears that the function of LaeA in *P. chrysogenum* is context dependent. It may vary under different nutrient limiting and environmental conditions (Veiga *et al.*, 2012), but also the strain background may determine the response as some of the *P. chrysogenum* strains currently studied were already subjected to classical strain improvement and this may have interfered with regulation.

Targeting of the histone acetylation status of the chromosome is another way to modulate the expression of genes. A balanced activity of histone acetylases (HAC) and histone deacetylases (HDAC) is required for the structuring of the chromatin (Brosch *et al.*, 2008). A number of successful attempts for deregulation of this equilibrium by genetic manipulations or enzymatic inhibition have been reported amongst others the production of calphostin B and cladochromes in *Cladosporium cladosporioides*. In *A. niger*, the production of nygerone A was induced when the growth medium was supplemented with the HDAC inhibitor suberoylanilide hydroxamic acid (SAHA) (Fisch *et al.*, 2009). An altered secondary metabolite production was also observed for *Alternaria alternata* and *P. expansum* treated with another HDAC inhibitor Trichostatin A (Shwab *et al.*, 2007). Direct targeting of the HDAC encoding genes has been performed for aspergilli reporting their regulatory role in secondary metabolism as well as germination and resistance to stress environmental conditions (Lee *et al.*, 2009; Shwab *et al.*, 2007; Tribus *et al.*, 2005). Our recent results suggest that an epigenomic approach can be successfully applied for the activation of secondary metabolism in *P. chrysogenum*. The Pc21g14570 gene of this fungus encodes an orthologue of a class 2 histone deacetylase termed HdaA which plays a role in epigenetic regulation of secondary metabolism. Deletion of the *hdaA* gene induces a significant pleiotropic effect on the expression of a set of PKS and NRPS encoding genes (M. Samol, unpublished data). The *hdaA* deletion mutant exhibits decreased conidial pigmentation that links to a reduced expression of the PKS gene Pc21g16000 responsible for the production of the conidial pigment precursor naphtha-γ-pyrone. Moreover, the lack of *hdaA* caused decreased levels of the yellow pigment chrysogine that could be associated with the down-regulation of the NRPS encoding gene Pc21g12630 and neighbouring genes that are part of a cluster. In addition, transcriptional activation of an unknown PKS gene cluster containing Pc21g05070 and Pc21g05080 occurred. However, these genes already acquired single nucleotide substitutions at early stages of the classical strain improvement of *P. chrysogenum* for β-lactam production that likely inactivated their activities and thus no specific metabolites could be detected.

Co-cultivation and other environmental constraints

One common method to active silent gene clusters is to grow the fungus on different types of media, both in submersed cultures and on surfaces. However, it is generally believed that this method mimics environmental conditions and thus are surrogate in reflecting native interactions in true ecosystems. Bacteria and fungi co-inhabit a wide variety of environments, from soil and food products to plants and mammals (Wargo and Hogan, 2006). Mutual interactions may have diverse effects on survival, colonization, and pathogenesis of these organisms, and often result in a vast diversity in chemical products (e.g. virulence determinants, pheromones, symbiotic factors, or unknown molecules). The concept of interspecies associations and their biosynthetic potential is an emerging field of research, though still largely unexplored. Natural communities of microorganisms such as soil-dwelling bacteria and fungi were already known to produce a multitude of molecules. By co-cultivating of a collection of 58 actinomycetes with *A. nidulans* sharing the same habitat, one species of *Streptomyces hygroscopicus* was found to induce fungal secondary metabolite production (Schroeckh *et al.*, 2009). By the use of a specific *A. nidulans* secondary metabolism gene array, the up-regulation of PKS and NRPS gene clusters could be demonstrated. Using metabolic profiling, the chemical structures of orsellinic acid, the typical lichen metabolite lecanoric acid, and the cathepsin K inhibitors F-9775A and F-9775B could be resolved. These compounds require the archetypal orsellinic acid polyketide synthase for biosynthesis as confirmed through gene inactivation experiments. To exclude the involvement of low molecular weight signalling molecules in the induction of fungal gene expression during the co-cultivation experiment, the fungal culture was also treated with supernatant of the bacterial culture and this did not result in fungal secondary metabolite production. Therefore, it was concluded that intimate contact is needed between *A. nidulans* and the *Streptomyces* strain (Schroeckh *et al.*, 2009), which results in the induction of otherwise silent biosynthesis genes.

The functional role of natural products likely confers biological advantages for the symbiotic associations. For example, peramine – a potent insect feeding deterrent – is a pyrrolopyrazine synthesized by *Epichloë/Neotyphodium* mutualistic endophytes in association with their grass hosts (Tanaka *et al.*, 2005) and thus protects the host plant from insect *Graminacious herbivore* (Rowan *et al.*, 1990). Its structure suggests that it is the product of a reaction catalysed by a two-module NRPS, derived from proline and arginine via a diketopiperazine intermediate (Tanaka *et al.*, 2005). Some fungal plant pathogens can also function in an antagonistic way. An example concerns isolates containing a host selective protein, which determine the pathogenicity of the fungus *Pyrenophora tritici-repentis* towards wheat (Ciuffetti *et al.*, 1997). The hybrid PKS–NRPS product 2-pyridone tenellin is released by the entomopathogenic fungus *Beauveria bassiana* upon contact with insects (Halo *et al.*, 2008). Simulating the natural microbial communication under laboratory conditions provides a largely unexplored method to activate biosynthetic gene clusters and may reveal the molecular (and chemical) context behind interspecies cross-talks. Co-cultivation experiments of *P. chrysogenum* with the bacterium *Bacillus subtilis* did not induce secondary metabolite formation, but instead caused the expression of two α-1,3 endoglucanases that are likely involved in the degradation of bacterial exopolysaccharides (Bajaj *et al.*, 2014).

Conclusions and future trends

P. chrysogenum is well known for industrial applications for the production of β-lactam antibiotics but its further potential for secondary metabolite formation has less well been studied. In recent years, by the use of combined metabolic profiling and gene inactivation or epigenetic interference, a vast potential of *P. chrysogenum* to produce secondary metabolites has been uncovered. However, many of the potential secondary metabolite associated genes are transcriptionally silent and not expressed under laboratory conditions. This may represent a largely untapped source for potential interesting bioactive compounds. Because of its industrial usage, and as a result of the extensive classical strain improvement program, current *P. chrysogenum* strains have superb characteristics for large-scale fermentation and, importantly, nitrogen metabolism is geared towards the production of nonribosomal peptides. This is evident when the multiple penicillin biosynthetic gene clusters are genetically removed, causing a redirection in the nitrogen flow and the high level production of other nonribosomal peptides. This implies that *P. chrysogenum* is an excellent host for the production of such compounds, which can be exploited in the heterologous expression of NRPS clusters. The potential for industrial secondary metabolite formation is also shows by the recent demonstration that a β-lactam negative strain of *P. chrysogenum* can be used for the heterologous production of the cholesterol-lowering drug pravastatin (McLean *et al.*, 2015). With the availability of more advanced genetic methods and procedures for genome editing, the development of *P. chrysogenum* into a generic host for peptide production may become a reality in the near future.

Acknowledgements

This work was supported by STW (Stichting Technische Wetenschappen) in the framework of the GenBiotics Perspectief Programme, and by the following public private networks: Integration of Biosynthesis and Organic Synthesis (IBOS) programme of Advanced Chemical Technologies for Sustainability (ACTS), and by the research program of the Bio-based Ecologically Balanced Sustainable Industrial Chemistry (BE-BASIC).

Web resources

AntiSMASH antibiotics and Secondary Metabolite Analysis SHell: http://antismash.secondarymetabolites.org

Smurf Secondary metabolite unique regions finder: http://jcvi.org/smurf/index.php

References

Abdo, A., Caboche, S., Leclere, V., Jacques, P., and Pupin, M. (2012). A new fingerprint to predict nonribosomal peptides activity. J. Comput. Aided Mol. Des. *26*, 1187–1194.

Abe, N., Arakawa, T., Yamamoto, K., and Hirota, A. (2002). Biosynthesis of bisorbicillinoid in Trichoderma sp. USF-2690; evidence for the biosynthetic pathway, via sorbicillinol, of sorbicillin, bisorbicillinol, bisorbibutenolide, and bisorbicillinolide. Biosci. Biotechnol. Biochem. *66*, 2090–2099.

Abe, N., Yamamoto, K., and Hirota, A. (2000). Novel fungal metabolites, demethylsorbicillin and oxosorbicillinol, isolated from Trichoderma sp. USF-2690. Biosci. Biotechnol. Biochem. *64*, 620–622.

Ali, H., Ries, M.I., Lankhorst, P.P., van der Hoeven, R.A., Schouten, O.L., Noga, M., Hankemeier, T., van Peij, N.N., Bovenberg, R.A., Vreeken, R.J., *et al.* (2014). A non-canonical NRPS is involved in the synthesis of fungisporin and related hydrophobic cyclic tetrapeptides in *Penicillium chrysogenum*. PLoS One *9*, e98212.

Ali, H., Ries, M.I., Nijland, J.G., Lankhorst, P.P., Hankemeier, T., Bovenberg, R.A., Vreeken, R.J., and Driessen, A.J. (2013). A branched biosynthetic pathway is involved in production of roquefortine and related compounds in *Penicillium chrysogenum*. PLoS One *8*, e65328.

Anslow, W.K., Breen, J., and Raistrick, H. (1940). Studies in the biochemistry of micro-organisms: Emodic acid (4:5:7-trihydroxyanthraquinone-2-carboxylic acid) and omega-hydroxyemodin (4:5:7-trihydroxy-2-(hydroxymethyl)-anthraquinone), metabolic products of a strain of *Penicillium cyclopium* Westling. Biochem. J. *34*, 159–168.

Bacha, N., Dao, H.P., Atoui, A., Mathieu, F., O'Callaghan, J., Puel, O., Liboz, T., Dobson, A.D., and Lebrihi, A. (2009). Cloning and characterization of novel methylsalicylic acid synthase gene involved in the biosynthesis of isoasperlactone and asperlactone in *Aspergillus westerdijkiae*. Fungal Genet. Biol. *46*, 742–749.

Bajaj, I., Veiga, T., van Dissel, D., Pronk, J.T., and Daran, J.M. (2014). Functional characterization of a *Penicillium chrysogenum* mutanase gene induced upon co-cultivation with *Bacillus subtilis*. BMC Microbiol. *14*, 114.

Basaran, P., and Demirbas, R.M. (2010). Spectroscopic detection of pharmaceutical compounds from an aflatoxigenic strain of *Aspergillus parasiticus*. Microbiol. Res. *165*, 516–522.

Bayram, O., Krappmann, S., Ni, M., Bok, J.W., Helmstaedt, K., Valerius, O., Braus-Stromeyer, S., Kwon, N.J., Keller, N.P., Yu, J.H., *et al.* (2008). VelB/VeA/LaeA complex coordinates light signal with fungal development and secondary metabolism. Science *320*, 1504–1506.

Bergmann, S., Funk, A.N., Scherlach, K., Schroeckh, V., Shelest, E., Horn, U., Hertweck, C., and Brakhage, A.A. (2010). Activation of a silent fungal polyketide biosynthesis pathway through regulatory cross talk with a cryptic nonribosomal peptide synthetase gene cluster. Appl. Environ. Microbiol. *76*, 8143–8149.

Boettger, D., and Hertweck, C. (2013). Molecular diversity sculpted by fungal PKS–NRPS hybrids. Chembiochem *14*, 28–42.

Bouhired, S., Weber, M., Kempf-Sontag, A., Keller, N.P., and Hoffmeister, D. (2007). Accurate prediction of the *Aspergillus nidulans* terrequinone gene cluster boundaries using the transcriptional regulator LaeA. Fungal Genet. Biol. *44*, 1134–1145.

Boukhalfa, H., and Crumbliss, A.L. (2002). Chemical aspects of siderophore mediated iron transport. Biometals *15*, 325–339.

Brakhage, A.A., and Schroeckh, V. (2011). Fungal secondary metabolites – strategies to activate silent gene clusters. Fungal Genet. Biol. *48*, 15–22.

Bringmann, G., Gulder, T.A., Lang, G., Schmitt, S., Stohr, R., Wiese, J., Nagel, K., and Imhoff, J.F. (2007). Large-scale biotechnological production of the antileukemic marine natural product sorbicillactone A. Mar. Drugs *5*, 23–30.

Bringmann, G., Lang, G., Gulder, T.A.M., Tsuruta, H., Muhlbacher, J., Maksimenka, K., Steffens, S., Schaumann, K., Stohr, R., Wiese, J., *et al.* (2005). The first sorbicillinoid alkaloids, the antileukemic sorbicillactones A and B, from a sponge-derived Penicillium chrysogenum strain. Tetrahedron *61*, 7252–7265.

Bringmann, G., Lang, G., Muhlbacher, J., Schaumann, K., Steffens, S., Rytik, P.G., Hentschel, U., Morschhauser, J., and Muller, W.E. (2003). Sorbicillactone A: a structurally unprecedented bioactive novel-type alkaloid from a sponge-derived fungus. Prog. Mol. Subcell. Biol. *37*, 231–253.

Brosch, G., Loidl, P., and Graessle, S. (2008). Histone modifications and chromatin dynamics: a focus on filamentous fungi. FEMS Microbiol. Rev. *32*, 409–439.

Brown, M.P., Brown-Jenco, C.S., and Payne, G.A. (1999). Genetic and molecular analysis of aflatoxin biosynthesis. Fungal Genet. Biol. *26*, 81–98.

Bushley, K.E., Ripoll, D.R., and Turgeon, B.G. (2008). Module evolution and substrate specificity of fungal nonribosomal peptide synthetases involved in siderophore biosynthesis. BMC Evol. Biol. *8*, 328.

Cane, D.E., and Walsh, C.T. (1999). The parallel and convergent universes of polyketide synthases and nonribosomal peptide synthetases. Chem. Biol. *6*, R319–325.

Challis, G.L., Ravel, J., and Townsend, C.A. (2000). Predictive, structure-based model of amino acid recognition by nonribosomal peptide synthetase adenylation domains. Chem. Biol. *7*, 211–224.

Chung, Y.J., Coates, N.H., Viana, M.E., Copeland, L., Vesper, S.J., Selgrade, M.K., and Ward, M.D. (2005). Dose-dependent allergic responses to an extract of *Penicillium chrysogenum* in BALB/c mice. Toxicology *209*, 77–89.

Ciuffetti, L.M., Tuori, R.P., and Gaventa, J.M. (1997). A single gene encodes a selective toxin causal to the development of tan spot of wheat. Plant Cell *9*, 135–144.

Clutterbuck, P.W., Lovell, R., and Raistrick, H. (1932). Studies in the biochemistry of micro-organisms: The formation from glucose by members of the *Penicillium chrysogenum* series of a pigment, an alkali-soluble protein and penicillin-the antibacterial substance of Fleming. Biochem. J. *26*, 1907–1918.

Conti, E., Stachelhaus, T., Marahiel, M.A., and Brick, P. (1997). Structural basis for the activation of phenylalanine in the non-ribosomal biosynthesis of gramicidin S. EMBO J. *16*, 4174–4183.

Corse, J.W., Jones, R.G., Soper, Q.F., Whitehead, C.W., and Behrens, O.K. (1948). Biosynthesis of penicillins; substituted phenylacetic acid derivatives as penicillin precursors. J. Am. Chem. Soc. *70*, 2837–2843.

Cox, R.J. (2007). Polyketides, proteins and genes in fungi: programmed nano-machines begin to reveal their secrets. Org. Biomol. Chem. *5*, 2010–2026.

Cox, R.J., and Simpson, T.J. (2009). Fungal type I polyketide synthases. Method Enzymol. *459*, 49–78.

Cram, D.J., and Tishler, M. (1948). Mold metabolites; isolation of several compounds from clinical penicillin. J. Am. Chem. Soc. *70*, 4238.

Dimroth, P., Walter, H., and Lynen, F. (1970). [Biosynthesis of 6-methylsalicylic acid]. Eur. J. Biochem. *13*, 98–110.

Evans, B.S., Robinson, S.J., and Kelleher, N.L. (2011). Surveys of non-ribosomal peptide and polyketide assembly lines in fungi and prospects for their analysis *in vitro* and *in vivo*. Fungal Genet. Biol. *48*, 49–61.

Fierro, F., Garcia-Estrada, C., Castillo, N.I., Rodriguez, R., Velasco-Conde, T., and Martin, J.F. (2006). Transcriptional and bioinformatic analysis of the 56.8 kb DNA region amplified in tandem repeats containing the penicillin gene cluster in Penicillium chrysogenum. Fungal Genet. Biol. *43*, 618–629.

Finking, R., and Marahiel, M.A. (2004). Biosynthesis of nonribosomal peptides1. Annu. Rev. Microbiol. *58*, 453–488.

Fisch, K.M., Gillaspy, A.F., Gipson, M., Henrikson, J.C., Hoover, A.R., Jackson, L., Najar, F.Z., Wagele, H., and Cichewicz, R.H. (2009). Chemical induction of silent biosynthetic pathway transcription in *Aspergillus niger*. J. Ind. Microbiol. Biotechnol. *36*, 1199–1213.

Fitzgerald, J.T., Ridley, C.P., and Khosla, C. (2011). Engineered biosynthesis of the antiparasitic agent frenolicin B and rationally Designed analogs in a heterologous host. J. Antibiot. (Tokyo) *64*, 759–762.

Fleming, A. (1955). The story of penicillin. Bull. Georgetown Univ. Med. Cent. *8*, 128–132.

Fleming, A. (2001). On the antibacterial action of cultures of a penicillium, with special reference to their use in the isolation of B. influenzae. 1929. Bull. World Health Organ. *79*, 780–790.

Foreman, C.M., Dieser, M., Greenwood, M., Cory, R.M., Laybourn-Parry, J., Lisle, J.T., Jaros, C., Miller, P.L., Chin, Y.P., and McKnight, D.M. (2011). When a habitat freezes solid: microorganisms over-winter within the ice column of a coastal Antarctic lake. FEMS Microbiol. Ecol. *76*, 401–412.

Fujii, I. (2010). Functional analysis of fungal polyketide biosynthesis genes. J. Antibiot. (Tokyo) *63*, 207–218.

Fujii, I., Mori, Y., Watanabe, A., Kubo, Y., Tsuji, G., and Ebizuka, Y. (2000). Enzymatic synthesis of 1,3,6,8-tetrahydroxynaphthalene solely from malonyl coenzyme A by a fungal iterative type I polyketide synthase PKS1. Biochemistry *39*, 8853–8858.

Fujii, I., Watanabe, A., Sankawa, U., and Ebizuka, Y. (2001). Identification of Claisen cyclase domain in fungal polyketide synthase WA, a naphthopyrone synthase of *Aspergillus nidulans*. Chem. Biol. *8*, 189–197.

Gao, S.S., Li, X.M., Li, C.S., Proksch, P., and Wang, B.G. (2011). Penicisteroids A and B, antifungal and cytotoxic polyoxygenated steroids from the marine alga-derived endophytic fungus *Penicillium chrysogenum* QEN-24S. Bioorg. Med. Chem. Lett. *21*, 2894–2897.

Gao, X., Haynes, S.W., Ames, B.D., Wang, P., Vien, L.P., Walsh, C.T., and Tang, Y. (2012). Cyclization of fungal nonribosomal peptides by a terminal condensation-like domain. Nat. Chem. Biol. *8*, 823–830.

Garcia-Estrada, C., Ullan, R.V., Albillos, S.M., Fernandez-Bodega, M.A., Durek, P., von Dohren, H., and Martin, J.F. (2011). A single cluster of coregulated genes encodes the biosynthesis of the mycotoxins roquefortine C and meleagrin in *Penicillium chrysogenum*. Chem. Biol. *18*, 1499–1512.

Garcia-Rico, R.O., Fierro, F., Mauriz, E., Gomez, A., Fernandez-Bodega, M.A., and Martin, J.F. (2008). The heterotrimeric Galpha protein pga1 regulates biosynthesis of penicillin, chrysogenin and roquefortine in *Penicillium chrysogenum*. Microbiology *154*, 3567–3578.

Geltner, C., Lass-Florl, C., Bonatti, H., Muller, L., and Stelzmuller, I. (2013). Invasive pulmonary mycosis due to *Penicillium chrysogenum*: a new invasive pathogen. Transplantation 95, e21–23.

Georges, K., Jayaprakasam, B., Dalavoy, S.S., and Nair, M.G. (2008). Pest-managing activities of plant extracts and anthraquinones from Cassia nigricans from Burkina Faso. Bioresour. Technol. *99*, 2037–2045.

Gressler, M., Zaehle, C., Scherlach, K., Hertweck, C., and Brock, M. (2011). Multifactorial induction of an orphan PKS-NRPS gene cluster in Aspergillus terreus. Chem. Biol. *18*, 198–209.

Guo, C.J., Sun, W.W., Bruno, K.S., and Wang, C.C. (2014). Molecular genetic characterization of terreic acid pathway in *Aspergillus terreus*. Org. Lett. *16*, 5250–5253.

Guo, W., Li, D., Peng, J., Zhu, T., Gu, Q., and Li, D. (2015). Penicitols A-C and penixanacid A from the mangrove-derived *Penicillium chrysogenum* HDN11–24. J. Nat. Prod. *78*, 306–310.

Hacker, C., Glinski, M., Hornbogen, T., Doller, A., and Zocher, R. (2000). Mutational analysis of the N-methyltransferase domain of the multifunctional enzyme enniatin synthetase. J. Biol. Chem. *275*, 30826–30832.

Halo, L.M., Marshall, J.W., Yakasai, A.A., Song, Z., Butts, C.P., Crump, M.P., Heneghan, M., Bailey, A.M., Simpson, T.J., Lazarus, C.M., *et al.* (2008). Authentic heterologous expression of the tenellin iterative polyketide synthase nonribosomal peptide synthetase requires coexpression with an enoyl reductase. Chembiochem 9, 585–594.

Harris, D.M., van der Krogt, Z.A., Klaassen, P., Raamsdonk, L.M., Hage, S., van den Berg, M.A., Bovenberg, R.A., Pronk, J.T., and Daran, J.M. (2009). Exploring and dissecting genome-wide gene expression responses of *Penicillium chrysogenum* to phenylacetic acid consumption and penicillinG production. BMC Genomics *10*, 75.

Henk, D.A., Eagle, C.E., Brown, K., M.A., V.D.B., Dyer, P.S., Peterson, S.W., and Fisher, M.C. (2011). Speciation despite globally overlapping distributions in *Penicillium chrysogenum*: the population genetics of Alexander Fleming's lucky fungus. Mol. Ecol. *20*, 4288–4301.

Hidalgo, P.I., Ullan, R.V., Albillos, S.M., Montero, O., Fernandez-Bodega, M.A., Garcia-Estrada, C., Fernandez-Aguado, M., and Martin, J.F. (2014). Molecular characterization of the PR-toxin gene cluster in Penicillium roqueforti and Penicillium chrysogenum: cross talk of secondary metabolite pathways. Fungal Genet. Biol. *62*, 11–24.

Hikino, H., Nabetani, S., and Takemoto, T. (1973). [Structure and biosynthesis of chrysogine, a metabolite of *Penicillium chrysogenum*]. Yakugaku Zasshi *93*, 619–623.

Hoff, B., Kamerewerd, J., Sigl, C., Mitterbauer, R., Zadra, I., Kurnsteiner, H., and Kuck, U. (2010). Two components of a velvet-like complex control hyphal morphogenesis, conidiophore development, and penicillin biosynthesis in *Penicillium chrysogenum*. Eukaryot. Cell *9*, 1236–1250.

Holm, D.K., Petersen, L.M., Klitgaard, A., Knudsen, P.B., Jarczynska, Z.D., Nielsen, K.F., Gotfredsen, C.H., Larsen, T.O., and Mortensen, U.H. (2014). Molecular and chemical characterization of the biosynthesis of the 6-MSA-derived meroterpenoid yanuthone D in *Aspergillus niger*. Chem. Biol. *21*, 519–529.

Hornbogen, T., Riechers, S.P., Prinz, B., Schultchen, J., Lang, C., Schmidt, S., Mugge, C., Turkanovic, S., Sussmuth, R.D., Tauberger, E., *et al.* (2007). Functional characterization of the recombinant N-methyltransferase domain from the multienzyme enniatin synthetase. Chembiochem *8*, 1048–1054.

Houbraken, J., Frisvad, J.C., and Samson, R.A. (2011). Fleming's penicillin producing strain is not *Penicillium chrysogenum* but *P. rubens*. IMA Fungus *2*, 87–95.

Houbraken, J., Frisvad, J.C., Seifert, K.A., Overy, D.P., Tuthill, D.M., Valdez, J.G., and Samson, R.A. (2012). New penicillin-producing *Penicillium* species and an overview of section Chrysogena. Persoonia *29*, 78–100.

Hsieh, K.P., Yu, S., Wei, Y.H., Chen, C.F., and Wei, R.D. (1986). Inhibitory effect *in vitro* of PR toxin, a mycotoxin from Penicillium roqueforti, on the mitochondrial HCO-3-ATPase of the rat brain, heart and kidney. Toxicon *24*, 153–160.

Hur, G.H., Vickery, C.R., and Burkart, M.D. (2012). Explorations of catalytic domains in non-ribosomal peptide synthetase enzymology. Nat. Prod. Rep. *29*, 1074–1098.

Jenke-Kodama, H., and Dittmann, E. (2009). Bioinformatic perspectives on NRPS/PKS megasynthases: advances and challenges. Nat. Prod. Rep. *26*, 874–883.

Kale, S.P., Milde, L., Trapp, M.K., Frisvad, J.C., Keller, N.P., and Bok, J.W. (2008). Requirement of LaeA for secondary metabolism and sclerotial production in *Aspergillus flavus*. Fungal Genet. Biol. *45*, 1422–1429.

Keating, T.A., Ehmann, D.E., Kohli, R.M., Marshall, C.G., Trauger, J.W., and Walsh, C.T. (2001). Chain termination steps in nonribosomal peptide synthetase assembly lines: directed acyl-S-enzyme breakdown in antibiotic and siderophore biosynthesis. Chembiochem *2*, 99–107.

Khaldi, N., Seifuddin, F.T., Turner, G., Haft, D., Nierman, W.C., Wolfe, K.H., and Fedorova, N.D. (2010). SMURF: Genomic mapping of fungal secondary metabolite clusters. Fungal Genet. Biol. *47*, 736–741.

Koglin, A., Mofid, M.R., Lohr, F., Schafer, B., Rogov, V.V., Blum, M.M., Mittag, T., Marahiel, M.A., Bernhard, F., and Dotsch, V. (2006). Conformational switches modulate protein interactions in peptide antibiotic synthetases. Science *312*, 273–276.

Kohli, R.M., Trauger, J.W., Schwarzer, D., Marahiel, M.A., and Walsh, C.T. (2001). Generality of peptide cyclization catalyzed by isolated thioesterase domains of nonribosomal peptide synthetases. Biochemistry *40*, 7099–7108.

Komagata, D., Fujita, S., Yamashita, N., Saito, S., and Morino, T. (1996). Novel neuritogenic activities of pseurotin A and penicillic acid. J. Antibiot. (Tokyo) *49*, 958–959.

Kosalkova, K., Dominguez-Santos, R., Coton, M., Coton, E., Garcia-Estrada, C., Liras, P., and Martin, J.F. (2015). A natural short pathway synthesizes roquefortine C but not meleagrin in three different *Penicillium roqueforti* strains. Appl. Microbiol. Biotechnol. *99*, 7601–7612.

Kosalkova, K., Garcia-Estrada, C., Ullan, R.V., Godio, R.P., Feltrer, R., Teijeira, F., Mauriz, E., and Martin, J.F. (2009). The global regulator LaeA controls penicillin biosynthesis, pigmentation and sporulation, but not roquefortine C synthesis in *Penicillium chrysogenum*. Biochimie *91*, 214–225.

Kroken, S., Glass, N.L., Taylor, J.W., Yoder, O.C., and Turgeon, B.G. (2003). Phylogenomic analysis of type I polyketide synthase genes in pathogenic and saprobic ascomycetes. Proc. Natl. Acad. Sci. U.S.A. *100*, 15670–15675.

Lee, I., Oh, J.H., Shwab, E.K., Dagenais, T.R., Andes, D., and Keller, N.P. (2009). HdaA, a class 2 histone deacetylase of *Aspergillus fumigatus*, affects germination and secondary metabolite production. Fungal Genet. Biol. *46*, 782–790.

Li, H., and Durbin, R. (2009). Fast and accurate long-read alignment with Burrows-Wheeler transform. Bioinformatics *25*, 589–595.

Li, H., Handsaker, B., Wysoker, A., Fennell, T., Ruan, J., Homer, N., Marth, G., Abecasis, G., Durbin, R., and 1000 Genome Project Data Processing Subgroup. (2009). The Sequence Alignment/Map format and SAMtools. Bioinformatics *25*, 2078–2079.

Linne, U., Doekel, S., and Marahiel, M.A. (2001). Portability of epimerization domain and role of peptidyl carrier protein on epimerization activity in nonribosomal peptide synthetases. Biochemistry *40*, 15824–15834.

Macdonald, K.D., Hutchinson, J.M., and Gillett, W.A. (1963). Heterokaryon studies and the genetic control of penicillin and chrysogenin production in *Penicillium chrysogenum*. J. Gen. Microbiol. *33*, 375–383.

Maskey, R.P., Grun-Wollny, I., and Laatsch, H. (2005). Sorbicillin analogues and related dimeric compounds from *Penicillium notatum*. J. Nat. Prod. *68*, 865–870.

McLean, K.J., Hans, M., Meijrink, B., van Scheppingen, W.B., Vollebregt, A., Tee, K.L., van der Laan, J.M., Leys, D., Munro, A.W., and van den Berg, M.A. (2015). Single-step fermentative production of the cholesterol-lowering drug pravastatin via reprogramming of Penicillium chrysogenum. Proc. Natl. Acad. Sci. U.S.A. *112*, 2847–2852.

McNeill, J., Barrie, F.R., Burdet, H.M., Demoulin, V., Hawksworth, D.L., Marhold, K., Nicolson, D.H., Prado, J., Silva, P.C., Skog, J.E., *et al.* (2006). International Code of Botanical Nomenclature (Vienna Code): adopted by the Seventeenth International Botanical Congress Vienna, Austria, July 2005. Regnum Vegetabile.

Meijer, W.H., Gidijala, L., Fekken, S., Kiel, J.A., van den Berg, M.A., Lascaris, R., Bovenberg, R.A., and van der Klei, I.J. (2010). Peroxisomes are required for efficient penicillin biosynthesis in Penicillium chrysogenum. Appl. Environ. Microbiol. *76*, 5702–5709.

Miller, R.F., and Huang, S. (1995). Isolation and structure of sorrentanone: a new tetrasubstituted quinone from *Penicillium chrysogenum*. J. Antibiot. (Tokyo) *48*, 520–521.

Moule, Y., Jemmali, M., and Rousseau, N. (1976). Mechanism of the inhibition of transcription by PR toxin, a mycotoxin from Penicillium roqueforti. Chem. Biol. Interact. *14*, 207–216.

Muller, W.H., van der Krift, T.P., Krouwer, A.J., Wosten, H.A., van der Voort, L.H., Smaal, E.B., and Verkleij, A.J. (1991). Localization of the pathway of the penicillin biosynthesis in *Penicillium chrysogenum*. EMBO J. *10*, 489–495.

Newbert, R.W., Barton, B., Greaves, P., Harper, J., and Turner, G. (1997). Analysis of a commercially improved *Penicillium chrysogenum* strain series: involvement of recombinogenic regions in amplification and deletion of the penicillin biosynthesis gene cluster. J. Ind. Microbiol. Biotechnol. *19*, 18–27.

Nicolaou, K.C., Simonsen, K.B., Vassilikogiannakis, G., Baran, P.S., Vidali, V.P., Pitsinos, E.N., and Couladouros, E.A. (1999). Biomimetic explorations towards the bisorbicillinoids: total synthesis of bisorbicillinol, bisorbibutenolide, and trichodimerol. Angew. Chem. Int. Ed. Engl. *38*, 3555–3559.

Otten, L.G., Schaffer, M.L., Villiers, B.R., Stachelhaus, T., and Hollfelder, F. (2007). An optimized ATP/PP(i)-exchange assay in 96-well format for screening of adenylation domains for applications in combinatorial biosynthesis. Biotechnol. J. *2*, 232–240.

Ouchaou, K., Maire, F., Salo, O., Ali, H., Hankemeier, T., van der Marel, G.A., Filippov, D.V., Bovenberg, R.A., Vreeken, R.J., Driessen, A.J., *et al.* (2015). A mutasynthesis approach with a *Penicillium chrysogenum* DeltaroqA strain yields new roquefortine D analogues. Chembiochem *16*, 915–923.

Perrin, R.M., Fedorova, N.D., Bok, J.W., Cramer, R.A., Wortman, J.R., Kim, H.S., Nierman, W.C., and Keller, N.P. (2007). Transcriptional regulation of chemical diversity in *Aspergillus fumigatus* by LaeA. PLoS Pathog. *3*, e50.

Petersen, L.M., Holm, D.K., Knudsen, P.B., Nielsen, K.F., Gotfredsen, C.H., Mortensen, U.H., and Larsen, T.O. (2014). Characterization of four new antifungal yanuthones from Aspergillus niger. J. Antibiot. (Tokyo) *68*, 201–205.

Priebe, S., Linde, J., Albrecht, D., Guthke, R., and Brakhage, A.A. (2011). FungiFun: a web-based application for functional categorization of fungal genes and proteins. Fungal Genet. Biol. *48*, 353–358.

Rahimipour, S., Bilkis, I., Peron, V., Gescheidt, G., Barbosa, F., Mazur, Y., Koch, Y., Weiner, L., and Fridkin, M. (2001). Generation of free radicals by emodic acid and its [D-Lys6]GnRH-conjugate. Photochem. Photobiol. *74*, 226–236.

Raper, K.B., Alexander, D.F., and Coghill, R.D. (1944). Penicillin: II. Natural variation and penicillin production in *Penicillium notatum* and allied species. J. Bacteriol. *48*, 639–659.

Rausch, C., Weber, T., Kohlbacher, O., Wohlleben, W., and Huson, D.H. (2005). Specificity prediction of adenylation domains in nonribosomal peptide synthetases (NRPS) using transductive support vector machines (TSVMs). Nucleic Acids Res. *33*, 5799–5808.

Reategui, R.F., Wicklow, D.T., and Gloer, J.B. (2006). Phaeofurans and sorbicillin analogues from a fungicolous *Phaeoacremonium* species (NRRL 32148). J. Nat. Prod. *69*, 113–117.

Ridley, C.P., and Khosla, C. (2007). Synthesis and biological activity of novel pyranopyrones derived from engineered aromatic polyketides. ACS Chem. Biol. *2*, 104–108.

Ries, M.I., Ali, H., Lankhorst, P.P., Hankemeier, T., Bovenberg, R.A., Driessen, A.J., and Vreeken, R.J. (2013). Novel key metabolites reveal further branching of the roquefortine/meleagrin biosynthetic pathway. J. Biol. Chem. *288*, 37289–37295.

Rowan, D.D., Dymock, J.J., and Brimble, M.A. (1990). Effect of fungal metabolite peramine and analogs on feeding and development of argentine stem weevil (*Listronotus bonariensis*). J. Chem. Ecol. *16*, 1683–1695.

Samson, R.A., and Frisvad, J.C. (2004). *Penicillium* subgenus *Penicillium*: new taxonomic schemes, mycotoxins and other extrolites. Stud. Mycol. *49*, 257.

Samson, R.A., Hadlok, R., and Stolk, A.C. (1977). A taxonomic study of the *Penicillium chrysogenum* series. Antonie Van Leeuwenhoek *43*, 169–175.

Sanchez, J.F., Somoza, A.D., Keller, N.P., and Wang, C.C. (2012). Advances in *Aspergillus* secondary metabolite research in the post-genomic era. Nat. Prod. Rep. *29*, 351–371.

Scheltema, R.A., Jankevics, A., Jansen, R.C., Swertz, M.A., and Breitling, R. (2011). PeakML/mzMatch: a file format, Java library, R library, and tool-chain for mass spectrometry data analysis. Anal. Chem. *83*, 2786–2793.

Scherlach, K., Boettger, D., Remme, N., and Hertweck, C. (2010). The chemistry and biology of cytochalasans. Nat. Prod. Rep. *27*, 869–886.

Schroeckh, V., Scherlach, K., Nutzmann, H.W., Shelest, E., Schmidt-Heck, W., Schuemann, J., Martin, K., Hertweck, C., and Brakhage, A.A. (2009). Intimate bacterial–fungal interaction triggers biosynthesis of archetypal polyketides in *Aspergillus nidulans*. Proc. Natl. Acad. Sci. U.S.A. *106*, 14558–14563.

Scott, P.M., Kennedy, B.P., Harwig, J., and Blanchfield, B.J. (1977). Study of conditions of production of roquefortine and other metabolites of *Penicillin roqueforti*. Appl. Environ. Microbiol. 33, 249–253.

Shwab, E.K., Bok, J.W., Tribus, M., Galehr, J., Graessle, S., and Keller, N.P. (2007). Histone deacetylase activity regulates chemical diversity in *Aspergillus*. Eukaryot. Cell 6, 1656–1664.

Sims, J.W., Fillmore, J.P., Warner, D.D., and Schmidt, E.W. (2005). Equisetin biosynthesis in *Fusarium heterosporum*. Chem. Commun. (Camb.), 186–188.

Sims, J.W., and Schmidt, E.W. (2008). Thioesterase-like role for fungal PKS–NRPS hybrid reductive domains. J. Am. Chem. Soc. *130*, 11149–11155.

Smith, K.M., Phatale, P.A., Bredeweg, E.L., Connolly, L.R., Pomraning, K.R., and Freitag, M. (2006). Epigenetics of filamentous fungi. In Encyclopedia of Molecular Cell Biology and Molecular Medicine, Meyers, R.A., ed. (Wiley-VCH Verlag GmbH & Co. KGaA, Corvallis, OR).

Song, Z., Cox, R.J., Lazarus, C.M., and Simpson, T.T. (2004). Fusarin C biosynthesis in *Fusarium moniliforme* and *Fusarium venenatum*. Chembiochem 5, 1196–1203.

Stachelhaus, T., and Marahiel, M.A. (1995). Modular structure of peptide synthetases revealed by dissection of the multifunctional enzyme GrsA. J. Biol. Chem. *270*, 6163–6169.

Stachelhaus, T., Mootz, H.D., and Marahiel, M.A. (1999). The specificity-conferring code of adenylation domains in nonribosomal peptide synthetases. Chem. Biol. 6, 493–505.

Stachelhaus, T., Schneider, A., and Marahiel, M.A. (1995). Rational design of peptide antibiotics by targeted replacement of bacterial and fungal domains. Science *269*, 69–72.

Staunton, J., and Weissman, K.J. (2001). Polyketide biosynthesis: a millennium review. Nat. Prod. Rep. *18*, 380–416.

Stocking, E.M., and Williams, R.M. (2003). Chemistry and biology of biosynthetic Diels-Alder reactions. Angew. Chem. Int. Ed. Engl. *42*, 3078–3115.

Stodola, F.H., Wachtel, J.L., Moyer, A.J., and Coghill, R.D. (1945). Penitrinic acid, a new pigment from *Penicillium notatum*. J. Biol. Chem. *159*, 67–70.

Sutherland, A., Auclair, K., and Vederas, J.C. (2001). Recent advances in the biosynthetic studies of lovastatin. Curr. Opin. Drug Discov. Devel. *4*, 229–236.

Tanaka, A., Tapper, B.A., Popay, A., Parker, E.J., and Scott, B. (2005). A symbiosis expressed non-ribosomal peptide synthetase from a mutualistic fungal endophyte of perennial ryegrass confers protection to the symbiotum from insect herbivory. Mol. Microbiol. *57*, 1036–1050.

Tokuoka, M., Seshime, Y., Fujii, I., Kitamoto, K., Takahashi, T., and Koyama, Y. (2008). Identification of a novel polyketide synthase-nonribosomal peptide synthetase (PKS-NRPS) gene required for the biosynthesis of cyclopiazonic acid in *Aspergillus oryzae*. Fungal Genet. Biol. *45*, 1608–1615.

Tribus, M., Galehr, J., Trojer, P., Brosch, G., Loidl, P., Marx, F., Haas, H., and Graessle, S. (2005). HdaA, a major class 2 histone deacetylase of Aspergillus nidulans, affects growth under conditions of oxidative stress. Eukaryot. Cell *4*, 1736–1745.

Turgay, K., Krause, M., and Marahiel, M.A. (1992). Four homologous domains in the primary structure of GrsB are related to domains in a superfamily of adenylate-forming enzymes. Mol. Microbiol. *6*, 2743–2744.

Udwary, D.W., Merski, M., and Townsend, C.A. (2002). A method for prediction of the locations of linker regions within large multifunctional proteins, and application to a type I polyketide synthase. J. Mol. Biol. *323*, 585–598.

van den Berg, M.A., Albang, R., Albermann, K., Badger, J.H., Daran, J.M., Driessen, A.J., Garcia-Estrada, C., Fedorova, N.D., Harris, D.M., Heijne, W.H., *et al.* (2008). Genome sequencing and analysis of the filamentous fungus *Penicillium chrysogenum*. Nat. Biotechnol. *26*, 1161–1168.

van den Berg, M.A., Westerlaken, I., Leeflang, C., Kerkman, R., and Bovenberg, R.A. (2007). Functional characterization of the penicillin biosynthetic gene cluster of *Penicillium chrysogenum* Wisconsin54-1255. Fungal Genet. Biol. *44*, 830–844.

Veiga, T., Nijland, J.G., Driessen, A.J., Bovenberg, R.A., Touw, H., van den Berg, M.A., Pronk, J.T., and Daran, J.M. (2012). Impact of velvet complex on transcriptome and penicillin G production in glucose-limited chemostat cultures of a beta-lactam high-producing *Penicillium chrysogenum* strain. OMICS *16*, 320–333.

Villiers, B.R., and Hollfelder, F. (2009). Mapping the limits of substrate specificity of the adenylation domain of TycA. Chembiochem *10*, 671–682.

von Dohren, H. (2004). Biochemistry and general genetics of nonribosomal peptide synthetases in fungi. Adv. Biochem. Eng. Biotechnol. *88*, 217–264.

von Dohren, H., Dieckmann, R., and Pavela-Vrancic, M. (1999). The nonribosomal code. Chem. Biol. *6*, R273–279.

Walzel, B., Riederer, B., and Keller, U. (1997). Mechanism of alkaloid cyclopeptide synthesis in the ergot fungus *Claviceps purpurea*. Chem. Biol. *4*, 223–230.

Wargo, M.J., and Hogan, D.A. (2006). Fungal–bacterial interactions: a mixed bag of mingling microbes. Curr. Opin. Microbiol. *9*, 359–364.

Watanabe, A., and Ebizuka, Y. (2004). Unprecedented mechanism of chain length determination in fungal aromatic polyketide synthases. Chem. Biol. *11*, 1101–1106.

Weber, G., and Leitner, E. (1994). Disruption of the cyclosporin synthetase gene of *Tolypocladium niveum*. Curr. Genet. *26*, 461–467.

Weber, G., Schorgendorfer, K., Schneider-Scherzer, E., and Leitner, E. (1994). The peptide synthetase catalyzing cyclosporine production in *Tolypocladium niveum* is encoded by a giant 45.8-kilobase open reading frame. Curr. Genet. *26*, 120–125.

Weber, S.S., Bovenberg, R.A., and Driessen, A.J. (2012). Biosynthetic concepts for the production of beta-lactam antibiotics in *Penicillium chrysogenum*. Biotechnol. J. *7*, 225–236.

Weber, T., Blin, K., Duddela, S., Krug, D., Kim, H.U., Bruccoleri, R., Lee, S.Y., Fischbach, M.A., Muller, R., Wohlleben, W., *et al.* (2015). antiSMASH 3.0-a comprehensive resource for the genome mining of biosynthetic gene clusters. Nucleic Acids Res. *43*, W237–243.

Westendorf, J., Marquardt, H., Poginsky, B., Dominiak, M., Schmidt, J., and Marquardt, H. (1990). Genotoxicity of naturally occurring hydroxyanthraquinones. Mutat. Res. *240*, 1–12.

Wiest, A., Grzegorski, D., Xu, B.W., Goulard, C., Rebuffat, S., Ebbole, D.J., Bodo, B., and Kenerley, C. (2002). Identification of peptaibols from *Trichoderma virens* and cloning of a peptaibol synthetase. J. Biol. Chem. *277*, 20862–20868.

Wolf, F.T., Kim, Y.T., and Jones, E.A. (1960). Spectral studies on chrysogeuin, a pigment produced by *Penicillium chrysogenum*. Physiologia Plantarum *13*, 621–627.

Yadav, G., Gokhale, R.S., and Mohanty, D. (2003a). Computational approach for prediction of domain organization and substrate specificity of modular polyketide synthases. J. Mol. Biol. *328*, 335–363.

Yadav, G., Gokhale, R.S., and Mohanty, D. (2003b). SEARCHPKS: A program for detection and analysis of polyketide synthase domains. Nucleic Acids Res. *31*, 3654–3658.

pH Modulation by Fungal Secreted Effecting Molecules: A Mechanism Affecting Pathogenicity and Mycotoxin Accumulation During Colonization by *Penicillium expansum* 10

Dov Prusky, Shiri Barad, Nofar Glam, Nancy Keller and Amir Sherman

Abstract

A postharvest pathogen as *Penicillium* can start its attack process as soon as spores land on wounded tissue; other pathogens as *Colletotrichum* breach the unripe fruit cuticle, remain quiescent for months until the fruit ripens and then colonize the host. Both postharvest fungal pathogens initiate their development by secreting non-proteinaceous pH modulating factors as organic acids or ammonia, which acidify or alkalinize the ambient host environment. These fungal secreted pH effecting molecules (SEM) modulate the host environment, activate secondary metabolism and regulate an arsenal of enzymes in order to increase fungal pathogenicity under any specific conditions. This arsenal includes genes and processes that compromise host defences, enhance programmed cell death (PCD), contribute to intracellular signalling, produce cell wall-degrading enzymes, regulate specific transporters, induce redox protectant systems and generate factors needed by the pathogen to cope effectively with the hostile environment found within the host. In *Penicillium*, genes that contribute to gluconic acid (GLA) and secondary metabolites accumulation found to be pH regulated by the specific transcription factor PacC, indicating the capability to respond under acidifying modes of attack. In this chapter we will analyse *Penicillium* capability to modulate the environment pH as a mechanism for colonization.

Introduction

The infection and colonization by postharvest pathogens, and the resistance of unripe fruit to infection are considered as a dynamic process that is activated during host maturation and ripening by the transformation of quiescent-biotrophic fungi to active necrotrophic organisms (Denison *et al.*, 1995; Calvo *et al.*, 2002; Caracuel *et al.*, 2003b; O'Meara *et al.*, 2010).

Transcriptome analysis of the biotrophic stage of *Colletotrichum higginsianum* revealed a large inventory of candidate effectors (Kunz *et al.*, 2006). These tagged effectors were found in appressoria penetration pores before host invasion, and revealed subtle early levels of functional complexity for this fungal organ (Kunz *et al.*, 2006). Their presence indicates that secreted effector proteins manipulate the host during penetration and quiescence, which highlights the fact that is not a passive interaction (Kunz *et al.*, 2006; O'Meara *et al.*, 2010). Although *C. higginsianum* do not display a quiescent phase, it likely reflects behaviour of a similar nature in other *Colletotrichum* species that do pause between their biotrophic and necrotrophic stages; for example *C. gloeosporioides* secretes a large variety of stage-specific effectors, non-ribosomal peptide- synthases, cutinases and others cell-killing factors during quiescence (Alkan *et al.*, 2013). Effectors are key factors in eliciting as well as subverting host-defence responses and inducing PCD; thus, they might contribute to activation of the quiescent infection, however, functional analysis of their contribution should be further studied.

In parallel to the secreted protein molecules, at very early stages of germination and during the later stages of necrotrophic development, non-proteinaceous secreted pH effecting molecules were shown to contribute to host pH modulation and modulate fungal morphogenesis, penetration and colonization (Prusky *et al.*, 2013). The pH-modulating molecules described were ammonia and organic acids (Prusky and Yakoby, 2003; Prusky *et al.*, 2004), and are less specific than the effector proteins that were detected at early stages of spore germination, penetration and host colonization (Kunz *et al.*, 2006).

Small secreted molecules affect fungal biology

Environmental pH is one of the most important physiochemical factors that affect cell growth and development. The capability to secrete pH effecting molecules, that can regulate their internal environmental conditions, possibly have significant importance in cellular processes, such as protein synthesis, growth and proliferation (Pouyssegur, 1985; Bravo and Macdonald-Bravo, 1986; Chambard and Pouyssegur, 1986; Musgrove *et al.*, 1987; Chiche *et al.*, 2010). These responses reflect the inability of the cells to function normally at suboptimal pH conditions.

The pathogen's ability to change the environment pH and activate signalling pathways enables it to control not only cell growth and development, but also colonization (Prusky and Yakoby, 2003). During the activation of the infection, changes in external pH rapidly alter plant gene expression and modulate host responses, similarly to effectors (Lager *et al.*, 2010). Fungal factors that modulate the host environmental pH will promote the secretion of the arsenal of pH-specific pathogenicity factors, which include: toxins, expression-regulating factors, genes that modulate host defence, specific transporters, redox-protective substances, cell-wall-degrading enzymes and proteinases (Espeso *et al.*, 1993; Calvo *et al.*, 2002; Spielvogel *et al.*, 2008).

Fungal small secreted molecules modulate environmental pH and pathogenicity

A large number of fungal plant pathogens modify host pH. These include host alkalization by, e.g. *C. gloeosporioides, C. coccodes, C. higginsianum* (Prusky *et al.*, 2001; Alkan *et al.*, 2008;

Miyara *et al.*, 2010; O'Connell *et al.*, 2012), *Alternaria alternata* (Eshel *et al.*, 2002a,b), and *F. oxysporum* (Caracuel *et al.*, 2003a) as well as host acidification by *Penicillium expansum* and *P. digitatum* (Lee and Dean, 1993; Prusky *et al.*, 2004), *Botrytis cinerea* (Wubben *et al.*, 2000; Ten Have *et al.*, 2001; Manteau *et al.*, 2003), *Sclerotinia sclerotiorum* (Rollins and Dickman, 2001), *Phomopsis mangiferae* (Davidzon *et al.*, 2010), and *Monilinia fructicola* (Perpetua *et al.*, 1996). The transition from germinating to necrotrophic colonization involves the accumulation of ammonia, gluconic and oxalic acids for *C. gloeosporioides, P. expansum* and *S. sclerotiorum*, respectively (Prusky *et al.*, 1989; Poussereau *et al.*, 2001; Reinehr *et al.*, 2007); these are present at the various stages of fungal attack: germination, appressoria formation, penetration and late colonization.

Ammonia production by fungi seems to result from activation of proteases, followed by deamination of amino acids, which leads to ambient alkalization (Jennings, 1989) (Fig. 10.1). Ammonium secretion by developing mycelia was detected in association with pathogenicity of many *Colletotrichum* species, including *C. gloeosporioides, C. acutatum, C. higginsianum, C. graminicola* and *C. coccodes* (Alkan *et al.*, 2008; Dieguez-Uribeondo *et al.*, 2008; Miyara *et al.*, 2010; O'Connell *et al.*, 2012), *A. alternata* (Eshel *et al.*, 2002a,b) and *Fusarium oxysporum* (Caracuel *et al.*, 2003a). The ammonium secreted by these fungi results in alkalization of the host tissue; its concentration can reach approximately 5 mM, as found in decayed avocado, tomato and persimmon fruits (Eshel *et al.*, 2002a,b; Alkan *et al.*, 2008; Miyara *et al.*, 2010). In every case, increased ammonium accumulation has been related to enhanced pathogenicity of *Colletotrichum* spp. (Alkan *et al.*, 2008, 2009; Miyara *et al.*, 2010). In *Colletotrichum*, glutamate dehydrogenase (*gdh2*) is involved in catabolism of glutamate and release of ammonia and is differentially expressed during colonization (Miyara *et al.*,

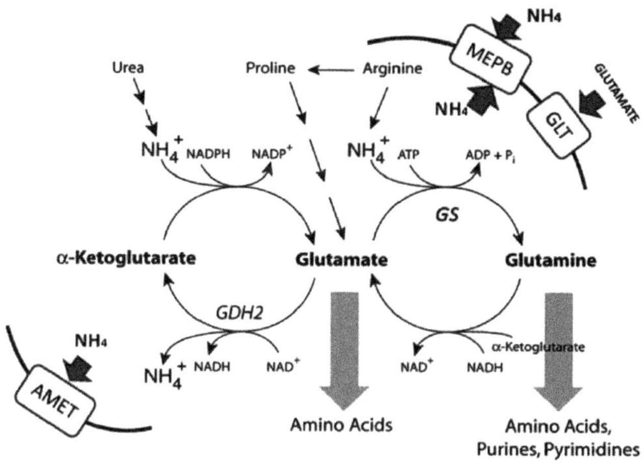

Figure 10.1 Central pathways for nitrogen metabolism. Nitrogenous compounds in the cell are synthesized from either glutamate or glutamine. The major pathway for glutamate synthesis is through combination of ammonia with α-ketoglutarate, which is synthesized in the early steps of the citric acid cycle. Glutamine is synthesized by the combination of ammonia with glutamate. Pathways for utilization of a variety of nitrogen sources, including urea, proline, and arginine, are shown. Production of ammonia by alkalizing fungi is mainly by activation of GDH2. The MEPB, AMET and GLT transporters have key roles in ammonia secretion (adapted from Miyara *et al.*, 2012).

2010). Development of the $\Delta gdh2$ strain reduces ammonia accumulation in freshly produced spores and inhibits appressoria formation and pathogenicity (Miyara *et al.*, 2010). Recent findings suggest that various fungal pathogens, such as *F. oxysporum*, *C. graminicola*, and *C. higginsianum* secrete ammonia in order to modulate the surrounding pH and create an optimal pH for expression and activity of their pathogenicity factors (Miyara *et al.*, 2012; O'Connell *et al.*, 2012).

On the contrary to alkalizing pathogens, pathogens such as *P. expansum* and *Phomopsis mangiferae* modify the host pH by secretion of gluconic acid (Davidzon *et al.*, 2010; Barad *et al.*, 2012, 2014), and its production is catalysed by oxidation of β-D-glucose to D-glucono-δ lactone and H_2O_2 by glucose oxidase (*gox2*) (Fig. 10.2). The relationship between gluconic acid and *P. expansum* pathogenicity was confirmed by determining the relationship between strain aggressiveness and gluconic acid production (Hadas *et al.*, 2007). Furthermore, *P. expansum gox*-RNAi mutants with reduced *gox2* expression showed inhibited pathogenicity during fruit colonization (Barad *et al.*, 2012), which indicates the importance of the gluconic acid/acidification process in pathogenicity of *Penicillium* (Table 10.1).

Given that *P. expansum* requires wound for successful penetration and colonization, reactive oxygen species (ROS) produced at early time points as a consequence of wound responses might be critical for *gox2* activation and fungal ROS production (Hadas *et al.*, 2007; Macarisin *et al.*, 2007). Therefore, generated superoxide and subsequently generated ROS may act as second messenger that regulate the transcription of hypoxic genes, such as *gox*, and activate its involvement in pathogenicity of *Penicillium* (Dutton and Evans, 1996; Kalisz *et al.*, 1997; Hadas *et al.*, 2007). Given that H_2O_2 generation by *P. expansum* is added to the massive ROS accumulation at the lesion site, ROS contribution to the regulation of pathogenicity is a viable possibility.

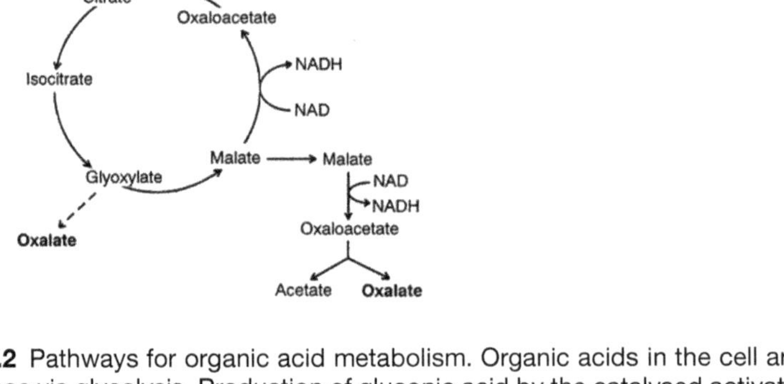

Figure 10.2 Pathways for organic acid metabolism. Organic acids in the cell are synthesized from glucose via glycolysis. Production of gluconic acid by the catalysed activation of glucose oxidase occurs in the early stages of glucose oxidation. The synthesis of oxalic acid occurs in the late stages of the citric acid cycle (adapted from Munir *et al.*, 2001; http://www.pnas.org/content/98/20/11126/F5.expansion.html).

Table 10.1 The importance of the gluconic acid/acidification process in pathogenicity of *Penicillium* (Barad *et al.*, 2012; adapted with the permission of the author)

Inhibition (%)	Gluconic acid (mg/g DW)	GOX2 Relative expression	Strain
0	6.91 ± 0.15	1.00 ± 0.42	Pe-21
7.0	4.22 ± 0.23	0.17 ± 0.05	TPe116
19.5	3.21 ± 0.53	0.19 ± 0.01	TPe114
47.9	2.77 ± 0.15	0.08 ± 0.01	TPe130
69.0	0.83 ± 0.08	0.07 ± 0.01	TPe141

GLA requirement for colonization and conidiation by *P. expansum*

Functional analysis of the *gox*-RNAi mutants showed that modulation of GLA level affected differently the pathogen's interaction during germination and colonization. The greater down-regulation of *gox2* in *gox*-RNAi mutants, the stronger impairment of GLA production, medium acidification and apple fruit infection (Table 10.1). Conditions reducing gluconic acid accumulation enhanced spore germination (Table 10.2) (Barad *et al.*, 2012). These results indicate that the observed increase in GLA production is a factor that differentially modulates surface sporulation and colonization. While the highest production of GLA occurs under neutral conditions, the functional effect on pathogenicity occurs at low pH conditions, however germination and biomass formation is promoted under acidic conditions in which GLA production is repressed (Barad *et al.*, 2012) (Table 10.2).

pH and gene activation

Ambient pH changes induced by the fungus are important in determining the capability of the pathogen to successfully invade, colonize and kill the targeted host with the aid of secreted pathogenicity factors. Although a steady intracellular pH must be maintained to ensure that enzymes have optimal conditions for their activity, the pH is a critical factor in fungal attack strategy for gene activation. Fungal pathogens have developed environmental sensing mechanisms to best suit their offensive arsenal and at the same time to compromise the host defences. They are using the PaL signal transduction pathway as a mechanism for

Table 10.2 Germination, total D-gluconic acid (GLA) and biomass production for *P. expansum* WT cultivated at a range of pH levels (Barad *et al.*, 2012; adapted with the permission of the author)

pH[a]	Germination (%)	Biomass (g DW)	Total D-gluconic acid (mg/g DW)	ΔpH (initial – final)
3	70 ± 4.2	0.3352 ± 0.09	0.54 ± 0.04	−1.09 ± 0.23
4	73 ± 1.3	0.2243 ± 0.08	0.58 ± 0.05	−0.41 ± 0.15
5	60 ± 1.2	0.2075 ± 0.03	5.25 ± 0.45	0.83 ± 0.12
6	31 ± 1.4	0.1535 ± 0.06	17.08 ± 1.01	2.37 ± 0.22
7	25 ± 1.2	0.0964 ± 0.04	27.2 ± 0.90	3.05 ± 0.21

response to pH modulating conditions (Arechiga-Carvajal and Ruiz-Herrera, 2005; Arst *et al.*, 1994; Arst and Penalva, 2003; Caracuel *et al.*, 2003a,b; Cervantes-Chavez *et al.*, 2010; Fonseca-Garcia *et al.*, 2012; Lamb *et al.*, 2001; Lamb and Mitchell, 2003; Negrete-Urtasun *et al.*, 1999; Penalva and Arst, 2002; Rollins, 2003; Tilburn *et al.*, 1995).The PaL signalling pathway promotes the proteolytic activation of a zinc finger transcription factor named PacC in filamentous fungi or RIM101 in yeasts (Penalva and Arst, 2002). This factor activates the transcription of some genes at alkaline pH and represses others at acidic pH (Su and Mitchell, 1993; Espeso and Arst, 2000; Errakhi *et al.*, 2008; Miyara *et al.*, 2010). Regulation of those genes has been assumed to be correlated with their optimal protein activity (Hadas *et al.*, 2007).

Under alkaline conditions, two proteolytic steps occur on the whole protein, $PacC^{72}$: first, $PacC^{72}$ is cut to $PacC^{53}$ – this proteolysis is pH dependent, based on a primary interaction with PalA through two specific motifs found at the C-terminal region of PacC (Vincent *et al.*, 2003) and is presumably carried out by the protease PalB (Denison *et al.*, 1995); second, independently of pH, $PacC^{53}$ is converted to $PacC^{27}$, which is transcriptionally active and is responsible for repression of acidic genes (Espeso and Arst, 2000) and for the activation of alkaline genes, including *pacC* itself, at alkaline pH (Orejas *et al.*, 1995; Suarez and Penalva, 1996; Mingot *et al.*, 1999; Caracuel *et al.*, 2003a,b). Under acidic ambient conditions, PacC remains cytosolic and inaccessible for proteolysis, but in response to pH signalling, the PaL apparatus activates PacC. In its active form, $PacC^{27}$ is translocated to the nucleus, where the zinc-finger region recognizes the consensus DNA site 5′-GCCARG-3′ (Tilburn *et al.*, 1995; Mingot *et al.*, 2001). PacC regulation has been most extensively studied in the genus *Aspergillus*, but it has also been found in many other fungi (Arst and Penalva, 2003; Penalva *et al.*, 2008). However, few publications have addressed all the genes that might be regulated by PacC specifically during fungal pathogenicity (Yakoby *et al.*, 2000; Eshel *et al.*, 2002b).

The expression of genes encoding enzymes that are secreted only at alkaline pH has been suggested to be a pathogenicity regulator under alkaline environments (Alkan *et al.*, 2013). In *C. gloeosporioides pel*B (encoded PL) is expressed and secreted at pH levels higher than 5.7, also similar to those present in decaying tissue (Prusky *et al.*, 1989; Yakoby *et al.*, 2000, 2001). In *C. gloeosporioides*, *pel*B expression during pH alkalization followed a pattern similar to that of *pac1*, suggesting the presence of a regulatory mechanism for control of enzyme secretion (Drori *et al.*, 2003). Miyara *et al.* (2008) used a *ΔpacC* mutant to demonstrate that genes encoding for pectolytic enzymes under alkaline conditions were activated by PacC in *C. gloeosporioides*. Similarly, for acidic environments, polygalacturonases (*pg*) expressed by *F. oxysporum*, were controlled by acidic pH (Caracuel *et al.*, 2003a,b); moreover, in *F. oxysporum* a *Δpac*C mutant exhibited significantly decreased virulence at acidic pH, suggesting that *pacC* is a negative regulator of genes that are important for fungal attack under acidic conditions. In another acidifying pathogen, *P. expansum*, analysis of the transcript levels of *pepg*1 (encodes for polygalacturonase) showed the highest accumulation at pH 4.0 and only minor expression at pH values higher than 5 (Prusky *et al.*, 2004). Transcription of the *pacC* homologue, *pac*1 declined during acidification concomitantly with increasing *pg1* expression, which suggests the existence of a complex virulence mechanism that controls *pg1* expression under acidic conditions that has not been described. This may indicate that the function of PacC is not only to switch on or off genes, but to assist in the fine-tuning of the expression of pH modulated systems (transporters and antioxidants). This may be in order to maintain homeostasis and expression of pathogenicity factors for orchestration of the genomic arsenal under changing pH. Therefore, in a fungus that acidifies its surroundings, PacC will probably transiently activate alkaline-expressed genes and repress acid-expressed genes during initial colonization, until the secretion of organic acids lowers the environmental pH. Reciprocally, in fungi that alkalinize the environment, *pacC* will be activated only after the fungi raise the surrounding pH. Since fungi are likely to encounter a broad range of initial environmental pH levels and responses to carbon and nitrogen nutritional balance, conservation of a broad range of pH responses is justified independently of the preferred pH induced for pathogenicity.

pH and secondary metabolism of patulin production

Patulin is a common secondary metabolite present in decayed tissue colonized by *P. expansum* and other *Penicillium* spp. (Puel *et al.*, 2010). Following the inoculation of *P. expansum*, a GLA accumulation and patulin production are easily detectable in the colonized tissue. Barad *et al.* (2014) suggested that pathogenicity of *P. expansum* is modulated by changes in pH mediated by GLA accumulation, as a factor for activation of polygalacturonase under acidic conditions. However, the joint accumulation in the colonized tissue of two fungal metabolites, GLA and patulin, raised the question of their contribution, either separately or synergistically, to *P. expansum* pathogenicity.

In order to analyse the contribution of the accumulated GLA and patulin to pathogenicity, the *gox*-RNAi mutants, previously described, were used. Using these mutants, Barad *et al.* (2012) demonstrated that down-regulation of *gox2* is accompanied by impairment in the ability to produce GLA and apple colonization. Moreover, the higher the impairment in GLA accumulation, the higher inhibition of the relative expression of isoepoxidon dehydrogenase (*idh*) (one of the last genes in patulin biosynthesis) and patulin accumulation (Barad *et al.*, 2014). These results suggest that the concurrent dynamic pH regulation by GLA accumulation is the critical factor in the induction of this secondary metabolism of patulin production and fungal colonization.

The metabolic capability for patulin accumulation by *P. expansum* under conditions of GLA accumulation and decreasing pH suggests that GLA might be one of the precursors for patulin's biosynthetic accumulation. This might explain McCallum *et al.*'s (2002) results, where without knowing the cause for the pH decrease, they found that more aggressive isolates of *P. expansum* greatly reduce the pH of the growth medium and produce detectable levels of patulin, whereas the weaker growers produce a smaller reduction in pH and no detectable patulin. Interestingly, few reports have referred to the sources of carbon metabolism for patulin biosynthesis (Puel *et al.*, 2010). Barad *et al.* (2014) confirmed that growth of *P. expansum* in the presence of sucrose, glucose or GLA as sole carbon source, were efficient precursors for patulin accumulation. Like aflatoxin, glucose is an optimal carbohydrate source for patulin production but it is unclear whether the carbon source directly induces patulin production or modulates its biosynthesis through general metabolism. Ramachandra *et al.* (2006) suggested that the GLA may be metabolized to patulin through glycolysis to pyruvate and from there probably via general metabolism to patulin (Fig. 10.2), confirming Barad's results indicating the importance of GLA accumulation during early stages of *P. expansum* colonization of fruits as a precursor for patulin accumulation (Hadas *et al.*, 2007; Barad *et al.*, 2012) (Fig. 10.3).

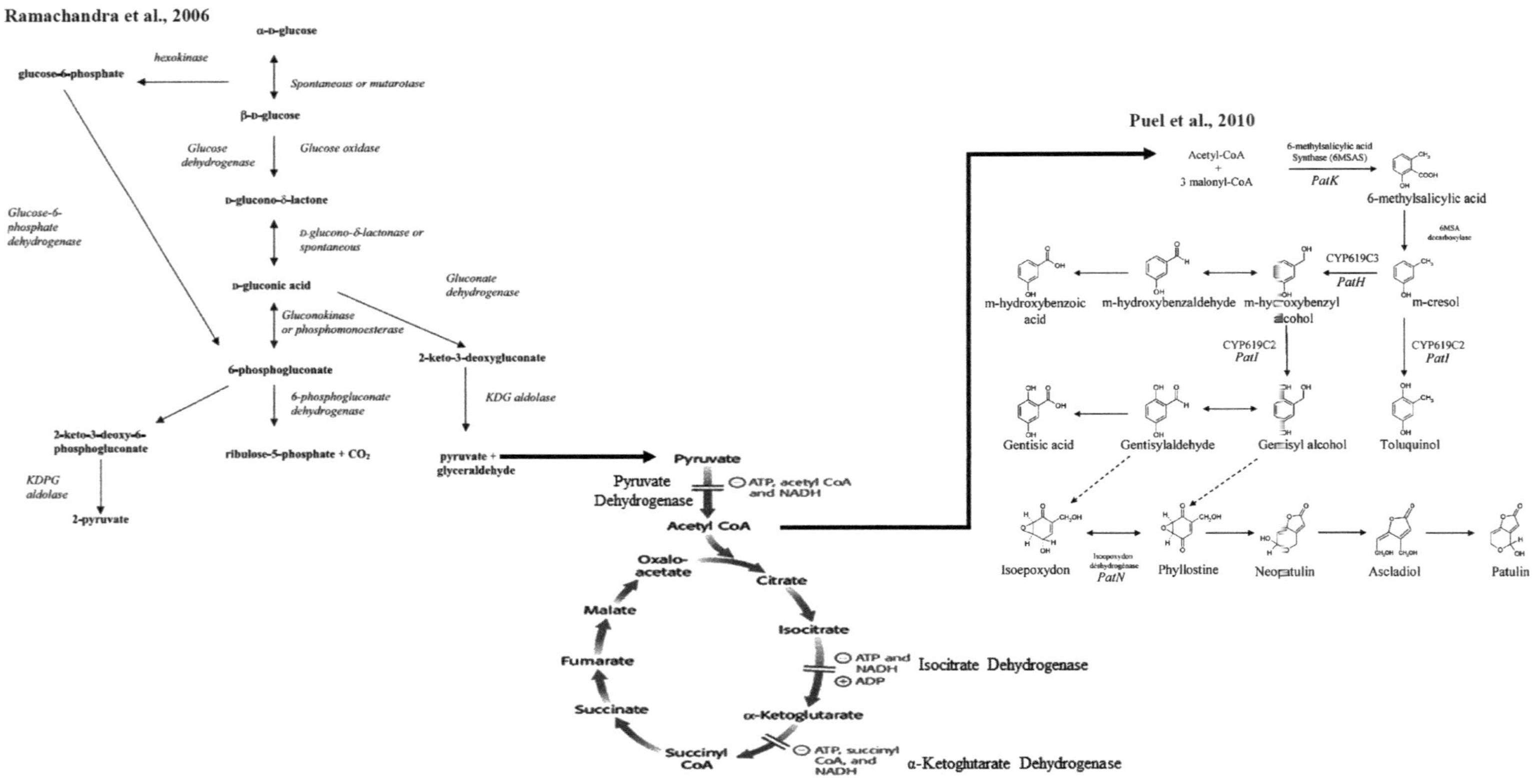

Figure 10.3 Synthesis of patulin through the gluconic acid (adapted from Ramachandra *et al.*, 2006, and Puel *et al.*, 2010).

The effect of acidification by GLA accumulation on patulin and pathogenicity of *P. expansum*

Several reports have already indicated that the production of secondary metabolites is determined by generic factors and associated cellular regulatory systems, such as the global nitrogen, carbon and pH regulatory systems common to most fungi (Yu and Keller, 2005). Nitrogen, carbon source and low pH play key roles in the regulation of sterigmatocystin/aflatoxin production in *Aspergillus* spp. (Kachholz and Demain, 1983; Keller *et al.*, 1997a; Woloshuk *et al.*, 1997). In *Aspergillus nidulans,* low pH is required for sterigmatocystin production, whereas alkaline pH is required for penicillin production (Shah *et al.*, 1991; Keller *et al.*, 1997a). Low pH is also conducive to the production of fumonisin by *Fusarium proliferatum* (Keller *et al.*, 1997b) and ochratoxin by *Aspergillus ochraceus* (O'Callaghan *et al.*, 2006). There are no published reports on pH modulation as a factor in *P. expansum* pathogenicity. However, Barad *et al.* (2014) showed that expression of *gox2* and *idh* is higher at pH 7.0 than at pH 4.0. This suggest that pathogenicity factors that modulate *P. expansum* colonization might be regulated by the transcription factor PacC. At acidic pH, the PacC protein is in an inactive conformation and is unable to bind to the promoter sites of the target genes, but, under alkaline growth conditions, activated PacC acts as both an activator of alkaline-expressed genes and a repressor of acid-expressed genes (Penalva *et al.*, 2008; Ment *et al.*, 2015). Loss-of-function mutations in *pacC* cause an acidity-mimicking phenotype and result in increased expression of 'acid' genes and reduced expression of 'alkaline' genes, as was found for *Colletotrichum gloeosporioides* (Miyara *et al.*, 2012). Under neutral/alkaline conditions, PacC has been reported to regulate many genes in *A. nidulans, P. chrysogenum* (Denison, 2000; Espeso and Arst, 2000; Eisendle *et al.*, 2004), *C. gloeosporioides* (Miyara *et al.*, 2008, 2010), and *Sclerotinia sclerotiorum* (Kim *et al.*, 2008), as found for *P. expansum.* In *S. sclerotiorum,* which regulates acidification of the environment with oxalic acid, site-directed mutagenesis of *pac1* created a dominant activating *pac1*^{c} mutation that resulted in the activation of *pac1,* leading to oxalic acid accumulation that was largely independent of ambient pH (Kim *et al.*, 2008). This indicated a positive role for *pac1* in oxalic acid regulation.

Functional analysis of *pacC*-RNAi mutants of *P. expansum* were used to demonstrate the regulation of GLA and patulin accumulation during pathogenicity by PacC. *pacC*-RNAi mutants showed reduced GLA and patulin accumulation and a 45% reduction in pathogenicity, while the pH decreases in the WT and mutants were similar. These results suggest that in *P. expansum, pacC* has a significant role in both factors that modulate pathogenicity, i.e. GLA and patulin accumulation. These results are similar to those obtained with *Sclerotinia,* whose oxalic acid production increased with increasing ambient pH (Maxwell, 1973; Rollins and Dickman, 2001; Rollins, 2003). In *P. expansum,* PacC also displayed pH-regulated expression, its steady-state levels increasing with growth of the medium pH (Barad *et al.*, 2014). How GLA and patulin accumulation is regulated during fungal pathogenicity in fruits is still not clear. However, aside from the effect of GLA on patulin accumulation, previous work has indicated that nitrogen level may regulate the expression of patulin metabolism. Grootwassink and Gaucher (1980) suggested that a distinct metabolic profile can be expressed in cultures containing ammonium as the nitrogen source. When ammonium ions were added to cultures producing patulin, rapid down-regulation of *idh* and *6-msas* transcripts and reduced patulin accumulation were observed (Rollins and Gaucher, 1994). Maceration of fruit tissue by pectolytic enzymes may lead to differential accumulation of nitrogen precursors that affect pathogen metabolism, as was reported for the down-regulation of sterigmatocystin and ochratoxin A (Feng and Leonard, 1998; Abbas *et al.*, 2009). These changes might explain in part the significant differences in patulin accumulation, given the differential nitrogen contents of host fruits. Further research is needed to fully understand the mechanism governing the down-regulation of genes modulating patulin biosynthesis.

In summary, GLA appears to play a central role in pH responsiveness. Its production is a pH-regulated process and its accumulation, by virtue of environmental acidification, may serve as a regulator of acid pH regulated processes such as *pg1* expression (Prusky *et al.*, 2004; Hadas *et al.*, 2007). On the other hand, *P. expansum* uses GLA accumulation during modification of the ambient pH environment as a regulatory cue for disease and accumulation of secondary metabolites such as patulin.

Modulating the activation of small secreted molecules

The capability of pathogens to secrete pH-regulating molecules, on one hand, and the transcriptome analysis of PacC modulated genes, on the other hand, revealed an arsenal of pathogenicity factors, transporters and antioxidants which are induced by either alkalization or acidification of the environment. Reports that, in most cases, a given pathogen have a single, specific way of life by which it modulates its host pH, do not indicate that the same pathogen may not act in the opposite direction (Table 10.3). The questions are: how specific are the pH-regulating patterns for each particular fungal species during pathogenicity, and what is the signal that may differentially activate the specific pH modulation during colonization?

Can different host nutritional conditions, such as increasing sugar levels during fruit ripening, modulate the type of fungal small secreted molecules that modulates the host pH? Fungi possess sensitive gene regulatory mechanisms for responding to nutrient fluctuations in the environment as occur in ripening fruits or growing plants. Such mechanisms must be essential in pathogens for three reasons: (1) they would signal the transition of the fungus from the nutrient-free surface to the sugar-rich interior of the host; (2) they would allow the

Table 10.3 Fungal pathogens and small secreted molecules that modulate pH for the activation of pathogenicity factors. *Penicillium* spp., *Botrytis cinerea*, *Sclerotinia sclerotiorum* and *Phomopsis mangiferae* secrete organic acids for acidification. On the other hand, *Colletotrichum* spp., *Alternaria alternata* and *Fusarium oxysporum* secrete ammonia for alkalization

Pathogens	Alkalizers	Acidifiers
Colletotrichum	Ammonia	
Alternaria	Ammonia	
Fusarium	Ammonia	
Penicillium		Gluconic acid
Phomopsis		Gluconic acid
Monilinia		Gluconic acid
Sclerotinia		Oxalic acid
Botrytis		Oxalic acid

fungus to respond rapidly to the nutritional status of the host; and (3) they would temper the voracious appetite of pathogen during the necrotrophic growth stage in the plant. High glucose/sucrose concentrations may be oxidized to CO_2 via tricarboxylic acid, with high rates of glycolysis and the production of organic acids that contribute to the secretion of metabolites that decrease the host pH (Fig. 10.2). In low sugar concentrations, the importance of glutaminolysis for cell energy supply is clearly evident, and ammonia is generated as a byproduct of the glutaminase and glutamate dehydrogenase synthesis reactions (Newland *et al.*, 1990). Understanding the genetic pathways that regulate the responses of pathogenic fungi to their environment is paramount to development of effective disease mitigation strategies. Pathogens use specific gene-induction pathways to metabolize a wide range of carbon and nitrogen compounds, but this voracity is moderated by two global regulatory systems that ensure the preferential utilization of a few favoured carbon and nitrogen sources. Nitrogen metabolite repression (NMR) ensures the preferential utilization of ammonium and L-glutamine over less favoured nitrogen sources such as nitrate (Wilson and Arst, 1998; Fernandez *et al.*, 2014). GATA-family transcription factors, such as *AreA*, play global roles in NMR by acting as wide-domain positive regulators of genes required for alternative nitrogen source utilization. Carbon catabolite repression (CCR) is a global regulatory mechanism found in a wide range of microbial organisms; it ensures the utilization of preferred carbon sources such as glucose over less favourable ones. However, little is known about the components of CCR that interact with other pH-modulating nitrogen systems: CCR operates via the negatively acting zinc finger repressor CreA to ensure that glucose is utilized preferentially, by preventing the expression of genes required for the metabolism of less preferred carbon sources; it promotes glucose metabolism, including glucose uptake and utilization, and represses the expression of alternative carbon-utilizing genes including cell wall degrading enzymes, genes involved in glucosamine, proline metabolism and gluconeogenic genes (Fernandez *et al.*, 2012, 2014). Interestingly, both CCR and NMR regulatory systems converge on genes required for metabolizing key compounds that can be used as both carbon and nitrogen sources for pH modulation. How are both systems activated? This question is of high importance for understanding the differential pH response and the consequent expression of genes that modulate pathogenicity under dynamic pH and colonization conditions.

Acknowledgement

We acknowledge the support by the Binational US-Israel Agricultural Research and Development Fund (BARD) during several stages of our work. A scholarship from the Agricultural Research Organization, the Volcani Centre was granted to SB.)

References

Abbas, A., Valez, H., and Dobson, A.D. (2009). Analysis of the effect of nutritional factors on OTA and OTB biosynthesis and polyketide synthase gene expression in *Aspergillus ochraceus*. Int. J. Food. Microbiol. *135*, 22–27.

Alkan, N., Fluhr, R., Sherman, A., and Prusky, D. (2008). Role of ammonia secretion and pH modulation on pathogenicity of *Colletotrichum coccodes* on tomato fruit. Mol. Plant Microbe Interact. *21*, 1058–1066.

Alkan, N., Davydov, O., Sagi, M., Fluhr, R., and Prusky, D. (2009). Ammonium secretion by *Colletotrichum coccodes* activates host NADPH oxidase activity enhancing host cell death and fungal virulence in tomato fruits. Mol. Plant Microbe Interact. 22, 1484–1491.

Alkan, N., Meng, X., Friedlander, G., Reuveni, E., Sukno, S., Sherman, A., Thon, M., Fluhr, R., and Prusky, D. (2013). Global aspects of pacC regulation of pathogenicity genes in *Colletotrichum gloeosporioides* as revealed by transcriptome analysis. Mol. Plant Microbe Interact. *26*, 1345–1358.

Arechiga-Carvajal, E.T., and Ruiz-Herrera, J. (2005). The RIM101/pacC homologue from the basidiomycete *Ustilago maydis* is functional in multiple pH-sensitive phenomena. Eukaryot. Cell *4*, 999–1008.

Arst, H.N., and Penalva, M.A. (2003). pH regulation in *Aspergillus* and parallels with higher eukaryotic regulatory systems. Trends Genet. *19*, 224–231.

Arst, H.N. Jr., Bignell, E., and Tilburn, J. (1994). Two new genes involved in signalling ambient pH in *Aspergillus nidulans*. Mol. Gen. Genet. *245*, 787–790.

Barad, S., Horowitz, S.B., Moscovitz, O., Lichter, A., Sherman, A., and Prusky, D. (2012). A *Penicillium expansum* glucose oxidase-encoding gene, GOX2, is essential for gluconic acid production and acidification during colonization of deciduous fruit. Mol. Plant Microbe Interact. *25*, 779–788.

Barad, S., Horowitz, S.B., Kobiler, I., Sherman, A., and Prusky, D. (2014). Accumulation of the mycotoxin patulin in the presence of gluconic acid contributes to pathogenicity of *Penicillium expansum*. Mol. Plant Microbe Interact. *27*, 66–77.

Bravo, R., and Macdonald-Bravo, H. (1986). Effect of pH on the induction of competence and progression to the S-phase in mouse fibroblasts. FEBS Lett. *195*, 309–312.

Calvo, A.M., Wilson, R.A., Bok, J.W., and Keller, N.P. (2002). Relationship between secondary metabolism and fungal development. Microbiol. Mol. Biol. Rev. *66*, 447–459.

Caracuel, Z., Casanova, C., Roncero, M.I., Di Pietro, A., and Ramos, J. (2003a). pH response transcription factor PacC controls salt stress tolerance and expression of the P-Type Na+ -ATPase Ena1 in *Fusarium oxysporum*. Eukaryot. Cell 2, 1246–1252.

Caracuel, Z., Roncero, M.I., Espeso, E.A., Gonzalez-Verdejo, C.I., Garcia-Maceira, F.I., and Di Pietro, A. (2003b). The pH signalling transcription factor PacC controls virulence in the plant pathogen *Fusarium oxysporum*. Mol. Microbiol. *48*, 765–779.

Cervantes-Chavez, J.A., Ortiz-Castellanos, L., Tejeda-Sartorius, M., Gold, S., and Ruiz-Herrera, J. (2010). Functional analysis of the pH responsive pathway Pal/Rim in the phytopathogenic basidiomycete *Ustilago maydis*. Fungal Genet. Biol. *47*, 446–457.

Chambard, J.C., and Pouyssegur, J. (1986). Intracellular pH controls growth factor-induced ribosomal protein S6 phosphorylation and protein synthesis in the G0-G1 transition of fibroblasts. Exp. Cell. Res. *164*, 282–294.

Chiche, J., Brahimi-Horn, M.C., and Pouyssegur, J. (2010). Tumour hypoxia induces a metabolic shift causing acidosis: a common feature in cancer. J. Cell. Mol. Med. *14*, 771–794.

Davidzon, M., Alkan, N., Kobiler, I., and Prusky, D. (2010). Acidification by gluconic acid of mango fruit tissue during colonization via stem end infection by *Phomopsis mangiferae*. Postharvest Biol. Technol. *55*, 71–77.

Denison, S.H. (2000). pH regulation of gene expression in fungi. Fungal Genet. Biol. *29*, 61–71.

Denison, S.H., Orejas, M., and Arst, H.N. Jr. (1995). Signaling of ambient pH in *Aspergillus* involves a cysteine protease. J. Biol. Chem. *270*, 28519–28522.

Dieguez-Uribeondo, J., Forster, H., and Adaskaveg, J.E. (2008). Visualization of localized pathogen-induced pH modulation in almond tissues infected by *Colletotrichum acutatum* using confocal scanning laser microscopy. Phytopathology *98*, 1171–1178.

Drori, N., Kramer-Haimovich, H., Rollins, J., Dinoor, A., Okon, Y., Pines, O., and Prusky, D. (2003). External pH and nitrogen source affect secretion of pectate lyase by *Colletotrichum gloeosporioides*. Appl. Environ. Microbiol. *69*, 3258–3262.

Dutton, M.V., and Evans, C.S. (1996). Oxalate production by fungi: Its role in pathogenicity and ecology in the soil environment. Can. J. Microbiol. *42*, 881–895.

Eisendle, M., Oberegger, H., Buttinger, R., Illmer, P., and Haas, H. (2004). Biosynthesis and uptake of siderophores is controlled by the PacC-mediated ambient-pH regulatory system in *Aspergillus nidulans*. Eukaryot. Cell 3, 561–563.

Errakhi, R., Meimoun, P., Lehner, A., Vidal, G., Briand, J., Corbineau, F., Rona, J.P., and Bouteau, F. (2008). Anion channel activity is necessary to induce ethylene synthesis and programmed cell death in response to oxalic acid. J. Exp. Bot. *59*, 3121–3129.

Eshel, D., Lichter, A., Dinoor, A., and Prusky, D. (2002a). Characterization of *Alternaria alternata* glucanase genes expressed during infection of resistant and susceptible persimmon fruits. Mol. Plant. Pathol. 3, 347–358.

Eshel, D., Miyara, I., Ailing, T., Dinoor, A., and Prusky, D. (2002b). pH regulates endoglucanase expression and virulence of *Alternaria alternata* in persimmon fruit. Mol. Plant Microbe Interact. *15*, 774–779.

Espeso, E.A., and Arst, H.N. (2000). On the mechanism by which alkaline pH prevents expression of an acid-expressed gene. Mol. Cell. Biol. *20*, 3355–3363.
Espeso, E.A., Tilburn, J., Arst, H.N. Jr., and Penalva, M.A. (1993). pH regulation is a major determinant in expression of a fungal penicillin biosynthetic gene. EMBO J. *12*, 3947–3956.
Feng, G.H., and Leonard, T.J. (1998). Culture conditions control expression of the genes for aflatoxin and sterigmatocystin biosynthesis in *Aspergillus parasiticus* and *A. nidulans*. Appl. Environ. Microbiol. *64*, 2275–2277.
Fernandez, J., Wright, J.D., Hartline, D., Quispe, C.F., Madayiputhiya, N., and Wilson, R.A. (2012). Principles of carbon catabolite repression in the rice blast fungus: Tps1, Nmr1-3, and a MATE-family pump regulate glucose metabolism during infection. PLoS Genet. *8*, e1002673.
Fernandez, J., Marroquin-Guzman, M., and Wilson, R.A. (2014). Mechanisms of nutrient acquisition and utilization during fungal infections of leaves. Annu. Rev. Phytopathol. *52*, 155–174.
Fonseca-Garcia, C., Leon-Ramirez, C.G., and Ruiz-Herrera, J. (2012). The regulation of different metabolic pathways through the Pal/Rim pathway in *Ustilago maydis*. FEMS Yeast Res. *12*, 547–556.
Grootwassink, J.W., and Gaucher, G.M. (1980). *De novo* biosynthesis of secondary metabolism enzymes in homogeneous cultures of *Penicillium urticae*. J. Bacteriol. *141*, 443–455.
Hadas, Y., Goldberg, I., Pines, O., and Prusky, D. 2007. Involvement of gluconic acid and glucose oxidase in the pathogenicity of *Penicillium expansum* in apples. Phytopathology *97*, 384–390.
Jennings, D.H. (1989). Some perspectives on nitrogen and phosphorus metabolism in fungi. In Nitrogen, Phosphorus and Sulphur Utilization by Fungi, Boddy, L., Marchant, R., and Read, D.J., eds. (Cambridge University Press, UK), pp. 1–13.
Kachholz, T., and Demain, A.L. (1983). Nitrate repression of averufin and aflatoxin biosynthesis. J. Nat. Prod. *46*, 499–506.
Kalisz, H.M., Hendle, J., and Schmid, R.D. (1997). Structural and biochemical properties of glycosylated and deglycosylated glucose oxidase from *Penicillium amagasakiense*. Appl. Microbiol. Biotechnol. *47*, 502–507.
Keller, N.P., Nesbitt, C., Sarr, B., Phillips, T.D., and Burow, G.B. (1997a). pH regulation of sterigmatocystin and aflatoxin biosynthesis in *Aspergillus* spp. Phytopathology *87*, 643–648.
Keller, S.E., Sullivan, T.M., and Chirtel, S. (1997b). Factors affecting the growth of *Fusarium proliferatum* and the production of fumonisin B1: oxygen and pH. J. Ind. Microbiol. Biotechnol. *19*, 305–309.
Kim, K.S., Min, J.Y., and Dickman, M.B. (2008). Oxalic acid is an elicitor of plant programmed cell death during *Sclerotinia sclerotiorum* disease development. Mol. Plant Microbe Interact. *21*, 605–612.
Kunz, C., Vandelle, E., Rolland, S., Poinssot, B., Bruel, C., Cimerman, A., Zotti, C., Moreau, E., Vedel, R., Pugin, A., and Boccara, M. (2006). Characterization of a new, nonpathogenic mutant of *Botrytis cinerea* with impaired plant colonization capacity. New Phytol. *170*, 537–550.
Lager, I., Andreasson, O., Dunbar, T.L., Andreasson, E., Escobar, M.A., and Rasmusson, A.G. (2010). Changes in external pH rapidly alter plant gene expression and modulate auxin and elicitor responses. Plant Cell Environ. *33*, 1513–1528.
Lamb, A.L., Torres, A.S., O'Halloran, T.V., and Rosenzweig, A.C. (2001). Heterodimeric structure of superoxide dismutase in complex with its metallochaperone. Nat. Struct. Biol. *8*, 751–755.
Lamb, T.M., and Mitchell, A.P. (2003). The transcription factor Rim101p governs ion tolerance and cell differentiation by direct repression of the regulatory genes NRG1 and SMP1 in *Saccharomyces cerevisiae*. Mol. Cell. Biol. *23*, 677–686.
Lee, Y.H., and Dean, R.A. (1993). cAMP regulates infection structure formation in the plant pathogenic fungus *Magnaporthe grisea*. Plant Cell *5*, 693–700.
Macarisin, D., Cohen, L., Eick, A., Rafael, G., Belausov, E., Wisniewski, M., and Droby, S. (2007). *Penicillium digitatum* suppresses production of hydrogen peroxide in host tissue during infection of citrus fruit. Phytopathology *97*, 1491–1500.
Manteau, S., Abouna, S., Lambert, B., and Legendre, L. (2003). Differential regulation by ambient pH of putative virulence factor secretion by the phytopathogenic fungus *Botrytis cinerea*. FEMS Microbiol. Ecol. *43*, 359–366.
Maxwell, D.P. (1973). Oxalate formation in *Whetzelinia sclerotiorum* by oxaloacetate acetylhydrolase. Physiol. Plant Pathol. *3*, 279–288.
McCallum, J., Tsao, R., and Zhou, T. (2002). Factors affecting patulin production by *Penicillium expansum*. J. Food Prot. *65*, 1937–1942.
Ment, D., Alkan, N., Luria, N., Bi, F.C., Reuveni, E., Fluhr, R., and Prusky, D. (2015). A role of AREB in the regulation of PACC-dependent acid-expressed-genes and pathogenicity of *Colletotrichum gloeosporioides*. Mol. Plant Microbe Interact. *28*, 154–166.

Mingot, J.M., Tilburn, J., Diez, E., Bignell, E., Orejas, M., Widdick, D.A., Sarkar, S., Brown, C.V., Caddick, M.X., Espeso, E.A., Arst, H.N. Jr., and Penalva, M.A. (1999). Specificity determinants of proteolytic processing of *Aspergillus* PacC transcription factor are remote from the processing site, and processing occurs in yeast if pH signalling is bypassed. Mol. Cell. Biol. *19*, 1390–1400.

Mingot, J.M., Espeso, E.A., Diez, E., and Penalva, M.A. (2001). Ambient pH signaling regulates nuclear localization of the *Aspergillus nidulans* PacC transcription factor. Mol. Cell. Biol. *21*, 1688–1699.

Miyara, I., Shafran, H., Kramer Haimovich, H., Rollins, J., Sherman, A., and Prusky, D. (2008). Multi-factor regulation of pectate lyase secretion by *Colletotrichum gloeosporioides* pathogenic on avocado fruits. Mol. Plant. Pathol. *9*, 281–291.

Miyara, I., Shafran, H., Davidzon, M., Sherman, A., and Prusky, D. (2010). pH regulation of ammonia secretion by *Colletotrichum gloeosporioides* and its effect on appressorium formation and pathogenicity. Mol. Plant Microbe Interact. *23*, 304–316.

Miyara, I., Shnaiderman, C., Meng, X., Vargas, W.A., Diaz-Minguez, J.M., Sherman, A., Thon, M., and Prusky, D. (2012). Role of nitrogen-metabolism genes expressed during pathogenicity of the alkalinizing *Colletotrichum gloeosporioides* and their differential expression in acidifying pathogens. Mol. Plant Microbe Interact. *25*, 1251–1263.

Musgrove, E., Seaman, M., and Hedley, D. (1987). Relationship between cytoplasmic pH and proliferation during exponential-growth and cellular quiescence. Exp. Cell. Res. *172*, 65–75.

Negrete-Urtasun, S., Reiter, W., Diez, E., Denison, S.H., Tilburn, J., Espeso, E.A., Penalva, M.A., and Arst, H.N. Jr. (1999). Ambient pH signal transduction in *Aspergillus*: completion of gene characterization. Mol. Microbiol. *33*, 994–1003.

Newland, M., Greenfield, P.F., and Reid, S. (1990). Hybridoma growth limitations: the roles of energy metabolism and ammonia production. Cytotechnology 3, 215–229.

O'Callaghan, J., Stapleton, P.C., and Dobson, A.D. (2006). Ochratoxin A biosynthetic genes in *Aspergillus ochraceus* are differentially regulated by pH and nutritional stimuli. Fungal Genet. Biol. *43*, 213–221.

O'Connell, R.J., Thon, M.R., Hacquard, S., Amyotte, S.G., Kleemann, J., Torres, M.F., Damm, U., Buiate, E.A., Epstein, L., Alkan, N., *et al.* (2012). Lifestyle transitions in plant pathogenic *Colletotrichum* fungi deciphered by genome and transcriptome analyses. Nat. Genet. *44*, 1060–1065.

O'Meara, T.R., Norton, D., Price, M.S., Hay, C., Clements, M.F., Nichols, C.B., and Alspaugh, J.A. (2010). Interaction of *Cryptococcus neoformans* Rim101 and protein kinase A regulates capsule. PLoS Pathog. *6*, e1000776.

Orejas, M., Espeso, E.A., Tilburn, J., Sarkar, S., Arst, H.N. Jr., and Penalva, M.A. (1995). Activation of the *Aspergillus* PacC transcription factor in response to alkaline ambient pH requires proteolysis of the carboxy-terminal moiety. Genes Dev. *9*, 1622–1632.

Penalva, M.A., and Arst, H.N. Jr. (2002). Regulation of gene expression by ambient pH in filamentous fungi and yeasts. Microbiol. Mol. Biol. Rev. *66*, 426–446.

Penalva, M.A., Tilburn, J., Bignell, E., and Arst, H.N. (2008). Ambient pH gene regulation in fungi: making connections. Trends Microbiol. *16*, 291–300.

Perpetua, N.S., Kubo, Y., Yasuda, N., Takano, Y., and Furusawa, I. (1996). Cloning and characterization of a melanin biosynthetic THR1 reductase gene essential for appressorial penetration of *Colletotrichum lagenarium*. Mol. Plant Microbe Interact. *9*, 323–329.

Poussereau, N., Gente, S., Rascle, C., Billon-Grand, G., and Fevre, M. (2001). aspS encoding an unusual aspartyl protease from *Sclerotinia sclerotiorum* is expressed during phytopathogenesis. FEMS Microbiol Lett. *194*, 27–32.

Pouyssegur, J. (1985). The growth factor-activatable Na+/H+ exchange system – a genetic approach. Trends Biochem. Sci. *10*, 453–455.

Prusky, D., and Yakoby, N. (2003). Pathogenic fungi: leading or led by ambient pH? Mol. Plant Pathol. *4*, 509–516.

Prusky, D., Gold, S., and Keen, N.T. (1989). Purification and characterization of an endopolygalacturonase produced by *Colletotrichum gloeosporioides*. Physiol. Mol. Plant. Pathol. *35*, 121–133.

Prusky, D., McEvoy, J.L., Leverentz, B., and Conway, W.S. (2001). Local modulation of host pH by *Colletotrichum* species as a mechanism to increase virulence. Mol. Plant Microbe Interact. *14*, 1105–1113.

Prusky, D., McEvoy, J.L., Saftner, R., Conway, W.S., and Jones, R. (2004). Relationship between host acidification and virulence of *Penicillium* spp. on apple and citrus fruit. Phytopathology *94*, 44–51.

Prusky, D., Alkan, N., Mengiste, T., and Fluhr, R. (2013). Quiescent and necrotrophic lifestyle choice during postharvest disease development. Annu. Rev. Phytopathol. *51*, 155–176.

Puel, O., Galtier, P., and Oswald, I.P. (2010). Biosynthesis and toxicological effects of patulin. Toxins *2*, 613–631.

Ramachandra, S., Fontanille, P., Pandey, A., and Larroche, C. (2006). Gluconic acid: properties, applications and microbial production. Food Technol. Biotechnol. *44*, 185–195.

Reinehr, R., Gorg, B., Becker, S., Qvartskhava, N., Bidmon, H.J., Selbach, O., Haas, H.L., Schliess, F., and Haussinger, D. (2007). Hypoosmotic swelling and ammonia increase oxidative stress by NADPH oxidase in cultured astrocytes and vital brain slices. Glia *55*, 758–771.

Rollins, J.A. (2003). The *Sclerotinia sclerotiorum* pac1 gene is required for sclerotial development and virulence. Mol. Plant Microbe Interact. *16*, 785–795.

Rollins, J.A., and Dickman, M.B. (2001). pH signaling in *Sclerotinia sclerotiorum*: identification of a pacC/RIM1 homolog. Appl. Environ. Microbiol. *67*, 75–81.

Rollins, M.J., and Gaucher, G.M. (1994). Ammonium repression of antibiotic and intracellular proteinase production in *Penicillium urticae*. Appl. Microbiol. Biotechnol. *41*, 447–455.

Shah, A.J., Tilburn, J., Adlard, M.W., and Arst, H.N. Jr. (1991). pH regulation of penicillin production in *Aspergillus nidulans*. FEMS Microbiol Lett. *61*, 209–212.

Spielvogel, A., Findon, H., Arst, H.N., Araujo-Bazan, L., Hernandez-Ortiz, P., Stahl, U., Meyer, V., and Espeso, E.A. (2008). Two zinc finger transcription factors, CrzA and SltA, are involved in cation homoeostasis and detoxification in *Aspergillus nidulans*. Biochem. J. *414*, 419–429.

Su, S.S.Y., and Mitchell, A.P. (1993). Molecular characterization of the yeast meiotic regulatory gene Rim1. Nucleic Acids Res. *21*, 3789–3797.

Suarez, T., and Penalva, M.A. (1996). Characterization of a *Penicillium chrysogenum* gene encoding a PacC transcription factor and its binding sites in the divergent pcbAB-pcbC promoter of the penicillin biosynthetic cluster. Mol. Microbiol. *20*, 529–540.

Ten Have, A., Breuil, W.O., Wubben, J.P., Visser, J., and van Kan, J.A.L. (2001). *Botrytis cinerea* endopolygalacturonase genes are differentially expressed in various plant tissues. Fungal Genet. Biol. *33*, 97–105.

Tilburn, J., Sarkar, S., Widdick, D.A., Espeso, E.A., Orejas, M., Mungroo, J., Penalva, M.A., and Arst, H.N. Jr. (1995). The *Aspergillus* PacC zinc finger transcription factor mediates regulation of both acid- and alkaline-expressed genes by ambient pH. EMBO J. *14*, 779–790.

Vincent, O., Rainbow, L., Tilburn, J., Arst, H.N., and Penalva, M.A. (2003). YPXL/I is a protein interaction motif recognized by *Aspergillus* PalA and its human homologue, AIP1/Alix. Mol. Cell. Biol. *23*, 1647–1655.

Wilson, R.A., and Arst, H.N. Jr. (1998). Mutational analysis of AREA, a transcriptional activator mediating nitrogen metabolite repression in *Aspergillus nidulans* and a member of the "streetwise" GATA family of transcription factors. Microbiol. Mol. Biol. Rev. *62*, 586–596.

Woloshuk, C.P., Cavaletto, J.R., and Cleveland, T.E. (1997). Inducers of aflatoxin biosynthesis from colonized maize kernels are generated by an amylase activity from *Aspergillus flavus*. Phytopathology *87*, 164–169.

Wubben, J.P., ten Have, A., van Kan, J.A., and Visser, J. (2000). Regulation of endopolygalacturonase gene expression in *Botrytis cinerea* by galacturonic acid, ambient pH and carbon catabolite repression. Curr Genet. *37*, 152–157.

Yakoby, N., Freeman, S., Dinoor, A., Keen, N.T., and Prusky, D. (2000). Expression of pectate lyase from *Colletotrichum gloesosporioides in C. magna* promotes pathogenicity. Mol. Plant Microbe Interact. *13*, 887–891.

Yakoby, N., Beno-Moualem, D., Keen, N.T., Dinoor, A., Pines, O., and Prusky, D. (2001). *Colletotrichum gloeosporioides* pelB is an important virulence factor in avocado fruit–fungus interaction. Mol Plant Microbe Inter. *14*, 988–995.

Yu, J.H., and Keller, N. (2005). Regulation of secondary metabolism in filamentous fungi. Annu. Rev. Phytopathol. *43*, 437–458.

Evolutionary Adaptation as a Tool to Generate Targeted Mutant Strains as Evidenced by Increased Inulinase Production in *Aspergillus oryzae*

11

Helena Culleton, Eline Majoor, Vincent A. McKie and Ronald P. de Vries

Abstract

Inulin is found widely distributed in nature as a storage polysaccharide and consists of a linear polymer of β-1,2-linked D-fructose molecules which can be hydrolysed by *endo-*/*exo*-inulinases and fructofuranosidase (invertase) to give D-fructose and fructooligosaccharides. In this study we aimed to improve the inulin degradation potential of *Aspergillus oryzae* through the up-regulation of *exo*-inulinase production using an evolutionary adaptation method. This method has the advantage over UV or chemical mutagenesis in that it is likely to have a lower number of random mutations as only beneficial mutations will provide a competitive advantage and will therefore become dominant in the culture. As an organism with no predicted *endo*-inulinase function, improved inulin degradation in *A. oryzae* would be largely dependent on the overproduction of this enzyme. Subsequent generation growth of *Aspergillus oryzae* (Rib40) on inulin for 9 weeks successfully resulted in *exo*-inulinase-overproducing mutants.

Introduction

Inulin is widely distributed in nature as a storage carbohydrate in the roots and tubers of several plants such as dandelion, chicory, dahlia, Jerusalem artichoke and burdock (Vandamme and Derycke, 1983). It contains a linear polymer of β-1,2-linked D-fructose, connected to a terminal sucrose residue (Pavis *et al.*, 2001) and can be hydrolysed by inulinases (*endo-*/*exo-*) (Derycke and Vandamme, 1984) and fructofuranosidase (invertase) (Wallis *et al.*, 1997), resulting mainly in the formation of D-fructose and fructooligosaccharides. Fructose is the sweetest naturally occurring sugar compound identified to date and consequently is emerging fast as an important ingredient in the food and pharmaceutical industry. Fructose is a more favourable alternative sweetener to sucrose due to its higher sweetening capacity, low viscosity and beneficial effects associated with diabetic patients and iron absorption in children (Pawan, 1973). Furthermore, fructooligosaccharides have many positive functional and nutritional properties i.e. reduced caloric value, stimulation of intestinal bifidobacteria and as a source of dietary fibre in food preparations (Niness, 1999; Singh and Gill, 2006). Conventional methods of fructose production involving the enzymatic degradation of starch followed by the conversion of D-glucose to D-fructose using glucose isomerase yields only ~42% D-fructose, with the remaining end-products consisting of glucose (~50%) and oligosaccharides (~8%) (Singh and Gill, 2006). Other techniques, such as ion exchange chromatography, have been developed for fortification of fructose but all add to the production cost (Zittan, 1981). The use of microbial inulinases has therefore been proposed as a more favourable and promising approach for obtaining optimum D-fructose yields.

In a recent study, the plant polysaccharide degrading potential was compared for three different *Aspergillus* species i.e. *A. nidulans, A. niger* and *A. oryzae* (Coutinho *et al.*, 2009). Comparison of the CAZy (Carbohydrate-Active enzyme database (CAZy-http://www.cazy.org) (Lombard *et al.*, 2014)) content of the genomes of these fungi demonstrated significant differences, with the number of genes related to particular polysaccharide degradation varying substantially between the three species. *A. niger* has the highest number of inulin-related genes in its genome with five open reading frames (ORFs), compared to *A. nidulans* and *A. oryzae*, which have two ORFs and four ORFs respectively. Growth profiles for these species on inulin, compared on the Fungal Growth Database website (www.fung-growth.org), showed a correlation with genome content, with growth of *A. niger* exceeding that of both *A. nidulans* and *A. oryzae* on this carbon source. While such poor growth could be expected for *A. nidulans*, the genome of which had less than half the number of ORFs of *A. niger*, the results were somewhat surprising for *A. oryzae*. Further analysis into the function predictions of these ORF's showed that of the three species, *A. niger* was the only one to contain an *endo*-inulinase in its genome, thus enhancing its ability for efficient inulin degradation. Based on this information, *A. nidulans* and *A. oryzae* would be required to over-produce *exo*-inulinase in order to improve their growth on inulin, thus resulting in an increased yield of D-fructose. This suggests that these aspergilli have a unique biological approach towards the enzymatic degradation of the same substrate.

Industrial enzyme overproductions strains are often produced using UV or chemical mutagenesis (Nikolaev *et al.*, 2013; Peterson and Nevalainen, 2012), but this approach has the disadvantage that random mutations are introduced. While the mutants selected are screened for the desired phenotype, often other mutations are also introduced, some of which could be potentially disadvantageous. Now that genomics tools have become very affordable, there is a growing interest in re-sequencing mutant strains to identify the mutations that are responsible for the phenotype. Here we present an alternative method for generating such mutants, which is less likely to introduce random (potentially disadvantageous) mutations. The approach is similar to what has been used for the yeast *Saccharomyces cerevisiae* (Wisselink *et al.*, 2009). The use of this type of evolutionary screen has been very effective in obtaining microorganisms (primarily industrial) with improved phenotypes, such as an expanded substrate range, increased stress tolerance and efficient substrate utilization (Patnaik, 2008; Sauer, 2001). In this study we have looked at an evolutionary approach to develop a strain with increased *exo*-inulinase activity. We have taken the filamentous fungi, *A. oryzae* and *A. nidulans* and generationally grown them on inulin for 9 weeks to obtain *exo*-inulinase-overproducing mutants.

Materials and methods

Strains, media and culture conditions

Aspergillus nidulans (FGSCA4) and *Aspergillus oryzae* (Rib40) were used in this study. Both strains were taken from glycerol stocks stored at −45°C and grown on MEA (malt extract agar) prior to the evolution experiments. All plates were grown at 30°C.

Aspergillus minimal medium (MM) was described previously (de Vries *et al.*, 2004). Agar was added at 2% (w/v) for solid medium. For the evolution experiments, MM solid agar was supplemented with 1% (w/v) inulin (dahlia tubers) (Sigma; Cat. No. I3754). For the analysis and characterization of enzyme activities, 400 ml liquid cultures (2.5 l baffled flasks) of MM + 1% (w/v) inulin (dahlia tubers) (Sigma; Cat. No. I3754)/inulin (chicory) (Sigma: Cat. No. I2255) were inoculated with 1×10^6 spores/ml (final) and incubated at 30°C in an orbital shaker at 250 rpm for 4 days.

Preparation of evolution strains

From the initial regeneration on MEA agar, spores from *A. nidulans* and *A. oryzae* strains were harvested in 10 ml ACES (*N*-(2-Acetamido)-2-aminoethanesulfonic acid) buffer (0.2% (w/v) ACES + 0.02% (w/v) TWEEN® 80, pH 6.0) and 200 µl was used to inoculate the evolution specific agar (MM + 1% (w/v) inulin) in triplicate for each strain. All plates were incubated at 30°C. After one week, growth was examined and 100–200 µl of spores was used to inoculate new inulin plates. This process was repeated for a period of 9 weeks after which time the strains were purified on inulin by diluted plating and selecting the best growing colony. The purified strains were stored in 30% (w/v) glycerol at −80°C.

Enzyme profile comparison by SDS-PAGE

400 ml liquid cultures were inoculated with 1×10^6 spores/ml (final) and were grown in 2.5 l baffled flasks at 30°C and 250 rpm for 4 days. Following incubation, the mycelia were removed from the culture filtrate by filtration over a Büchner funnel with nylon gauze. SDS-PAGE was done using a 12% polyacrylamide gel containing 0.1% (w/v) SDS with 15 µl culture filtrate samples being loaded for each expression strain. Protein bands were detected by Coomassie Blue staining (Laemmli, 1970).

Enzyme activity measurements

For the measurement of enzyme activities from the *A. oryzae* generation strains, an initial 1:10 dilution of the culture filtrates was carried out in 100 mM sodium acetate buffer, pH 4.5 including BSA (1 mg/ml). Para-nitrophenol (*p*NP) assays were used for the measurement of α-/β-galactosidase and α-/β-glucosidase activities. Initial *α*-/*β*-galactosidase assays were performed in duplicate by the addition of 0.2 ml of diluted enzyme to 0.2 ml of 10 mM *p*-nitrophenyl *α*-D-galactopyranoside (pH 4.5)/β-D-galactopyranoside (pH 4.5) respectively and incubated at 40°C for 60 min. Similarly, for the initial measurement of α-/β-glucosidase activities, 0.2 ml of diluted enzyme was added in duplicate to 0.2 ml of 10 mM *p*-nitrophenyl α-D-glucopyranoside (pH 4.5)/β-D-glucopyranoside (pH 4.5) respectively and incubated at 40°C for 60 min. The reactions were stopped with the addition of 3 ml of 2% (w/v) tri-sodium orthophosphate, pH 12.0 and the absorbance was measured at 400 nm. Activities were expressed as units/ml, where 1 unit is defined as 1 micromole of *p*-nitrophenol liberated per minute per millilitre of culture filtrate. Depending on results obtained, further time-points were conducted to ensure that activity rates were linear.

Nelson–Somogyi reducing sugar assays were used for the measurement of *exo*-inulinase activities. Two different substrates were used in these experiments for comparison purposes: high MW (molecular weight) inulin (Raftiline; HP Inulin) and kestose (Megazyme; Cat. No. O-KTR). Initial *exo*-inulinase assays were performed in duplicate by the addition of 0.2 ml of diluted enzyme to 0.2 ml of 20 mg/ml (final) inulin (pH 4.5)/5 mg/ml (final) kestose (pH 4.5) respectively and incubated at 40°C for 60 minutes. The reactions were stopped by the addition of 0.5 ml of Stopping Solution (25 ml of Solution A (2.5% (w/v) sodium carbonate anhydrous, 2.5% (w/v) potassium sodium tartrate and 20% (w/v) sodium sulfate), to which 1 ml of Solution B (3% (w/v) copper sulfate pentahydrate) was added). Enzyme reactions were then boiled for 20 min and allowed to cool to room temperature for 5 minutes before the addition of 3.0 ml of a 1:5 dilution of Solution C (5% (w/v) ammonium molybdate, 4.2% (v/v) concentrated sulfuric acid and 0.6% (w/v) sodium arsenate heptahydrate). Depending on results obtained, further time-points were conducted to ensure that activity rates were linear.

Results and discussion

Development of evolution strains

Aspergillus nidulans (FGSCA4) and *Aspergillus oryzae* (Rib40) were grown on inulin for 9 weeks with weekly re-inoculation, over which period the growth rate of *A. oryzae* increased (data not shown). This effect was not seen for *A. nidulans*, indicating that the basal levels of enzymes required to degrade this carbon source had not increased to the same degree as they had in *A. oryzae*.

To further demonstrate this hypothesis, the *exo*-inulinase activity levels for both strains at the beginning and end of the generation experiment were tested. *A. oryzae* week 1 (*Ao*1.1) and 9 (*Ao*9.1) and *A. nidulans* week 1 (*An*1.1) and 9 (*An*9.1) generation strains were chosen, with the same biological clone being used for both time points to give an accurate comparison. Strains were grown in media containing inulin from dahlia tubers and incubated at 30°C for 4 days. Initial *exo*-inulinase activity levels for both generations of each strain were compared (Table 11.1), with *Ao*9.1 giving higher than double the activity levels (~125 mU/min/ml) of that obtained for *Ao*1.1 (~60 mU/min/ml). The activity levels measured for *An*1.1 and *An*9.1 remained unchanged and consistent with that obtained for *Ao*1.1, providing a strong correlation with the growth profiles.

Table 11.1 Initial *exo*-inulinase activity values obtained for *A. oryzae* week 1 (*Ao*1.1) and 9 (*Ao*9.1) and *A. nidulans* week 1 (*An*1.1) and 9 (*An*9.1) generation strains. Strains were grown in MM + 1% (w/v) inulin (dahlia tubers) and incubated at 30°C, 250 rpm for 4 days. *Exo*-inulinase activity assays conducted as described in 'Materials and methods'

Activities (mU/min/ml)	*A. oryzae* generation strains		*A. nidulans* generation strains	
	*Ao*1.1	*Ao*9.1	*An*1.1	*An*9.1
exo-inulinase	60.0	129.4	48.7	62.7

Extracellular enzyme profile and activity comparison

Based on the results obtained from the initial enzyme activity measurements, the extracellular activities in the *A. oryzae* mutant strains were examined in more detail. *A. oryzae* (Rib40) and the biological triplicates generated from week 9 (*Ao*9.1, *Ao*9.2 and *Ao*9.3) were chosen for this study, with *A. oryzae* (Rib40) acting as a control and indicative of the basal levels of enzyme activities expected when *A. oryzae* wild type (WT) is grown on inulin as a carbon source. Inulin from both dahlia tubers and chicory were examined to determine whether the precise structure of the inulin affects the enzyme mixtures produced by the mutants.

After 4 days of growth, the extracellular enzyme profiles generated for each strain were visualized by SDS-PAGE (Fig. 11.1). The *A. oryzae* mutant strains all demonstrated a clear increase in enzyme production levels when grown on inulin compared to the wild type (Rib40) with notable differences also being present between the three mutants. This again, was in correlation with the growth profiles of these strains with *Ao*9.3 generating more dense mycelia and a more concentrated enzyme profile in inulin rich liquid cultures than the other strains. Further evaluation of the enzyme profiles showed an intense band at approximately 52 kDa (Fig. 11.1), which likely corresponds to the extracellular *exo*-inulinase (InuE), for which the calculated molecular mass in *A. oryzae* Rib40 (AO090701000400) is ~55 kDa. The other visible protein bands are consequently expected to be multiple fructofuranosidases (invertases) i.e. AO090020000640, AO090701000038 and AO090103000043, the molecular weights which are ~64 kDa, 69 kDa and 121 kDa respectively (Coutinho *et al.*, 2009). Interestingly, higher enzyme levels were generated on inulin from chicory than from dahlia tubers, the latter being the source of inulin used for the generation studies. Previous studies using analytical HPLC to determine the chain lengths of both these commercial preparations of inulin showed that inulin from dahlia contains a higher molecular weight range which in turn has been known to affect solubility (Hincha *et al.*, 2000). This indicates

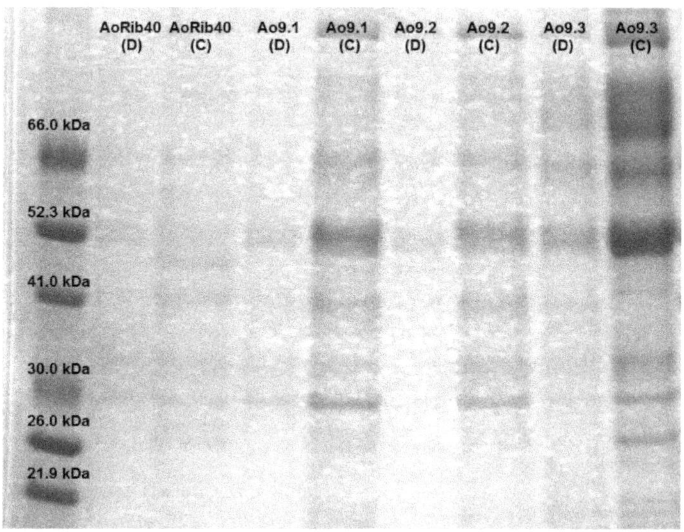

Figure 11.1 SDS-PAGE analysis of *A. oryzae* (Rib40) and week 9 evolved biological triplicate strains (*Ao*9.1, *Ao*9.2 and *Ao*9.3) when grown on MM + 1% inulin (dahlia tubers (D)/chicory (C)) at 30°C, 250 rpm for 4 days. Low molecular weight (LMW) ladder was produced in house.

that while the basic structure of both carbon sources are identical and require the same enzyme actions for degradation, the inulin from chicory may be more easily utilized by the fungal cell.

In combination with the visual profiles of extracellular enzyme production during growth on inulin, the levels of *exo*-inulinase activity were measured in duplicate and compared for the different strains. The activities of *α*- and *β*-galactosidase and *α*- and *β*-glucosidase were also measured in duplicate as a control and indicative of the basal level of change in production of un-associated enzyme activities. In correlation to what was discovered with the enzyme profiles, the overall levels of all enzyme activities were higher for the *A. oryzae* mutant strains than for the wild type (Fig. 11.2). This was especially the case for *Ao*9.3,

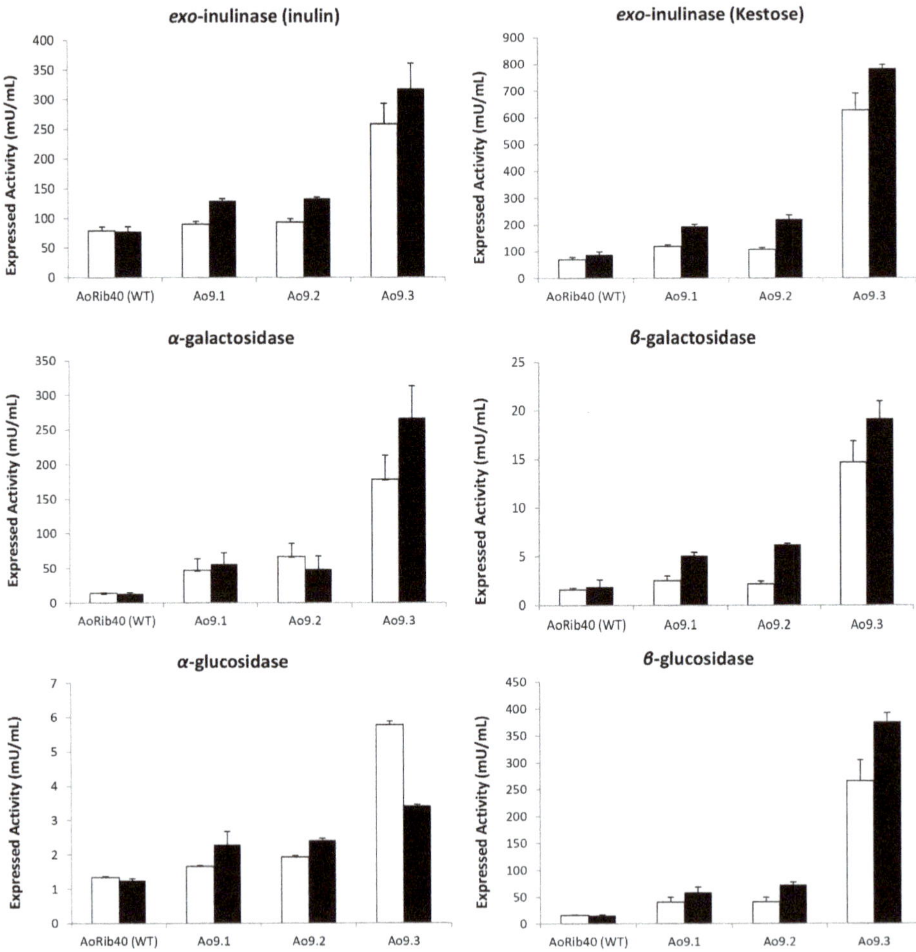

Figure 11.2 Measured activity values obtained from Day 4 culture filtrate samples from the indicated week 9 *A. oryzae* generation strains. *Ao*Rib40 was used as a control and indicative of the basal levels of activities generated when *A. oryzae* wild type (WT) is grown on inulin as a carbon source. Error bars calculated on standard deviations between technical duplicates. All polysaccharide substrates and *p*-nitrophenol substrates were tested at 40°C, pH 4.5 under conditions as described in Materials and Methods. White = Dahlia inulin, Black = chicory inulin.

which gave a 4- and 9-fold increase in *exo*-inulinase activity on inulin and kestose, respectively. Kestose contains more terminal ends than inulin, thus confirming the *exo*-acting properties of the enzyme being over-produced. An average of a 12-fold increase in activities for the control enzymes was also noted indicating that the evolution screening not only increased the levels of the required enzyme activities i.e. *exo*-inulinase, but also the overall growth rate and secretion levels of these strains when grown on inulin. This could indicate that instead of a specific *exo*-inulinase increase in the mutants, enzyme secretion as whole has been improved in these strains.

Another hypothesis involves the up-regulation of genes associated with AmyR in response to the complete degradation of inulin to fructose and glucose. AmyR controls genes involved in starch degradation, but it has been noted in *A. niger* that it also regulates genes encoding *beta*-glucosidases and *alpha*- and *beta*-galactosidases (vanKuyk *et al.*, 2012). While starch and maltose are the most common inducers of this regulator, glucose also induces AmyR activation in *A. niger* and *A. oryzae* (Carlsen and Nielsen, 2001; vanKuyk *et al.*, 2012). If, like in *A. niger*, AmyR is also responsible for the regulation of these otherwise un-associated genes in *A. oryzae*, the complete degradation of inulin may then also cause an increase in the associated enzyme activities. In *A. niger*, fructose is a relatively weak repressing carbon source (de Vries *et al.*, 2002), while glucose at higher concentrations can cause significant repression of many genes encoding poly- and oligosaccharide degrading enzymes. Degradation of inulin would only result in very low glucose levels that may result in AmyR activation.

The mutants were maintained for several generations on MEA and then tested again on inulin without losing their improved properties, signifying that the changes were genetically stable and not dependent on a continuous selection. Future studies will analyse the nature of the mutations in more detail.

The results here demonstrate that adaptive evolution is a powerful tool to make targeted mutant strains with improved properties. In future studies, the genome or transcriptome of these mutants will be analysed for the molecular mechanism underlying this phenotype. This has recently been done for a cellulase overproducing mutant of *A. niger* that was constructed by adaptative evolution and in which case the origin of the phenotype could be assigned to down-regulation of a repressor protein (Patyshakuliyeva *et al.*, unpublished results).

References

Carlsen, M., and Nielsen, J. (2001). Influence of carbon source on alpha-amylase production by *Aspergillus oryzae*. Appl. Microbiol. Biotechnol. *57*, 346–349.

Coutinho, P.M., Andersen, M.R., Kolenova, K., vanKuyk, P.A., Benoit, I., Gruben, B.S., Trejo-Aguilar, B., Visser, H., van Solingen, P., Pakula, T., *et al.* (2009). Post-genomic insights into the plant polysaccharide degradation potential of *Aspergillus nidulans* and comparison to *Aspergillus niger* and *Aspergillus oryzae*. Fungal. Genet. Biol. *46*(Suppl. 1), S161-S169.

de Vries, R.P., Burgers, K., van de Vondervoort, P.J., Frisvad, J.C., Samson, R.A., and Visser, J. (2004). A new black *Aspergillus* species, *A. vadensis*, is a promising host for homologous and heterologous protein production. Appl. Environ. Microbiol. *70*, 3954–3959.

de Vries, R.P., vanKuyk, P.A., Kester, H.C., and Visser, J. (2002). The *Aspergillus niger faeB* gene encodes a second feruloyl esterase involved in pectin and xylan degradation and is specifically induced in the presence of aromatic compounds. Biochem. J. *363*, 377–386.

Derycke, D.G., and Vandamme, E.J. (1984). Production and properties of *Aspergillus niger* inulinase. J. Chem. Technol. Biotechnol. *34*, 45–51.

Hincha, D.K., Hellwege, E.M., Heyer, A.G., and Crowe, J.H. (2000). Plant fructans stabilize phosphatidylcholine liposomes during freeze-drying. Eur. J. Biochem. *267*, 535–540.

Laemmli, U.K. (1970). Cleavage of structural proteins during the assembly of the head of bacteriophage *T4*. Nature *227*, 680–685.

Lombard, V., Golaconda Ramulu, H., Drula, E., Coutinho, P.M., and Henrissat, B. (2014). The carbohydrate-active enzymes database (CAZy) in 2013. Nucleic Acids Res. *42*, 21.

Nikolaev, I., Farmer Hansen, S., Madrid, S., and de Vries, R.P. (2013). Disruption of the L-arabitol dehydrogenase encoding gene in *Aspergillus tubingensis* results in increased xylanase production. Biotechnol. J. *8*, 905–911.

Niness, K.R. (1999). Inulin and oligofructose: What are they? J. Nutrit. *129*, 1402s–1406s.

Patnaik, R. (2008). Engineering complex phenotypes in industrial strains. Biotechnol. Progr. *24*, 38–47.

Pavis, N., Chatterton, N.J., Harrison, P.A., Baumgartner, S., Praznik, W., Boucaud, J., and Prud'homme, M.P. (2001). Structure of fructans in roots and leaf tissues of *Lolium perenne*. New Phytologist *150*, 83–95.

Pawan, G. (1973). Fructose. In Molecular Structure and Function of Food Carbohydrates, Birch, G.G., and Green, L.F., eds. (Applied Science Publishers, Barking, Essex, UK), pp. 65–80.

Peterson, R., and Nevalainen, H. (2012). *Trichoderma reesei* RUT-C30-thirty years of strain improvement. Microbiology *158*, 58–68.

Sauer, U. (2001). Evolutionary engineering of industrially important microbial phenotypes. Adv. Biochem. Eng. Biotechnol. *73*, 129–169.

Singh, P., and Gill, P.K. (2006). Production of inulinases: recent advances. Food Technol. Biotechnol. *44*, 151–162.

Vandamme, E.J., and Derycke, D.G. (1983). Microbial inulinases: fermentation process, properties, and applications. Adv. Appl. Microbiol. *29*, 983.

vanKuyk, P.A., Benen, J.A., Wosten, H.A., Visser, J., and de Vries, R.P. (2012). A broader role for AmyR in *Aspergillus niger*: regulation of the utilisation of D-glucose or D-galactose containing oligo- and polysaccharides. Appl. Microbiol. Biotechnol. *93*, 285–293.

Wallis, G.L., Hemming, F.W., and Peberdy, J.F. (1997). Secretion of two beta-fructofuranosidases by *Aspergillus niger* growing in sucrose. Arch. Biochem. Biophys. *345*, 214–222.

Wisselink, H.W., Toirkens, M.J., Wu, Q., Pronk, J.T., and van Maris, A.J. (2009). Novel evolutionary engineering approach for accelerated utilization of glucose, xylose, and arabinose mixtures by engineered Saccharomyces cerevisiae strains. Appl. Environ. Microbiol. *75*, 907–914.

Zittan, L. (1981). Enzymatic hydrolysis of inulin – An alternative way to fructose production. Starch – Stärke *33*, 373–377.

Index

6-Methylsalicylate *see* 6-MSA
6-Methylsalicylate synthase 150, 153, 154
6-MSA 150, 153, 160–161, 163, 181

A

abaA 100
abfA 99
abfB 99
abfC 99
abnA 99
abnC 99
Acetyl xylan esterase 140
Acremonium cellulolyticus 7
see also Talaromyces pinophilus
Actin 5
Actinomycetes 165
ACV 149
Acyltransferase 149
Aflatoxin 18, 59, 181
Aflatoxin B 150
Aflatrem 59
agdB 103
AgnB 96
agsB 96
Allantoate permease 118
α-Aminoadipic acid 154
Alternaria 182
alternata 164, 175
Ammonia 174–176, 181, 183
Ammonia production 175
Amoxicillin 155
Ampicillin 155
AmyR 195
AN1263 124
AN4384 124
Anthranilic acid 160
Antibiotic 79
Antioxidants 182
AntiSMASH 60–61, 66, 82
Apex 9
Apples 75, 78–79, 179
Arabinose 100
Arabinoxylan 129–130
AreA 183
Arylsulfatase 122
Ascomycota 3
Asexual development 90, 93, 104
Aspergillaceae 10
Aspergillus 1–3, 5–11, 17, 20, 43, 57–58, 63, 81, 89, 95, 121–122, 125, 129, 133, 135, 139–140, 160, 163–164, 178, 189
acidus 7, 45
see also Aspergillus luchuensis
aculeatus 45
awamori sensu *Nakazawa* 7
awamori sensu *Perrone* 7
brasiliensis 45
campestris 45
candidus 4
carbonarius 2, 21, 45
citricus 7
see also Aspergillus niger
clavatus 45, 62, 80–82, 133, 136–139, 157, 160
cultivation 191
ficuum 7
see also Aspergillus welwitschiae
fischeri see Aspergillus fischerianus
fischerianus 11, 45, 56, 62, 118, 133, 136, 137, 138, 139, 157
flavus 1, 4, 18, 45, 59, 161
foetidus 7
foetidus var. *acidus see Aspergillus luchuensis*
fumigatus 1, 9, 20, 57, 59, 61–62, 64, 95, 97–98, 115- 117, 162, 163
fumigatus A1163 20, 44–45
fumigatus Af293 20, 44–45, 62
genomes 44–46
giganteus 80
glaucus 4, 9, 11, 45
see also Eurotium herbariorum
kawachii 45, 140
kawachii nom. inval. *see Aspergillus luchuensis*
longivesica 80

luchuensis 7
nidulans 9, 18, 20, 43, 45, 56, 57, 59, 61–63, 83, 92–100, 102–103, 114–121, 123–125, 133, 135–138, 151, 162–163, 165, 181, 190
nidulans FGSC A4 191, 192
niger 2, 7, 9, 18, 20, 57, 59, 61, 63, 90–93, 95–100, 102, 115–117, 132–134, 139, 160, 162, 164, 190, 195
niger ATCC 1015 20, 45, 57, 136–138
niger ATCC 11414 21
niger ATCC 13496 45
niger CBS 120.49 138, 140
niger CBS 513.88 20, 21, 44, 45, 57, 136, 137, 138
niger N402 136, 137
niger NRRL3 45
novofumigatus 45
ochraceoroseus 45
ochraceus 80, 181
oryzae 9, 20–21, 43, 45, 57, 59, 61, 63, 93, 115–117, 133, 136–138, 161, 189–190, 193–195
oryzae RIB326 21
oryzae RIB40 21, 191, 192, 193
parasiticus 18, 45, 158
patulum 160–161
phoenicis 7, 45
rambelli 45
ruber 45
see also Eurotium rubrum
saitoi nom. inval. *see Aspergillus tubingensis*
sect. *Nigri* 2
steynii 45
sydowii 45, 118
terreus 1, 45, 57, 59, 133, 136–139, 160, 162–163
tubingensis 7, 45
usamii 7
ustus 45, 56
versicolor 45, 64, 118
violaceofuscus 7
welwitschiae 2, 7
wentii 45
westerdikiaen 160
zonatus 45
Aspergillus Genome Database *see* AspGD
AspGD 47, 48
Aspire™ 77
astA 118
atg 95
deletion 95
atg1 95
Atg1 94–95
Atg13 94–95
atg4 95
atg7 95
atg8 95
atg9 95
AtmA 103
ATP 104
ATP sulfurylase 122
Autolysis 104
Autophagy 94–95, 103–105
Avogreen™ 78

B

Bacillus subtilis 149, 159, 165
Beauveria bassiana 165
BenA 6, 12
see also Tubulin, beta
Beta-lactams 145–146, 162, 165, 166
BgtD 98
Bioreactor cultivation 90
BioSave™ 77
Biotechnology 1, 3, 17
Biseriate 9
Bisvertinolone 159
BLAST 47, 49, 56, 151
BLASTP 57
Botrytis 182
cinerea 175
Brevianamide A 79
Brie 37
BrlA 93, 103
brlA 100–102
Byssochlamys 3

C

CADRE 48
Caffeic acid 130, 133, 140
Calmodulin 5–6
see also CaM
CaM 6, 12
see also Calmodulin
Carbohydrate esterase 132
Carbon catabolite repression 99, 102, 134, 181, 183
Carbon recycling 98
Carbon starvation 89–94, 96–97, 100, 102–105, Figure 6.1
Carbon stress 101
Cathepsin K 165
CAZy 48, 62, 132, 190
CAZymes 47, 62, 98–99, 103, 138
cbhB 99
CCR 183
Cell death 104–105
Cell wall 95, 96–99, 102, 129
biosynthesis 97
composition 98, 98
degradation 99, 100, 134
integrity 97
stress response 102
Cellulase 134
cfcA 97
CfcI 98

Chaetoglobosins 79
Chaetomium globosum 56
Chaetosartorya 9
Cheese culture 29, 37
CheesyTer 36
chiA 96
chiB 102
ChiB 103
chiB1 97, 100
ChiB1 97
Chicory 194
Chitin 95, 98
Choline 119
Chromocleista 8, 10
  see also Penicillium
Chrysogena 146
Chrysogenin 159–160
Chrysogine 160, 164
chsB 96
chsF 96
ChyA 162
chyA 153
ChyC 160
ChyD 160
ChyE 160
ChyH 160
ChyM 160
ChyR 160
Cinnamic acid 131
Cinnamoyl esterase 132
  see also FAE
Circumdati 9
Citric acid 9, 20
Citrinin 79–80
Citrus 81
Cladosporium cladosporioides 164
ClbR 99
Cleistothecia 4
ClrA 99
ClrB 99
Coccidioides immitis 162
Colletotrichum 174–175, 182
  lagenarium 151
  acutatum 175
  coccodes 174–175
  fioriniae 162
  gloeosporioides 174–175, 178, 181
  graminicola 175–176
  higginsianum 174–176
Colonization 177
Communesins 79
Comparative genomics 43–44, 47, 49, 57–59, 62, 65, 84
Comparative genomics approaches Table 4.3
Comparative proteomics 63–64
Comparative transcriptomics 63
Conidiation 94
Conidiogenesis 101
Conidiophore 9
Conidiospores 105
Coprogen 157, 162
CreA 99, 102–103, 134, 183
Cristaspora 8
  see also Aspergillus
CtcB 98
Cullin 121
Cutinase 174
Cyclopiazonic acid 79, 151
Cyclosporine 146, 148
Cyclotetrapeptides 158
Cycopenin 79
cys-3 122–123
Cys-3 125
cys4 124
cysA 119
cysB 118–119, 122, 124
CysB 121
cysB mutants 124
CysD 118
Cystathionine beta-synthase 124
Cysteine 114, 120–121, 123, 154
  biosynthesis 114, 118
Cysteine synthase 120
Cytochalasins 151
Cytochrome P450 160

D

D-Fructose 190, 195
D-Glucose 195
Dehydratase 151
Diferulates 129–131
Diketopiperazines 155
DNA barcode 5
DNA damage response 103
DNA sequencing 5, 17–18
DNA sequencing, next-generation 19, 22

E

Effectors 174
eglC 96
Elaphomyces 3–4
Elaphomycetaceae 2–4
Emericella 8, 9
  nidulans see Aspergillus nidulans
  see also Aspergillus
Emodin 160
EngA 96, 103
engA 100, 102
Enoyl reductase 151
Enterococcus 159
Enzyme profiling 193
Epigenetics 163, 164
Euchromatin 163
EuPathDB 48, 64
Eupenicillium 8, 10
  see also Penicillium

Eurotiales 1–2, 123, 125
Eurotiales 4
Eurotium 9
 herbariorum 11
 rubrum 45
Evolution 56
Expansolides 79
Extrolite 4
 see also Secondary metabolite

F

F-9775A 165
F-9775B 165
FA 140
FadA 101
FAE Table 8.1, 132–135, 137, 139
 activity 135
 production 134
 see also Feruloyl esterase
faeA 139, 140
FaeA 132, 134, 137, 138
FaeB 132, 133, 137, 138
faeB 140
Ferrichrome biosynthesis 157
Ferulic acid 129–131, 134–135, 139
Ferulic acid esterases 132
 see also FAE
Feruloyl esterase 129, 132
 see also FAE
Feruloylation 130
fksA 96
Flavipucine 151
Flb 101
FlbA 101, 103
Fludioxonil 79
FluG 101, 103
Folate metabolism 119
FphA 163
Fructose 189–190
Fumonisin 2
Functional genomics 83–84
Fungal Genetics Stock Center 19
Fungal Growth Database 190
FungiDB 47–48, 64
Fungisporin 158
Fusarine 162
Fusarium 36, 81, 182
 heterosporum 162
 oxysporum 175–176, 178
 proliferatum 181
 sambucinum 118

G

G-protein signalling 101–103
GABA shunt 122
Galactomannan 95
Galactosaminogalactan 95
Galactosidase, alpha 194–195
Galactosidase, beta 194–195
GanB 101
GCPSR
 see Genealogical concordance phylogenetic species recognition
gdh2 175, 176
Gel7 98
GelG 98
Gene cluster 19
 see also Secondary metabolite cluster
Genealogical concordance phylogenetic species recognition 5
Genetic mapping 18
Genetics 18
 forward 18
Genome databases 46–47
Genome sequencing 20, 66, 145
Genome-scale metabolic models 58
Genotype 18
Geosmithia 8, 11
 argillacea 7
  see also Rasamsonia argillacea
 eburnea 10
Geotrichum 37
 candidum 37
Germination 177
GHs 94–95, 98–99, 103
 see also Glycosyl hydrolases
GLA 173, 177, 179–182
Glandicoline A 155
Glandicoline B 155
GlkA 99
Glucoamylase 20
Gluconic acid 175–177
Gluconic acid 173
 see also GLA
Glucosamine 183
Glucose 138, 183, 190
Glucose depletion 102
Glucose oxidase 176
Glucose sucrose 179
Glucosidase, alpha 194
Glucosidase, beta 194
Glucuronoarabinoxylan 129
Glutamate dehydrogenase 183
Glutaminase 183
Glutaminolysis 183
Glycolysis 183
Glycosyl hydrolases 94
 see also GHs
Glyoxylate shunt 122
gox 177
gox2 176, 179
Graminacious herbivore 165

H

HAC 164
 see also Histone acetylases

Hamigera 3, 9
Hansenula polymorpha 118
hcpA 153
HcpA 158
HdaA 164
HDAC 164
  see also Histone deacetylases
Helvolic acid 59
Hemicarpenteles 8, 10
  see also Penicillium
Heterochromatin 163
HGT 35–36
Histidine kinase 122
Histidyl-tryptophanyldiketopiperazine 155
  see also HTD
Histone acetylases 164
  see also HAC
Histone deacetylases 164
  see also HDAC
Homocysteine 118, 124
  biosynthesis 124
  synthase 122
Horizontal gene transfer *see* HGT
Horizontally transferred regions 36
  see also HTR
HTD 155–156
HTR 36, 37
HxkA 99
HxkC 103
HxkD 103
Hybrid 165
Hydrogenperoxide 176
Hydrolase 102
Hydrophobins 95, 101
Hydroxycinnamic acid 129–130, 133
Hydroxycinnamic acid esterase 132
  see also FAE

I

ICN 5, 10
  see also International Code of Nomenclature for Algae, Fungi and Plants
ICPA 9
International Code of Nomenclature for Algae, Fungi and Plants 5
InuE 193
inuE 99
Inulin 189–193
Inulinase 189
  endo 190
  exo 190, 192, 194–195
Inulinases 190
Invasive human infections 146
Invertase 189
Ircinia fasciculata 146, 159
Iron 157
Isopenicillin N 154–155
Isopenicillin N acyltransferase 154
Itaconic acid 19, 57
ITS 5, 6, 12, 77

J

JGI 18–19, 21, 44, 46–49
Joint BioEnergy Institute 19
Joint Genome Institute *see* JGI

K

KapA 163
Kappa number 134
Kestose 192, 195
Ketoreductase 151
Ketosynthase 149
Kluyveromyces lactis 118
Koji 17, 21

L

L-Alanine 160
L-Cysteine 155
L-Glutamine 183
L-Histidine 155
L-Proline metabolism 183
L-Tryptophan 155
L-Valine 155
laeA 83
LaeA 163–164
Laurencia 146
Lecanoric acid 165
Leiothecium 3
Lignin 131
Lignocellulose 129, Figure 8.1
Lipase 133, 140
Lovastatin 161, 162
LovB 161
LovF 162
LreA 163
LreB 163

M

Magnaporthe grisea 162
Maize bran 138–139
MC 135, 139
  see also Methyl caffeate
MEA medium 195
mecA 118–119, 124
mecB 118, 124
mecC 124
Melanin 98
Meleagrin 155–156
  biosynthesis 156
Merimbla 11
Meroterpenoids 161
metA 119, 124
metabolic profiling 155
metB 118–119, 123–124
metE 119, 124–125
metF 119, 124

metG 118–119, 124
metH 119
Methionine 114, 120, 122–124
Methionine permease 122–123
Methionine synthase 119
Methionine, biosynthesis 114, 118–120
Methyl caffeate 132, 135–136
  see also MC
Methyl ferulate 132, 135–136
  see also MF
Methyl p-coumarate 132, 135–136
  see also MpC
Methyl sinapate 132, 135–136
  see also MS
metR 120, 122–123, 125
MetR 121
Metulae 9
metZ 123
MetZ 121, 125
MF 135, 139
  see also Methyl ferulate
MIC 79–80
Minimum inhibitory concentration (MIC) 79
Monascus 2, 3, 9–10
Monascus purpureus 162
Monilinia 182
Monilinia fructicola 175
Morphology 89
MpC 135, 139
  see also Methyl p-coumarate
mpkA 97
MpkB 101
MS 135, 139
  see also Methyl sinapate
Mutagenesis 19
  chemical 190
  UV 190
MycoCosm 48
Mycotoxins 27, 79–81, 174

N

nagA 97
Naphtha-γ-pyrone 164
NCBI 48, 56
Neosartorya 8, 9
  fischeri 45, 118
  see also Aspergillus
Neurospora 121
  crassa 19, 92, 100, 115–117, 122–123
  crassa 114
niaD 22
Nitrate 183
Nitrogen metabolism 166, Figure 10.1
Nitrogen metabolite repression 183
Nitrogen regulation 181
Nomenclature 5
Non-ribosomal peptide synthetase *see* NRPS
Norsolorinic acid 150
NRPS 61, 146–147, 151–152, 154, 160–161, 164–166, 174
  adenylation 147, 152
  condensation 147–148
  epimerization 148
  iterative 147
  linear 147
  nonlinear 147
  specificity 148–149
  thiolation 147
NRPS-like 61, 153
Nygerone A 164

O

Ochratoxin 79
Ochratoxin A 2, 21, 32, 79, 182
Oligoferulates 130
ω-Hydroxyemodin 159
Onychocola sclerotic 158
Organic acid metabolism 176, Figure 10.2
Organic acid production 183
Organic acids 174
Orsellinic acid 150, 165
Orthologues 47
Orthologues 49, 56
Oxalic acid 175, 181
Oxidative stress 103
Oxosorbicillinol 159

P

p-Coumaric acid 129, 130, 133–134, 140
pac1 178, 181
pacC 182
PacC 178–181
Paecilomyces 8, 11
PaL signalling pathway 178
Paracoccidiodes brasiliensis 160
Paralogues 49
Pathogenicity 176–177, 182, 183
Patulin 27, 35, 76, 79–80, 82–83, 160, 179–182
  biosynthesis 179, Figure 10.3
pbcC 163
pcbAB 152–153, 155
pcbC 152, 155
PCD 173–174
Pears 75, 78
Pectin 138–139
Pectionolytic enzymes 181
pelB 178
Pellet formation 90
Penbotec™ 79
  see also Fludioxonil
penDE 152, 155
Penicillic acid 79
Penicillin 17, 22, 29, 35, 80, 145, 152, 166
  biosynthesis 155
Penicillin G 152, 155
Penicillin V 152

Penicilliopsis 3
Penicillium 1–11, 17, 20–21, 27–28, 37–38, 43, 47, 57, 75–77, 79, 84, 95, 160, 173, 176–177, 182
aethiopicum 45
aurantiogriseum 28, 45, 76, 81
see also Penicillium expansum
biforme 28, 29, 32
see also Penicillium commune
bilaiae 45
brevicompactum 10, 45, 76
camemberti 10, 27–29, 32, 37, 45, 81
candidum 77
canescens 45
capsulatum 32, 45
carneum 28–29, 76, 80
chrysogenum 7, 17–18, 22, 37, 43–44, 46, 57–59, 81–82, 145–146, 149, 152, 155–156, 158–166, 181
chrysogenum genome 146, 152, 157
chrysogenum NRRL1951 158
chrysogenum P2niaD18 81
chrysogenum Wisconsin 54–1255 22, 46, 80–81, 158
citreoroseum 159
citrinum 80
clavigerum 80
commune 29, 37, 76
concentricum 80
coprobium 80
crustosum 76
cyaneofulvum 145
see also Penicillium chrysogenum
decumbens 22, 28, 32, 46
see also Penicilllium oxalicium
digitatum 27–29, 32, 34–35, 37, 81, 175
digitatum Pd1 46
digitatum PHI26 46
dipodomyicola 80
dupontii 11
see also Thermomyces dupontii
expansum 4, 28–29, 32, 35, 37, 46, 64, 75–78, 80–81, 83–84, 164, 175–182
expansum PEX1 46
expansum PEX2 46
expansum PEXP 46
expansum R19 46, 82
expansum T01 81, 82
fellatanum 46
funiculosum 7
see also Talaromyces funiculosus
fuscoglaucum 28–29, 32
see also Penicillium commune
genomes 28, 30–31, 35, 44–46, 80, 82
glabrum 46, 76
gladioli 80
glandicola 80
glaucum 77
griseofulvum 79–80, 160
griseoroseum 7, 145
see also Penicillium chrysogenum
italicum 27–29, 37, 46, 81, 83
italicum B3 81
italicum PITC 46
janthinellum 46
lanosocoeruleum 46
marinum 80
marneffei 46
meleagrinum 145
nalgiovense 27–29
nordicum 32, 46, 80–81
notatum 145–145, 158–159
see also Penicillium chrysogenum
oxalicium 46, 81
see also Penicillium decumbens
paneum 28–29, 80
paxilli 32, 46, 59
phylogeny 32, 33
purpurogenum 7
see also Talaromyces purpurogenus
raistrickii 46
roqueforti 10, 27–29, 35–38, 46, 76, 81, 155, 161
rubens 10, 17, 22, 27–29, 35, 37–38, 46, 115–117, 145, 160
rubrum 145
rugulosum 76
sclerotigenum 80
solitum 29, 76, 78, 82
stipitatum 46
see also Talaromyces stipitatus
transcriptomics 36, 37
verrucosum 76, 80
vulpinum 80
verrucosum 32
Penicitols 161
Penitrem A 79
Penitrinic acid 158, 161
Penixanacid 161
pepA 102
pepB 102
pepJ 102
PepJ 103
Pezizomycotina 118
PG 78
see also Polygalcturonases
pgaX 99
pgxV 99
pH 93, 103, 173–174, 177–182
regulation 182
signalling 178
Phaeoacremonium 158
Phenotype 18
Phialides 9, 10
Phialomyces 3
Phomopsis mangiferae 175, 176

Phomosis 182
Phosphatase 103
Phylogeny 5, 56, 77
Phytopathogens 84
Pichia pastoris 19
Pigmentation 164
PKS 61, 146, 149, 151–154, 158–162, 164–165
  dehydratase 150
  enoyl reductase 150
  highly reducing 150
  ketoreductase 150
  non-reducing 150
  partially reducing 150
  thiolation 149
  type I 149
  type I, iterative 149, 150
  type II 149
  type III 149
PKS-like 61
PKS-NRPS 151–152, 162–163, 165
  see also Hybrid
Plant biomass 134–135, 139
  degradation 134
Plant cell wall 129, 131–132
Plant polysaccharide 100
Podospora anserina 63
Polygalacturonases 78
  see also PG
Polyketide synthase *see* PKS
Polypaecilum 8
Pome 75–76, 81
PPTases 147
PR-toxin 161
Pravastatin 166
Primary metabolism 57, 58
Programmed cell death 173
  see also PCD
Proteases 62, 64, 102–103, 174, 178
Proteasome degradation 121
Protein synthesis 174
Proteolysis 178
Proteomics 62
prtA 102
PrtA 103
PrtT 102
Prx1 161
Prx2 161
Prx3 161
Prx4 161
Pseudotulostoma 3, 4
Pseurotin A 151
pssA 153
PssA 157, 162
pssB 153
PssB 157
pssC 153
Pyrenophora tritici-repentis 165
Pyrrolopyrazine 165

R

Rasamsonia 3, 7, 9–10
  aegroticola 10
  argillacea 7, 10
    see also Geosmithia argillacea
  brevistipitata 10
  byssochlamydoides 10
    see also Talaromyces byssichlamydoides
  composticola 10
  cylindrospora 10
  eburnea 10
    see also Talaromyces eburneus and Geosmithia eburnea
  emersonii 7, 10
  piperina 10
  pulvericola 10
Re-sequencing 190
Reactive oxygen species 124, 159, 176
  see also ROS
Reference genome sequences 18
Resequencing 19
Restriction length fragment polymorphism 77
  see also RFLP
Retentostat 91
Rezishanones 159
RFLP 77
RIM101 178
RIP 38
RlmA 102
rlmA 97
RNA polymerase 161
RNA polymerase II 5
  see also RPB2
RNAi 177
roqA 153
RoqA 156
RoqM 156
RoqN 156
RoqO 156
Roquefortine A 79
Roquefortine biosynthesis 156
Roquefortine C 79, 155–156
Roquefortine D 155–156
Roquefortines 155–156
ROS 124, 159, 176
RPB2 6, 12
  see also RNA polymerase II
rRNA 5
Rubratoxin B 79

S

Saccharomyces Genome Database 47
Saccharomyces cerevisiae 28, 43, 44, 58, 63, 99, 118, 124, 125, 190
Sagenomella 3, 8
Salmonella typhimurium 159
SBP 137, 139

SCF^SconB complex 120, 125
Schizosaccharomyces pombe 118
Scholar™ 79
Sclerocleista 3
Sclerotinia 182
    sclerotiorum 175, 181
sconB 120, 121, 122, 124
sconC 120, 121, 122
Screening 190
SDS-PAGE 191, 193
Secondary metabolism 59, 62, 101, 113, 145, 165, 179
    gene cluster 153
        see also Secondary metabolite cluster
Secondary metabolite 4, 35
Secondary metabolite cluster 20–21, 35, 35, 47, 60–62, 163
    activation 165
    prediction 60, 66, 151
    see also Gene cluster
Secretome 62
Septa 89, 105
Septal pores 90, 105
Sesquiterpenes 161
Sexual development 90, 101
SfgA 101
Shemer™ 78
SidE 157
Siderophore 157
    biosynthesis 157
Silent gene clusters 162, 165
Sinapic acid 129–130, 133
SMR 123–125
SMURF 60–61, 81–82, 152
SnowyOwl 67
SNPs 49
Sohirnones 159
Sorbicillinoids 158
Sorrentanone 159
srfA 149
Staphylococcus 159
Statins 161
steA 101
Sterigmatocystin 18, 59, 181–182
Stipe 10
Streptomyces hygroscopicus 165
Stress signaling 93, 97
Submerged cultivation 90
Sucrose 183
Sugar beet pulp 135
Sulfate 113
    assimilation 118
    permease 118–119, 123
Sulfide:quinone reductase 122
Sulfite reductase 122
Sulfur metabolism 113–117, 121–124, Figure 7.1
    repression 119–120, Figure 7.3, 125
        see also SMR
SulP 118
Superoxide 176
Synteny 57

T

Talaromyces 1–11
    atroroseus 11
    byssichlamydoides 10
    dupontii 11
        see also Thermomyces dupontii
    eburneus 10
    emersonii 7, 10
        see also Rasamsonia emersonii
    flavus 11
    funiculosus 7, 11
    leycettanus 9
    marneffei 1, 11
        see also Penicillium marneffei
    pinophilus 7, 11
    purpurogenus 7
    rugulosus 11
    stipitatus 46
        see also Penicillium stipitatum
    thermophilus 7, 11
        see also Thermomyces dupontii
Tannase 133, 140
Taxonomy 2–4, 77
    polyphasic 4
TE 32, 34, 36
Teleomorph 10
Terpene cyclase 61
Tetrapeptides 158
Thermoascus 3
Thermomyces 3, 7, 9, 11
    dupontii 7, 11
    lanuginosus 11
Thiabendazole 79
Tolypocladium niveum 148
Transcriptome 92, 174
    sequencing 83
    Penicillium 36–37
Translation elongation factor 1 alpha 5
Transporters 174
Transposable elements 32, 34
    see also TE
Transposons 32
    see also Transposable elements *and* TE
Triacetylfusarine 157
Trichocoma 3–4
Trichocomaceae 2–3
Trichoderma 158
    reesei 19, 134
    virens 147
    rubrum 162
    cutaneum 118
Trichostatin A 164
Tubulin, beta 5–6, 77
    see also BenA

U

Ubiquitin-protein ligase 120
Uniseriate 9

V

Valine 154
Vanillic acid 134
Vanillin 133, 134
VeA 101, 163
VelB 163
Velvet complex 163
Verticillium interextum 159
Vesicle 9
Viomellein 79
Viridicatin 79
VosA 163

W

wA 151
Wallaby 36
Warcupiella 3, 9
WB 138, 139
 see also Wheat bran
wetA 100
Wheat bran 135, 138
Woronin bodies 90

X

Xanthomegnin 79
Xerochrysium 3
Xeromyces 3
XlnR 63, 99, 134
XprG 102, 103, 103
Xylan 138, 140
Xylanase 134
Xyloglucan 130

Y

Yanuthones 161
Yarrowia lipolytica 118
YieldPlus™ 78

www.ingramcontent.com/pod-product-compliance
Ingram Content Group UK Ltd.
Pitfield, Milton Keynes, MK11 3LW, UK
UKHW022034041026
15591UKWH00012B/672